ELECTROMAGNETIC NONDESTRUCTIVE EVALUATION (III)

Studies in

Applied Electromagnetics and Mechanics

Editors

K. Miya, A.J. Moses, Y. Uchikawa, A. Bossavit, R. Collins, T. Honma,
G.A. Maugin, F.C. Moon, G. Rubinacci, P. Silvester, H. Troger and S.-A. Zhou

Volume 15

Previously published in this series:

Vol. 14. R. Albanese, G. Rubinacci, T. Takagi and S.S. Udpa (Eds.), Electromagnetic Nondestructive Evaluation (II)
Vol. 13. V. Kose and J. Sievert (Eds.), Non-Linear Electromagnetic Systems
Vol. 12. T. Takagi, J.R. Bowler and Y. Yoshida (Eds.), Electromagnetic Nondestructive Evaluation
Vol. 11. H. Tsuboi and I. Sebestyen (Eds.), Applied Electromagnetics and Computational Technology
Vol. 10. A.J. Moses and A. Basak (Eds.), Nonlinear Electromagnetic Systems
Vol. 9. T. Honma (Ed.), Advanced Computational Electromagnetics
Vol. 8. R. Collins, W.D. Dover, J.R. Bowler and K. Miya (Eds.), Nondestructive Testing of Materials
Vol. 7. C. Baumgartner, L. Deecke, G. Stroink and S.J. Williamson (Eds.), Biomagnetism: Fundamental Research and Clinical Applications

Volumes 1-6 have been published by Elsevier Science under the series title "Elsevier Studies in Applied Electromagnetics in Materials".

ISSN: 1383-7281

Electromagnetic Nondestructive Evaluation (III)

Edited by

D. Lesselier

Département de Recherche en Électromagnétisme — Laboratoire des Signaux et Systèmes, CNRS-SUPÉLEC, Gif-sur-Yvette, France

and

A. Razek

Laboratoire de Génie Électrique de Paris, CNRS-SUPÉLEC, Universités Paris XI et Paris VI, Gif-sur-Yvette, France

Amsterdam • Berlin • Oxford • Tokyo • Washington, DC

ISBN 90 5199 444 3 (IOS Press)
ISBN 4 274 90275 7 C3053 (Ohmsha)

Publisher
IOS Press
Van Diemenstraat 94
1013 CN Amsterdam
Netherlands
fax: +31 20 620 3419
e-mail: order@iospress.nl

Distributor in the UK and Ireland
IOS Press/Lavis Marketing
73 Lime Walk
Headington
Oxford OX3 7AD
England
fax: +44 1865 75 0079

Distributor in the USA and Canada
IOS Press, Inc.
5795-G Burke Center Parkway
Burke, VA 22015
USA
fax: +1 703 323 3668
e-mail: iosbooks@iospress.com

Distributor in Germany
IOS Press
Spandauer Strasse 2
D-10178 Berlin
Germany
fax: +49 30 242 3113

Distributor in Japan
Ohmsha, Ltd.
3-1 Kanda Nishiki-cho
Chiyoda-ku, Tokyo 101
Japan
fax: +81 3 3233 2426

PRINTED IN THE NETHERLANDS

Foreword

The E'NDE'98 Workshop was the fourth of its kind, following the previously held E'NDE'95 University College London, UK, E'NDE'96 University of Tokyo, Japan, and E'NDE'97 Create Reggio Calabria, Italy, and it is preceding the 5th E'NDE'98, Des Moines, USA, in August 1999.

It was organized—under the auspices of an international Standing Committee chaired by Professor Miya, University of Tokyo, since its creation— by the Laboratoire des Signaux et Systèmes – LSS (CNRS-SUPÉLEC-UPS) & Laboratoire de Génie Électrique de Paris – LGEP (CNRS-SUPÉLEC-UPS-UPMC) and by Électricité de France – Research and Development Division (EDF-DER), in cooperation with the Japan Society of Applied Electromagnetics and Mechanics (JSAEM), and with a great variety of French and international sponsorships: Centre National de la Recherche Scientifique – Département Sciences pour l'Ingénieur; Confédération Francaise pour les Essais Non-Destructifs (COFREND); International Compumag Society; Ministère de l'Éducation Nationale, de la Recherche et de la Technologie; École Supérieure d'Électricité (SUPÉLEC); Université Paris-Sud (Département de Physique & Direction de la Recherche).

Held in the excellent facilities of EDF-DER that are beautifully set in the island of Chatou, "l'Île des Impressionistes", located in the midst of the Seine river, 10km West of Paris, the Workshop was blessed by a sunny and warm September weather. Notice that a web site (with many useful electronic links) "http://supelec.supelec.fr/invi/ende98" was set up in February 1998 and quite actively maintained since then so as to allow the smooth and fast sharing of information during the preparation of the Workshop and afterwards, with a minimum of paperwork also.

Topics of interest included 5 key items, "Improved and novel methods of electromagnetic nondestructive testing", "Solution of direct and inverse problems", "Advanced sensing technology - design and optimization", "Proposal and results of benchmark problems", and "Industrial applications".

65 participants were officially registered, from Europe, the United States and Japan, and 32 multi-authored (and most often multi-institution) papers were presented. Short versions of these contributions have been published in the Workshop Digest which was provided to every registered participant and the reviewed and accordingly revised full papers are found herein, in addition to a summary of the keynote address.

After brief welcoming remarks by the two co-chairs, D. Lesselier and A. Razek, and by P. Meurgey in charge of the organization at EDF-DER, a 40' keynote address on "Non Destructive Evaluation at Électricité de France" was given in introduction by J. Samman from EDF. Then the contributions of 20' each were given in smooth succession, a comfortable pause schedule enabling a number of discussions "off-line".

6 sessions had been organized overall: "Probe design and related issues" (6 papers) chaired by D. Placko, "Defect evaluation with emphasis on stochastic methods (6 papers) chaired by S. S. Udpa, "Innovative modalities and applications" (3 papers) chaired by J. Pávó, on the first day; "Direct modeling with emphasis on numerical techniques" (8 papers) chaired half by G. Rubinacci and half by D. Ioan, "Defect evaluation with emphasis on deterministic methods" (7 papers) chaired by J. Bowler, and "Benchmark problems" (2 papers, and a round-robin examination) chaired and convened by T. Takagi, on the second day. Closing remarks were then made by K. Miya, and by S. S. Udpa, in charge of the forthcoming E'NDE'99.

This Workshop intended to allow a fruitful and well-focused discussion on a topic, Electromagnetic Non-Destructive Evaluation, which is crucial to a score of engineering activities and environmental evaluation and protection issues, and which needs all together deep theoretical insight, efficient modeling and data interpretation tools, and well-designed measurement systems and probes. It is the belief of the Workshop organizers that E'NDE'98 indeed allowed a better understanding of such intricate problems and fair advances in some of them, through confrontation of ideas and techniques, in an excellent cooperative spirit as highlighted by the great number of comprehensive investigations presently led in multinational, multidisciplinary frameworks. It is also their belief that these edited proceedings, "Electromagnetic Non-Destructive Evaluation (III)", published by IOS Press in the series "Studies in Applied Electromagnetics and Mechanics", will give a faithful image of these many endeavors, and should be of great interest to both theoreticians and practitioners in the field.

To conclude, the editors acknowledge the contribution of all members of the Organizing and Administrative Committees, and of the referees, without whom E'NDE'98 and the volume issued from it would not have been as such. Special thanks should go to C. Marchand and L. Pichon, who handled the registration of the participants and the submission of the abstracts and full papers, and to our co-organizer at EDF, P. Meurgey.

D. Lesselier and A. Razek
Co-Editors

A summary of the keynote address: E'NDE and the stakes of EDF

The Main Objectives of Nuclear Operators

In opening this Workshop, allow me to describe the position of Electromagnetic Non-Destructive Evaluation in the maintenance of Nuclear Power Plants (NPP) and its share in the stakes of EDF.

For the first time in its history, nuclear energy in Europe is on the verge of a decisive turning point. Several parameters have recently come into play to modify the scenario, such as the deregulation of the electricity market, the appearance of new means of production (fluidised bed coal boilers, small unit co-generation, etc.), but the essential parameter is the drop in the price of fossil fuels, particularly of natural gas.

Figure 1 shows the reference cost curves used to forecast the price of 1 kWh produced by an up-to-date power plant, burning different types of fuels.

One can clearly see that the benefit drawn between 1975 and 1990 from the low price of nuclear energy is going to disappear. Therefore, the main objectives of nuclear operators for remaining competitive are:

1. to reduce the maintenance costs, which represent 53% of direct operating expenses,
2. to obtain maximum availability in the European energy network,
3. to extend the life of financially amortised plants.

E'NDE Applications and NPP Maintenance

Non-Destructive inspection of NPP components forms part of the maintenance work. By the timely diagnosis of degradation capable of implicating the safety or the availability of the plant, it enables the replacement of the necessary components to be programmed and anticipated. It represents 10% of maintenance expenditure and has been estimated for 1996 at more than 900 million French Francs. E'NDE techniques, essentially those which employ Eddy Currents (ECT) represent more than half this expenditure, which is in the main devoted to two applications:

- The examination of Steam Generator (SG) tubes.
- The examination of Rod Cluster Control Assemblies (RCCA).

The strategy employed for these two examples illustrates the procedure applied also to other components.

The SG Tubes

These tubes are the place of transfer of all the energy from the reactor core to the turbine through the boiling of the water in the secondary circuit.

Substantial degradation in these tubes is due to the choice of the Inconel 600 alloy as the tube material. This choice was made for the great majority of PWR plants throughout the world. The stress corrosion cracks (IGSCC), appear from the primary side in the deformed zones like the U bends or the transition at the top of the tube sheet. But also it was initiated from the secondary side in the areas of concentration of pollutants, which are the deposits above the tube sheet or in the interstices between the tube and the tube support sheet. Generalised micro-cracking (Intergranular Attack IGA) due to corrosion has also been found under the deposits which are accompanying the IGSCC. Metal losses through wear have been observed at the anti-vibration bars (AVB) level in the top of the bundle and on the periphery in contact with loosing objects fallen into the SG. Figure 2 shows the zone where each type of defects appears.

Inspection of the tubes is carried out with an axial probe with two coils over the whole length of the tube. 320 000 tubes were examined in this fashion in 1996. In addition, so as to measure the length of cracks and be able to characterise them more readily, several rotating probes with two differential coils with ferrite cores are employed locally. More than 300 000 tubes were examined in the same year.

Identification of the defects from the ECT probe signals is difficult. As can be seen above, the signals are always mixed because of the presence either of two simultaneous defects, or of an interference signal linked to the environment (such as the magnetite or copper deposit, the presence of steel sheet or bars, local deformation such as rolled expansion, bending or denting).

The policy followed by EDF has been not to plug the tubes on a defect signal associated with a theoretical depth without knowing the nature of that defect and assessing its behaviour and its harmfulness. Hence, besides the very numerous laboratory simulation studies that have been conducted in order to learn how to recognise the type of defect, some tubes of the SG in operation have been pulled out for checking and validating the diagnosis made by END. For example, up to the end of 1996, 375 tubes were pulled out which represented approximately 2 billion French Francs for such investigations, without counting the loss of production due to the duration of the removal operation.

This whole procedure has enabled EDF to define a plugging policy suited to each degradation and closely linked to the ECT signal identified.

The Rod Cluster Control Assemblies

These are the fine rodlets, 10 mm in diameter, whose 0.5 mm cladding encloses a neutron absorber which enables the fission reaction to be controlled and stopped. They are grouped in clusters of 24 rods and inserted into different zones of the core.

Wear and tear, due to vibrations induced by the circulation of the fluid in the core, appears in the areas of friction by the rod against its guide. Locally it can lead to the puncturing or indeed the fatigue rupturing of the rod. For certain clusters, the end of which is inserted into a dense neutron flux, the irradiation causes swelling of the absorbent which expands the cladding until it cracks. The swelling also leads to a slowing down in the descent of the cluster or its jamming. Figure 3 shows the location of the defects.

The examination of each rodlet by an encircling ECT probe allows the presence of defects to be discerned. 24 coils enable the simultaneous inspection of a complete cluster. In 1996, the number of clusters examined was 32 000.

The hot laboratory investigation of several rods has enabled a relationship to be established between the mean section of wear and the ECT signal. However, some kinds of wear in the form of a V were poorly estimated. Similarly, in the swollen area, the cracking may be masked by the swelling signal. Solution to these two problems could not be found by the eddy current method. Therefore, profile measurement by a rotating ultrasonic probe is used each time the ECT signal in any area of the rod exceeds a predetermined threshold.

The establishment of criteria linked to the ECT signal has enabled a policy to be defined for scrapping a cluster or for using it for one or two cycles after its neutral position has been modified.

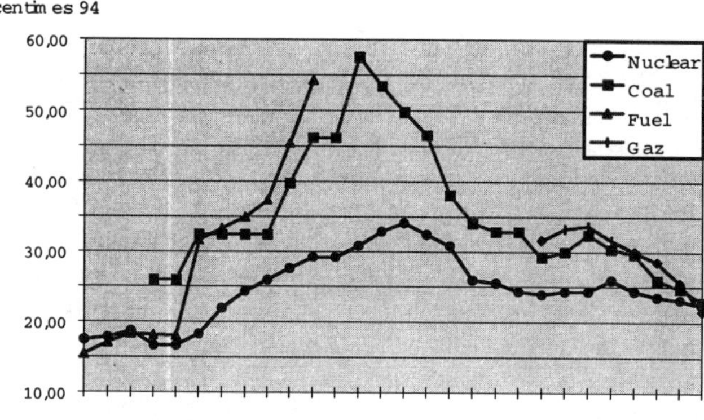

Figure 1 : Forecast reference cost per kWh

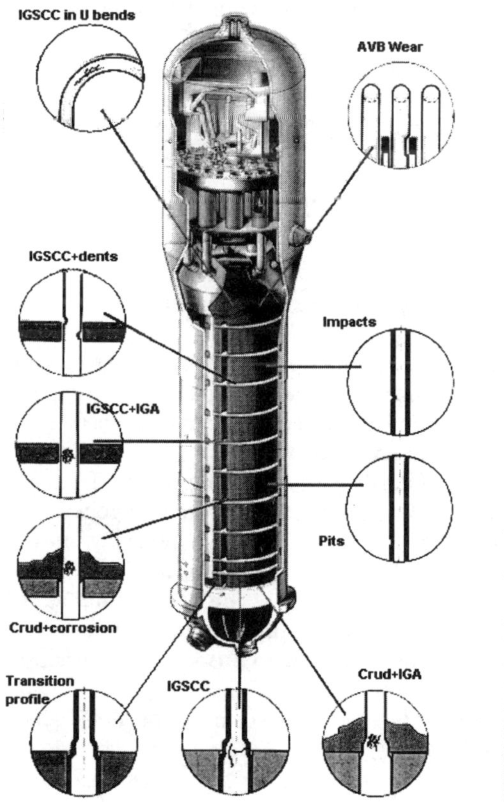

Figure 2 : Location of main degradations in Steam Generator tubes

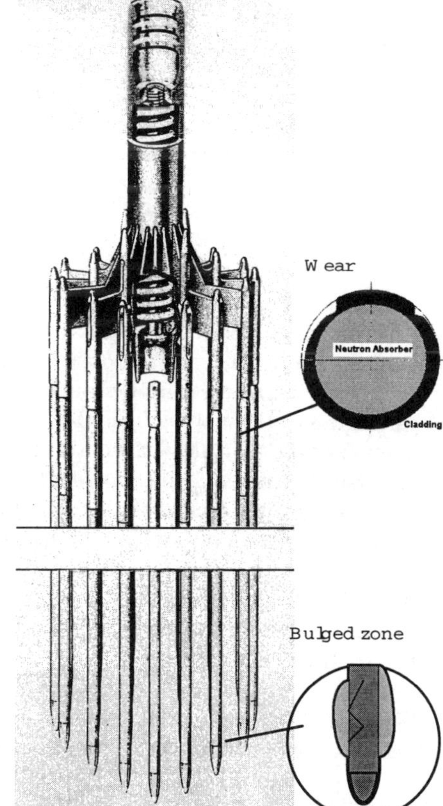

Figure 3 : location of main degradations in RCCA rodlets

Safety/Competitiveness Strategy

The efforts which have been spent in finding out the true mechanisms of degradation and the soundness of the END diagnoses, demonstrated in numerous cases of verification by analysis of the parts removed, have enabled the safety of these components to be established. This policy necessitates a perfect reproducibility of the measurements and their interpretation over the years so as to guarantee rigorous monitoring of the kinetics of evolution of degradation. For example, the accuracy of the length of the cracks in the tubes is of the same order as their lengthening in one cycle; this enables them to be monitored over several cycles before reaching the size required for plugging, which, that too, leaves a margin in relation to the critical size defined by the safety studies.

The strategy therefore has consisted in:

– stabilising the type of probe (with stringent requirements on manufacturing acceptance),
– EDF developing algorithms for processing the signal and for diagnosing the defects,
– enforcing service companies utilising these measures in their own industrial tooling.

Thanks to this strategy, productivity gains have been obtained by placing performances in competition on the basis of quality and rapidity in data acquisition.

Similarly the reliability of the diagnosis over a long period has enabled expense-savings, for example:

– by establishing a programme for replacing the SG in the medium term with anticipation of orders
– or of eliminating a succession of cluster inspection campaigns in order to arrive directly at their replacement.

Prospects

The development of the policy in matters of safety is pushing the regulators to demand that processes be qualified by an independent organisation. This is going to lead to formalising requirements and to conducting demonstrations of performances. So as to limit the tests, there will increasingly be recourse to mathematical modelling in order to explore the parameters of the methods. Similarly, the common use of recognised mock-ups (via the ENIQ for example) will allow the costs of creation to be shared.

As it has been noted above, in order to reach a final diagnosis, two or more ECT probes are employed, or indeed another method, that of ultrasonics, on the same component. To reduce the examination times, and therefore costs, we are going to try using **multi-technique** "probes" simultaneously. The data processing systems should be available commercially and should enable a diagnosis to be produced in **real time**. Despite the use of different techniques and different service providers, it must be possible to easily compare the results from one campaign to another, hence the obligation to work with a **common data format**.

These three notions (shown in bold) constitute the main lines of progress along which EDF is counting on moving in the future with its partners.

J. Samman
EDF – Groupe des Laboratoires

List of Referees

ALBANESE R. (Ass. EURATOM/ENEA/CREATE, Università degli Studi di Reggio Calabria)
BOSSAVIT A. (EDF–DER)
BOWLER J. R.(Department of Physics, University of Surrey)
DUCHÊNE B. (DRÉ – Laboratoire des Signaux et Systèmes, CNRS-SUPÉLEC)
HARFIELD N. (Department of Physics, University of Surrey)
IOAN D. ("Politehnica" University of Bucharest)
MARCHAND C. (Laboratoire de Génie Électrique de Paris, CNRS-SUPÉLEC-UPS-UPMC)
MASTORCHIO S. (EDF–DER)
MAYOS M. (EDF–DER)
MIYA K. (Nuclear Engineering Research Laboratory, The University of Tokyo)
MONEBHURRUN V. (DRÉ – Laboratoire des Signaux et Systèmes, CNRS-SUPÉLEC)
PÁVÓ J. (Department of Electromagnetic Theory, Technical University of Budapest)
PICHON L. (Laboratoire de Génie Électrique de Paris, CNRS-SUPÉLEC-UPS-UPMC)
PLACKO D. (LESIR, ENS Cachan)
RUBINACCI G. (Ass. EURATOM/ENEA/CREATE, Università degli Studi di Cassino)
TAKAGI T. (Institute of Fluid Science, Tohoku University)
UDPA L. (Department of Electrical and Computer Engineering, Iowa State University)
UDPA S. S. (Department of Electrical and Computer Engineering, Iowa State University)

E'NDE'98

The 4th International Workshop on Electromagnetic Non-Destructive Evaluation
Chatou, France
September 17-18, 1998

following previously held E'NDE'95 University College London, UK, E'NDE'96 University of Tokyo, Japan, and E'NDE'97 Create Reggio Calabria, Italy

Organized by:

- Laboratoire des Signaux et Systèmes - LSS (CNRS-SUPÉLEC-UPS) & Laboratoire de Génie Électrique de Paris - LGEP (CNRS-SUPÉLEC-UPS-UPMC)
- EDF - Research and Development Division

In cooperation with:

- Japan Society of Applied Electromagnetics and Mechanics, Japan

Sponsorships:

- CNRS - Département Sciences pour l'Ingénieur
- COFREND
- International Compumag Society
- Ministère de l'Éducation Nationale, de la Recherche et de la Technologie
- SUPÉLEC
- Université Paris-Sud Département de Physique & Direction de la Recherche

Standing Committee: K. Miya, Univ. Tokyo, Chairman,
R. Albanese, Univ. Reggio Calabria – J. R. Bowler, Univ. Surrey – R. Collins, Univ. College London – D. C. Ioan, Polytech. Inst. Bucharest – D. Lesselier, LSS – P. Meurgey, EDF-DER – M. S. Morioka, JAPEIC – N. Nakagawa, Iowa State Univ. – G. Z. Ni, Zhejing Univ. – J. Pávó, Tech. Univ. Budapest – A. Razek, LGEP – G. Rubinacci, Univ. Cassino – J. N. Sheng, X'ian Jiatong Univ. – T. Takagi, Tohoku Univ. – H. Takamatsu, KEPCO – S. S. Udpa, Iowa State Univ. – R. Zorgati, EDF-DER

Organizing Committee: D. Lesselier, LSS, and A. Razek, LGEP, co-chairs,
A. Bossavit, EDF-DER – B. Duchêne, LSS – C. Marchand, LGEP – S. Mastorchio, EDF-DER – M. Mayos, EDF-DER – V. Monebhurrun, LSS – L. Pichon, LGEP – R. Zorgati, EDF-DER

Administrative Committee:
E. Faure, LGEP – F. Mondesir, LGEP – H. Pieranska, LSS

E'NDE' 98 – List of Participants

ALBANESE R.
albanese@disna.dis.unina.it
Associazione EURATOM/ENEA/CREATE, DIMET, Università degli Studi di Reggio Calabria, Via Graziella, Loc. Feo di Vito, I-89128 Reggio Calabria, Italy

BELTRAME P.
Philippe.Beltrame@trotek.ec-lyon.fr
CEGELY - Ecole Centrale de Lyon, BP 163, 69131 Ecully, France

BERTHIAU G.
gberthiau@cea.fr
CEA / CEREM, STA / LCME, Bat. 611, Centre de Saclay, 91191 Gif sur- Yvette Cedex, France

BOSSAVIT A.
Alain.Bossavit@edfgdf.fr
EDF/DER, 1 Av du Général de Gaulle, 92141 Clamart Cedex, France

BOWLER J. R.
j.bowler@surrey.ac.uk
Department of Physics, University of Surrey, Guildford, Surrey GU2 5XH,UK

CARDELLI E.
ermanno@turbo.isten.ing.unipg.it
Institute of Energetics, University of Perugia, Via G. Duranti 1/A-4, 06125 Perugia, Italy

CHAMONINE M.
MChamonine@RosenGermany.de
H. Rosen Engineering GmbH, Research & Development Dept., Am Seitenkanal 8 D-49811 Lingen (Ems), Germany

CHAUSSECOURTE P.
Pierre.Chaussecourte@edfgdf.fr
EDF/DER, 1 Av du Général de Gaulle, 92141 Clamart Cedex, France

CHEN Z.
chn@oec.pnc.go.jp
Structure Safety Engineering Group, OEC/JNC 4002 Narida-chio, Oarai-machi, Ibaraki, 311-1393, Japan

CHENG W.
cheng@tokai.t.u-tokyo.ac.jp
Nuclear Engineering Research Laboratory Graduate School of Engineering, The University of Tokyo, 2-22 Shirakata-shirane, Tokai-mura, Naka-gun, Ibaraki, 319-1106, Japan

CORVAISIER-RICHE A.
corvaisi@etca.fr
Société de Calcul Mathématique 111, rue du Fbg Saint-Honoré, 75008 Paris, France

DE BARMON B.
bdebarmon@cea.fr
CEA / CEREM, STA / LCME, Bat. 611, Centre de Saclay, 91191 Gif sur Yvette Cedex, France

DUCHENE B.
Duchene@supelec.fr
Département de Recherche en Électromagnétisme - Laboratoire des Signaux et Systèmes, CNRS-SUPÉLEC, Plateau de Moulon, 91192 Gif-sur-Yvette Cedex, France

ENAMI K.
enami@wt.trdc.mhi.co.jp
Takasago R&D Center, Mitsubishi Heavy Industries, LtD
Shinhama 2-1-1, Arai-cho, Takasago, Hyogo, 676-8686, Japan, Japan

ENOKIZONO M.
enoki@cc.oita-u.ac.jp
Faculty of Engineering, Oita University, 700 Dannoharu, Oita, 870-1192, Japan

FAURE E.
Faure@lgep.supelec.fr
Laboratoire de Génie Électrique de Paris, CNRS-SUPÉLEC-Universités Paris XI et Paris VI, Plateau de Moulon, 91192 Gif-sur-Yvette Cedex, France

FORMISANO A.
formisano@kaos.dis.unina.it
Associazione EURATOM/ENEA/CREATE, Dip. Di Ing. dell'Informazione, Seconda Università degli Studi di Napoli, Via Roma 29, I-81031 Aversa (CE), Italy

FRANÇOIS N.
Francois@edfgdf.fr
EDF/DER, 1 Av du Général de Gaulle, 92141 Clamart Cedex, France

FRESA R.
fresa@diiie.unisa.it
DIIIE, Università degli Studi di Salerno, Via Ponte Don Melillo, I-84084, Fisciano (SA), Italy

FUKUTOMI H.
fukutomi@ohm.ifs.tohoku.ac.jp
Institute of Fluid Science, Tohoku University, Katahira 2-1-1, Aoba-ku, Sendai 980-8577, Japan

GASPARICS A.
ganti@rl.atki.kfki.hu
Research Institute for Technical Physics and Materials Sciences, H-1525 Budapest-114, P.O. Box 49, Hungary

HARFIELD N.
N.Harfield@surrey.ac.uk
Department of Physics, University of Surrey, Guildford, Surrey GU2 5XH, UK

HINKEL L.
LHinkel@RosenGermany.de
H. Rosen Engineering GmbH, Research & Development Dept., Am Seitenkanal 8, D-49811 Lingen (Ems), Germany

IOAN D.
daniel@lmn.pub.ro
"Politehnica" University of Bucharest, Numerical Methods Lab., Spl. Independentei 313, 77206, Romania

ISHIBASHI K.
isibasi@keyaki.cc.u-tokai.ac.jp
Mechanical Engineering, Faculty of Engineering, Tokai University, 2-28-4 Tomigaya, Shibuya, Tokyou
151-0063 Japan

JOUBERT P.-Y.
joubert@lesir.ens-cachan.fr
LESIR / ENS de Cachan, 61 Avenue du Président Wilson, 94 235 Cachan Cèdex, France

LESSELIER D.
lesselier@supelec.fr
Département de Recherche en Électromagnétisme - Laboratoire des Signaux et Systèmes, CNRS-SUPÉLEC, Plateau de Moulon, 91192 Gif-sur-Yvette Cedex, France

MARCHAND C.
Marchand@lgep.supelec.fr
Laboratoire de Génie Electrique de Paris (LGEP) CNRS-SUPÉLEC-Universités Paris XI et Paris VI, Plateau de Moulon, 91192 Gif-sur-Yvette Cedex, France

MASTORCHIO S.
Stephanie.Mastorchio@edfgdf.fr
EDF/DER, 6 Quai Watier, 78401 CHATOU Cedex, France

MATSUMOTO Y.
y-matsu@nfi.co.jp
Nuclear Fuel Industries, Ltd., 950 Ohaza-noda, Kumatori-cho, Sennan-gun, Osaka-fu 590-0451, Japan

MAYOS M.
Michel.Mayos@edfgdf.fr
EDF/DER, 6 Quai Watier, 78401 Chatou Cedex, France

MEURGEY P.
Patrick.Meurgey@edfgdf.fr
EDF/DER, 6 Quai Watier, 78401 Chatou Cedex, France

MIDROIT
Midroit@sollac.usinor.com
IRSID, Voie romaine, BP30320, 57283 Maizières-les-Metz Cedex, France

MINKOV D.
dminkov@genesis.rift.mech.tohoku.ac.jp
Research Institute for Fracture Technology, Tohoku University, Aoba-Ku, Sendaï, Miyagi 980, Japan

MIYA K.
miya@quest.gen.u-tokyo.ac.jp
Nuclear Engineering Research Laboratory, The University of Tokyo, 2-22 Shirakata Shirane, Tokai-mura, Naka-gun, Ibaraki 319-1106, Japan

MONEBHURRUN V.
Monebhurrun@supelec.fr
Département de Recherche en Électromagnétisme - Laboratoire des Signaux et Systèmes, CNRS-SUPÉLEC, Plateau de Moulon, 91192 Gif-sur-Yvette Cedex, France

NAKAGAWA N.
nnakagaw@cnde.iastate.edu
Center for Nondestructive Evaluation, Iowa State University, 1915 Scholl Rd, Bldg II, Ames, Iowa 50011, U.S.A.

NOVOTNY P.
Pavel.Novotny@vscht.cz
Institute of Chemical Technology, Technická 5, 166 28 Prague 6, Czech Rep.

PASTORINO M.
pastorino@dibe.unige.it
Department of Biophysical and Electronic Engineering, University of Genoa, Via Opera Pia 11A, I-16145 Genova, Italy

PÁVÓ J.
pavo@evtsz.bme.hu
Department of Electromagnetic Theory, Technical University of Budapest, H-1521 Budapest, Egry J. u. 18, Hungary

PICHON L.
Pichon@lgep.supelec.fr
Laboratoire de Génie Électrique de Paris, CNRS-SUPÉLEC-Universités Paris XI et Paris VI, Plateau de Moulon, 91192 Gif-sur-Yvette Cedex, France

PLACKO D.
placko@lesir.ens-cachan.fr
LESIR, ENS Cachan, 61 Avenue du Président Wilson, 94235 Cachan Cedex, France

POPA R. C.
rcpopa@tokai.t.u-tokyo.ac.jp
Nuclear Engineering Research Laboratory, The University of Tokyo, 2-22 Shirakata Shirane, Tokai-mura, Naka-gun, Ibaraki 319-1106, Japan

RAZEK A.
Razek@lgep.supelec.fr
Laboratoire de Génie Électrique de Paris, CNRS-SUPÉLEC-Universités Paris XI et Paris VI, Plateau de Moulon, 91192 Gif-sur-Yvette Cedex, France

RODGER D.
d.rodger@bath.ac.uk
University of Bath, Claverton Down, Bath, BA2 7AY, UK

ROUVE L.-L.
Laure-Line.Rouve@leg.ensieg.inpg.fr
Laboratoire de Magnétisme du Navire Domaine Universitaire ENSIEG, B.P. 46, 38402 St Martin d'Hères Cedex, France

RUBINACCI G.
rubinacci@ing.unicas.it
Associazione EURATOM/ENEA/CREATE, Dip. di Ing. Industriale, Università degli Studi di Cassino, Via Di Biasio 43, I-03043 Cassino (FR), Italy

RUOSI A.
adele@axpna1.na.infn.it
Instituto Nazionale per la Fisica della Materia (INFM), Unita di Napoli, Università di Napoli "Federico II", Piazzale Tecchio 80, 80125 Napoli, Italy

SAMMAN J.
Samman@edfgdf.fr
EDF/GDL, 21 Allée Privée, Carrefour Pleyel, 93206 Saint-Denis Cedex, France

SARTRE B.
bsartre@framatome.fr
Framatome BP13, 71380 Saint-Marcel, France

SHIMONE J.
jshimone@neltd.co.jp
Plant dept. Nuclear Engineering Ltd. (NEL), 1-3-7 Tosabori Nishi-ku Osaka 550-0001 Japan

SIKORA R.
rs@main.tuniv.szczecin.pl
Department of Theoretical Electrotechnics, Technical University of Szczecin, al. Piastow 19, 70-310, Szczecin, Poland

SUZUMA T.
suzuma@cstd.amaken.sumikin.co.jp
System Engineering Division, Sumitomo Metal Industries Ltd., 1-8 Fuso-cho Amagasaki 660-0891, Japan

TAKAGI T.
takagi@ifs.tohoku.ac.jp
Institute of Fluid Science, Tohoku University, Katahira 2-1-1, Aoba-ku, Sendai 980-8577, Japan

TALVARD M.
talvard@roseau.saclay.cea.fr
CEA/CEREM, STA/LCME, Bat. 611, Centre de Saclay, 91191 Gif-sur-Yvette Cedex, France

THEVENOT F.
DASSAULT_CDE@compuserve.com
Dassault Aviation CDE, Z.A. Louis Bréguet, BP 12, 78141 Vélizy-Villacoublay Cedex, France

TREVISAN F.
francesco.trevisan@diegm.uniud.it
Dipartimento di Ingegneria Elettrica Gest. e Mecc., Universita' di Udine, Via delle Scienze 208, 33100, Udine, Italy

TSUBOI H.
tsuboi@fuip.fukuyama-u.ac.jp
Dept. of Information Engineering, Fukuyama University, Gakuencho, Fukuyama 729-0292, Japan

UDPA S. S.
udpa@ee.iastate.edu
Materials Characterization Research Group, Department of Electrical and Computer Engineering, Iowa State University, 2215 Coover Hall, Ames, IA, 50011-3060 USA

VALENTINO M.
Valentin@cds.unina.it
Dipartimento di Ingegneria dei Materiali per la Produzione, Istituto Nazionale per la Fisica della Materia (INFM), Università di Napoli "Federico II", Piazzale Tecchio 80, 80125 Napoli, Italy

VÉRTESY G.
vert@r1.atki.kfki.hu
Research Institute for Technical Physics and Materials Sciences, H-1525 Budapest-114, PO. Box 49, Hungary

VÉRTESY Z.
zvert@r1.atki.kfki.hu
Research Institute for Technical Physics and Materials Science, H-1525 Budapest, PO. Box 49, Hungary

YAMADA K.
yamasan@fms.saitama-u.ac.jp
Graduate School of Saitama University, Urawa, Saitama 338-8570, Japan

ZAOUI F.
Zaoui@lgep.supelec.fr
Laboratoire de Génie Électrique de Paris, CNRS-SUPÉLEC-Universités Paris XI et Paris VI, Plateau de Moulon, 91192 Gif-sur-Yvette Cedex, France

ZORGATI R.
riadh.zorgati@edfgdf.fr
EDF/DER, 1 Av du Général de Gaulle, 92141 Clamart Cedex, France

Contents

Direct Modeling with Emphasis on Numerical Techniques

Defect Evaluation with Emphasis on Deterministic Methods

Benchmark Problems

Probe Design and Related Issues

Electromagnetic Nondestructive Evaluation (III)
D. Lesselier and A. Razek (Eds.)
IOS Press, 1999

Eddy Current Effects in the Core of a Magnetic Field Sensor

Raffaele ALBANESE
Associazione EURATOM/ENEA/CREATE, DIMET, Università degli Studi di Reggio Calabria, Via Graziella, Loc. Feo di Vito, I-89128 Reggio Calabria, Italy

Marco FEDERICO and Guglielmo RUBINACCI
Associazione EURATOM/ENEA/CREATE, Dip. di Ing. Industriale, Università degli Studi di Cassino, Via Di Biasio 43, I-03043 Cassino (FR), Italy

Alessandro FORMISANO
Associazione EURATOM/ENEA/CREATE, Dip. di Ing. dell'Informazione, Seconda Università degli Studi di Napoli, Via Roma 29, I-81031 Aversa (CE), Italy

Raffaele FRESA
DIIIE, Università degli Studi di Salerno, Via Ponte Don Melillo, I-84084 Fisciano (SA), Italy

Abstract. Aim of this paper is to assess the impact of the eddy currents induced in the core of the FLUXSET probe, a high sensitivity magnetic field sensor. Due to the conductivity of the magnetic core, at high frequencies the skin effect may in fact delay the penetration of the flux generated by the driving solenoid, with a consequent alteration of the output signal. Numerical calculations presented in previous papers indicated that the effects of the eddy currents on the output signal were negligible even at frequencies much higher than the inverse of the penetration time of the magnetic field. In this paper we propose an explanation of this phenomenon, which occurs in both linear and nonlinear cases. We also present 3D numerical calculations with both integral and differential formulations, in comparison with an approximate analytical solution for 1D models.

1. Introduction

In common magnetic field sensors based on Faraday's law (passive pick-up coils) the output signal is the time integral of the voltage measured at the terminals of an air core solenoid, which is proportional to the average field in the region occupied by the solenoid. The main limits of these sensors are the accuracy, the sensitivity, the calibration and the range of amplitude and frequency of the measured field.

To overcome these problems, and meet the stringent requirements needed for the NDT probes, other sensors have been proposed on the basis of different concepts [1, 2].

The common features of these active sensors are the presence of a driving solenoid, a pick-up coil, and a ferromagnetic core. In a *FLUXSET* sensor [1] the core is a thin sheet of annealed metallic glass, a soft magnetic material whose magnetic constitutive relationship, approximately piecewise linear with μ_r=85000 for B < B_{sat}=.65 T and μ_r=1 otherwise, allows to transform a field intensity measurement into a time delay measurement, by employing a triangular current waveform (Figure 1). In [2] the field intensity gauge is obtained by measuring the second harmonic of the output signal, by employing a sinusoidal current waveform.

Aim of this paper is to assess the impact of the eddy currents induced in the *FLUXSET* core by the driving solenoid. Due to the conductivity of the magnetic core, at high frequencies the skin effect may in fact delay the penetration of the flux generated by the driving solenoid, with a consequent alteration of the output signal.

In *FLUXSET Probe 97_1* design the ribbon thickness h is 20 μm, its conductivity is estimated to be $1.0 \cdot 10^5$ S/m. We analyze the eddy current effects when the frequency of the triangular waveform ranges from 200 Hz to 66.67 kHz.

In [3] a simple 1D analysis was carried out to estimate the penetration time of the magnetic field. For the unsaturated material, assuming a differential permeability $\mu_r = B_s/\mu_o H_s \approx 85000$, the skin depth for the fundamental harmonic of the triangular current waveform, given by $\delta = 1/\sqrt{\pi f \sigma \mu_r \mu_o}$, resulted to span from 21 μm at 66.7 kHz to 0.39 mm at 200 Hz, and the critical value $\delta = h/2$ was found to correspond to a frequency of about 35 kHz and a penetration time of 30 μs.

Nevertheless, numerical experiments found the growth-time of the magnetic flux to be 3-4 orders of magnitude faster [4].

The paper is organized as follows. Section 2 presents an explanation for the two apparently clashing results given in [3] and [4]. In Section 3 we recall the main characteristics of the numerical methods used in the analysis as well as the way of obtaining approximate analytical solutions. In Section 4, we report a detailed 3D numerical analysis for the penetration for both linear and nonlinear cases. Finally, in Section 5, we present our conclusions.

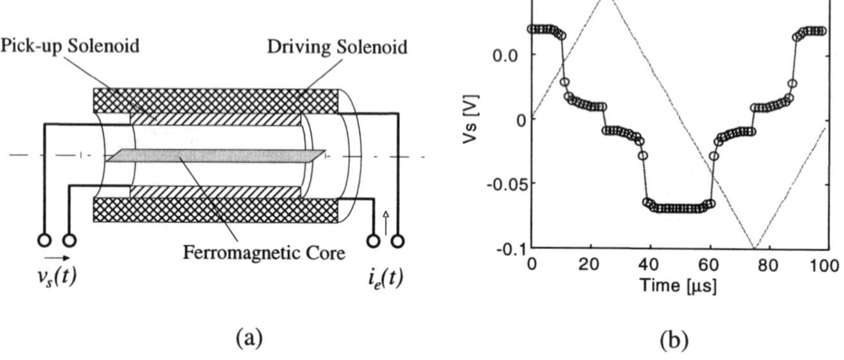

Figure 1. FLUXSET sensor : a) schematics; b) time behaviour of the output signal $v_s(t)$ for a triangular waveform of the driving current in the absence of external field.

2. Time scales for flux and field penetration

The two apparently clashing results of [3] and [4] found an explanation that is presented here. To clarify the point, let us consider two magnetostatic problems, in which the source is provided by a DC current flowing in the exciting solenoid and the magnetic core is replaced by a finite length rectangular block of perfectly magnetic material ($\mu \rightarrow \infty$). The block is massive in the first problem and hollow in the second case.

It is clear that the magnetic field outside the block is exactly the same in the two problems, because the problem in the air region is solved in both cases by imposing **H×n=0** at the interface between core and air region.

As a consequence, the magnetic fluxes linked with the solenoid and at any cross-section of the block are exactly the same in the two cases. The difference is in the distribution of the flux density inside the block.

In the equatorial b×h cross-section of the massive block, the flux Ψ is almost uniformly distributed, with an average value of the flux density $B_z = \Psi / bh$.

In the second case the flux density is zero in the internal vacuum region, and larger in the shell, where its average value is $B_z = \Psi / [bh-(b-\Delta)(h-\Delta)]$, where Δ is the shell thickness.

According to [3], during a transient phase of the order of .02 μs, the magnetic field has not yet penetrated along the whole thickness, and is confined in the skin depth $\delta=\pi(t/\sigma\mu_r\mu_0)^{1/2}$. Nevertheless, taking into account the solution of the second magnetostatic problem, this is not in contrast with the rapid penetration of the magnetic flux foreseen in [4], provided that the skin depth is wide enough to keep the value of B_z given before with $\delta\approx\Delta$ below the saturation value B_s.

On the other hand, as shown in Section 3.3, in the nonlinear case the penetration time of the magnetic flux practically coincides with that of the magnetic field.

3. Computational methods

3.1 3D non linear edge element integral formulation

The 3D non linear integral formulation [5-6] used in the analysis solves the integral equations in the conducting domain V_c and in the ferromagnetic domain V_f, in terms of the sources of the magnetic field, namely the current density and magnetization vectors.

The current density **J** is expressed in terms of edge shape functions $\mathbf{T}_k$:

$$\mathbf{J}(\mathbf{x},t) = \sum_{k=1}^{n} I_k(t)\nabla\times\mathbf{T}_k(\mathbf{x}) \tag{1}$$

The gauge based on the tree-cotree decomposition of the mesh [6] assures the uniqueness of $\mathbf{T}_k$.

The magnetization vector **M** is assumed to be uniform in each finite element within the iron and it can therefore be expressed in terms of elementary pulse functions $\mathbf{P}_k(\mathbf{x})$ as:

$$\mathbf{M}(\mathbf{x},t) = \sum_{1}^{m} M_k(t)\mathbf{P}_k(\mathbf{x}) \tag{2}$$

Applying the Galerkin's approach to the electric and magnetic constitutive equations, the following nonlinear system of equations is obtained:

$$\int_{V_c} \mathbf{T}_k \cdot (\eta\mathbf{J} - \mathbf{E})dV = 0 \qquad \forall\mathbf{T}_k \tag{3}$$

$$\int_{V_f} \mathbf{P}_k \cdot [\mathbf{M} - G(\mathbf{B})]dV = 0 \qquad \forall\mathbf{P}_k \tag{4}$$

in which $G(\mathbf{B})=\mu_0^{-1}\mathbf{B}-\mathbf{H}(\mathbf{B})$, the electric field **E** is expressed in terms of the magnetic and electric potentials as $\mathbf{E}=-\partial\mathbf{A}/\partial t-\nabla\partial\phi/\partial t$ and the flux density **B** can be obtained in terms of **J** and **M** using the Biot-Savart law.

The presence of a magnetic material requires the use of iterative methods. The nonlinear equation is then solved using an iterative method that can be proved to be a convergent Pichard-Banach procedure if the nonlinear mapping **H(B)** relating **B** to **H** verifies Lipschitz condition and is uniformly monotonic [5-6].

3.2 3D non linear edge element A, ϕ differential formulation

The non linear differential formulation used in the analysis solves the following weak formulation, in terms of the electric field $\mathbf{E}=-\partial\mathbf{A}/\partial t-\nabla\partial\phi/\partial t$:

$$\int_V \left\{\nabla\times\mathbf{N}_k\cdot\mathbf{H}[\nabla\times\mathbf{A}]+\mathbf{N}_k\cdot\sigma\frac{\partial}{\partial t}(\mathbf{A}+\nabla\phi)\right\}dV=\int_V \mathbf{N}_k\cdot\mathbf{J}_s dV+\int_{\partial V_h}\mathbf{N}_k\cdot\mathbf{h}dS,\ \forall\mathbf{N}_k \quad (5)$$

$$\int_V \nabla\varphi_k\cdot\sigma\frac{\partial}{\partial t}(\mathbf{A}+\nabla\phi)dV=\int_V \nabla\varphi_k\cdot\mathbf{J}_s dV+\int_{\partial V_h}\nabla\varphi_k\cdot\mathbf{h}dS,\quad \forall\varphi_k \quad (6)$$

where **A** is expressed in terms of edge element shape functions gauged using a tree-cotree decomposition of the finite element mesh as $\mathbf{A}(\mathbf{x},t)=\sum_1^n A_k(t)\mathbf{N}_k(\mathbf{x})$ and ϕ is expressed in terms of isoparametric shape functions φ_k. Vector **h** denotes the prescribed value of the tangential component of **H** on the boundary $\partial V_h\subseteq\partial V$.

3.3 1D analytical solutions: the linear and non linear cases

Analytical solutions are available for the one dimensional slab problem (infinite along y and z directions), in both linear and non-linear cases.

The linear case might be of interest for the simulation of the sensor behaviour in the unsaturated conditions ($B < B_{sat}$=.65 T everywhere), in which the magnetic permeability is quite high (μ_r=85000). The non-linear case is much more interesting, because the sensor is designed in such a way that the core is brought to saturation, which is essential for the FLUXSET working principle. In this case, the assumption of a step-like magnetic constitutive equation can be used to get an estimation of the penetration time.

In the linear case, the peculiarity of the problem under examination is that the presence of a magnetic material of finite extension in the direction of the field does not allow the straightforward specification of a boundary condition in terms of tangential **H**, namely H_z in our system of co-ordinates. As already anticipated in the introduction and discussed below, the correct statement of the 1D boundary value problem is closer to a flux driven problem even if the forcing term is specified in terms of current. It is then of interest in this case to considering a step change for the flux; it results that the flux density on the boundary is infinite at $t=0^+$ and approaches the stationary value in a way that cannot be stated a priori. It is then impossible to explicitly specify the boundary conditions for B_z or H_z. Therefore we solve the problem in terms of A_y, being $E_y=-\partial A_y/\partial t$. Accordingly, the diffusion equation in the linear case reduces to $\partial^2 A_y(x, t)/\partial x^2=\mu\sigma\partial A_y(x, t)/\partial t$. Being the flux linked with the slab constant for t>0, the boundary conditions are $A_y(0, t)=0$, $A_y(h/2, t)=A_0=\Psi/2b$. The initial conditions require $A_y(x, 0)=0$ everywhere but on the boundary. This is a classic diffusion problem whose analytical solution is well known.

In the non-linear case, an analytical solution can be obtained, following the classical approach presented in [7] in the limit of a step-like magnetic constitutive equation, by assuming

$$B=\begin{cases} B_s & \forall H>0 \\ -B_s & \forall H<0 \end{cases} \tag{7}$$

Following [7], at any given instant t, the magnetic induction is distributed in (0,h/2) as a square wave (mirrored in the other half of the slab) travelling with velocity $\frac{du}{dt}$:

$$B_z(x,t)=\begin{cases} B_s & x\in[0,u(t)[\\ 0 & x\in]u(t),h/2] \end{cases} \tag{8}$$

Applying Faraday's law

$$E_y=B_s\frac{du}{dt} \tag{9}$$

the electric constitutive equation:

$$J_y=\sigma E_y \tag{10}$$

and Ampère's law:

$$H_y(h/2,t)=\frac{NI}{\Delta z}=J_y(x,t)u(t) \tag{11}$$

where $NI(t)$ and Δz are the Ampere-turns and the height of the inducing solenoid, respectively, we get:

$$\frac{NI}{\Delta z}=\sigma B_s u\frac{du}{dt} \tag{12}$$

Therefore, taking $I(t)=I_M$ for t>0 and assuming that the applied field is able to saturate the slab, the position u(t) of the wave-front is given by:

$$u=\sqrt{\frac{2NI_M}{\sigma B_s\Delta z}t} \tag{13}$$

and the time needed to reach the flux saturation ($\Psi=\Psi_{max}$) is

$$t_{sat}=\frac{(h/2)^2}{2NI_M}\sigma B_s\Delta z \tag{14}$$

where Ψ is the flux linked with the slab:

$$\Psi = 2\int_0^{h/2} B_s\, b\, dx = 2B_s\, u\, b \tag{15}$$

The dynamic characteristic of the slab is finally given by:

$$I(t) = \frac{\sigma\Delta z}{8NB_s b^2}\frac{d}{dt}\left(\Psi^2\right), \qquad |\Psi| \le \Psi_{max} = B_s hb \tag{16}$$

4. Results of the analysis

A 3D numerical analysis of the field diffusion has been carried out using the integral and differential formulations briefly summarized in Section 3. In both cases, only one eighth of the slab has been discretized taking advantage of the symmetry of the problem. Of course, when using the integral formulation, only the ferromagnetic slab was discretized. In this case, the mesh is made of 12×7×8 not equally spaced hexahedral elements giving 1260 unknowns.

In the differential approach also the air region and the external solenoid have been discretized, so that the finite element mesh is made of 4928 elements, 5848 nodes giving 15144 unknowns. In this case the slab is discretized in 11×4×8 unevenly spaced hexahedral elements. To impose the boundary conditions at infinity, the mesh has been truncated at x = 9 mm, y = 9 mm, z = 11 mm.

The 3D numerical results have then been compared to the 1D analytical solutions.

4.1 The linear case

Figure 2 shows the average flux linked with the slab as a function of time for a step current, as computed with the differential formulation.

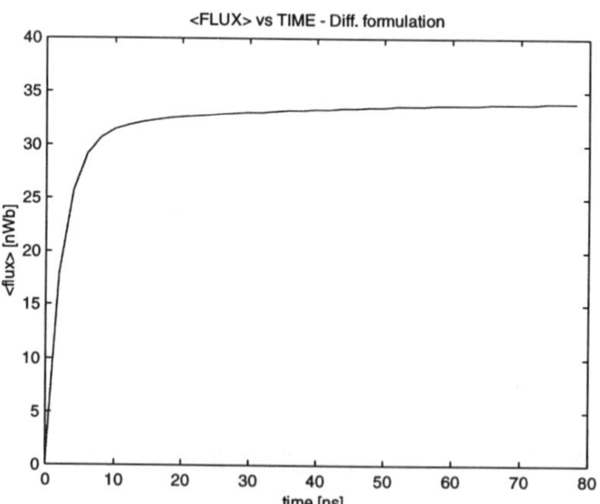

Figure 2. Linear case: 3D differential formulation. Time behavior of the average flux for a step input $NI=NI_M \cdot 1(t)$, $NI_M=5.64$ Ampereturns.

Figure 3 shows the corresponding results obtained for the flux density B_z at z=0 in the magnetic core of the sensor at various time instants.

Examining these results we can see that the diffusion process is characterised by two different time scales that clearly show up. While the flux reaches the final value after a few tens of ns, the flux density is at that time far away from its uniform distribution. Field and flux have very different time constants. The explanation is related to the magnetic nature of the slab and to its finite height. Taking into account what already discussed in Section 2, the flux linked with the core is mainly determined by the distribution of the normal component of the flux density at the iron-air interface at and around the extremities of the core.

We also computed the analytical solution imposing the instantaneous rise of the flux linked with the core. The behaviour of the flux density shown in Figures 3 and 4 in comparison with the numerical results obtained using the differential formulation are in an excellent agreement each with the other confirming both the validity of the assumptions made and the reliability of the numerical approach.

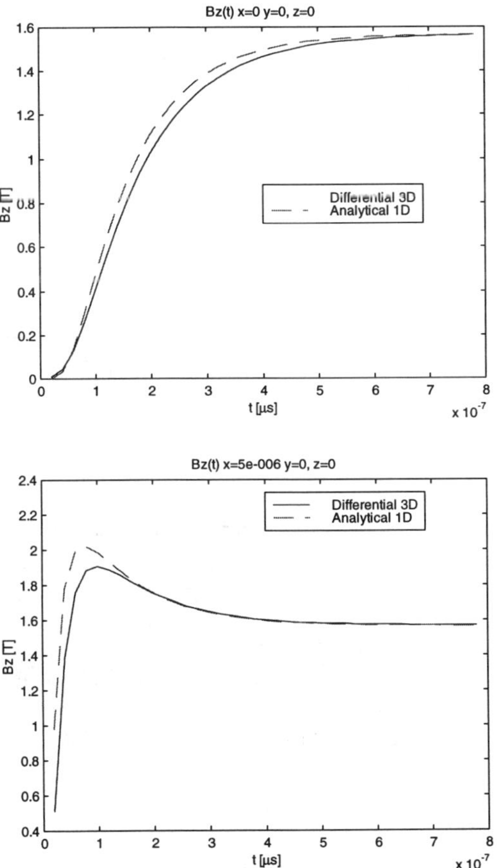

Figure 3. Linear case: 3D differential formulation and 1D analytical solution. Field at two different locations in the equatorial plane z=0 at for a step input input $NI=NI_M \cdot 1(t)$, NI_M=5.64 Ampereturns.

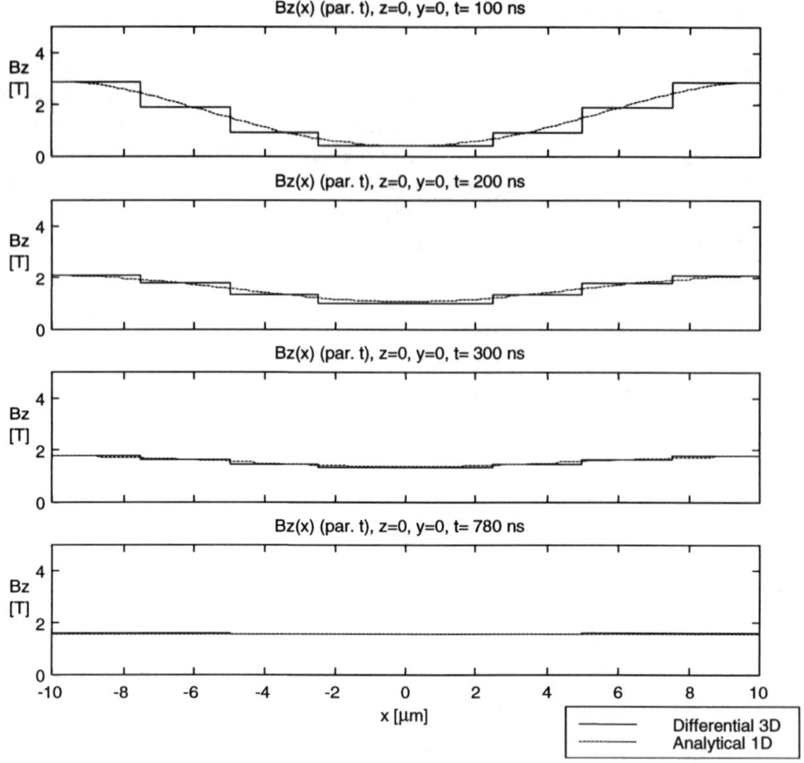

Figure 4. Linear analysis. Bz field at z=0 in the magnetic core of the sensor at various times. A step current of 5.64 Ampereturns flows in the exciting conductor.

4.2 *The non linear case*

A numerical analysis has also been carried out for the non linear case using a piecewise linear magnetic curve with μ_r=85000 for B < B_{sat}=.65 T and μ_r=1 otherwise, again for a NI step of 5.64 Ampereturns flowing in the 9 mm long exciting solenoid.

Figure 5 reports the time behavior of the flux inside the slab as obtained using the integral and the differential approach. In spite of the small number of finite elements along the thickness, both formulation provide a penetration time that is in excellent agreement with the approximate analytical value t_{sat}=5.2 ns predicted by (14).

A satisfactory agreement is also shown for the penetration of the wavefront (Figure 6), the spatial distribution (Figure 7) and the time behavior of the flux density (Figure 8).

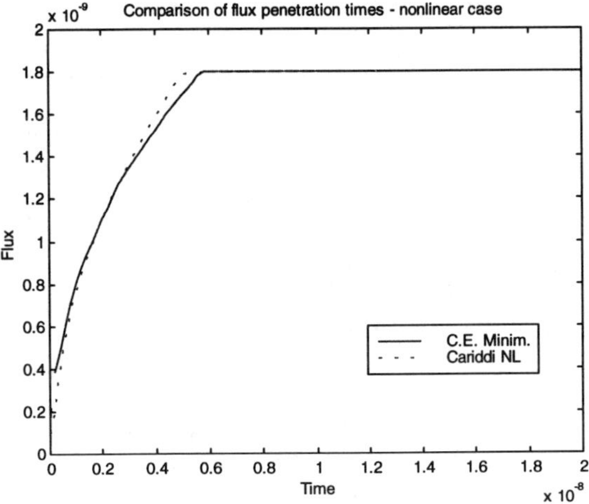

Figure 5. Nonlinear analysis. Average flux (Vs) vs time (s) for a step input $NI=NI_M \cdot 1(t)$, $NI_M=5.64$ Ampereturns: differential (continuous line) vs differential (dashed line) formulations.

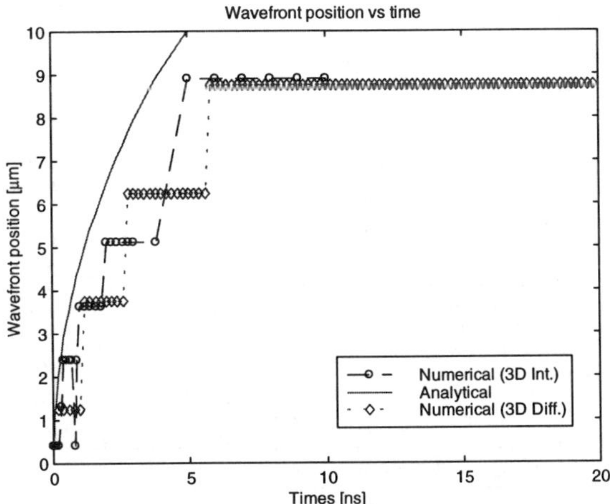

Figure 6. Nonlinear analysis. Wavefront penetration at z=0 for a step input $NI=NI_M \cdot 1(t)$, $NI_M=5.64$ Ampereturns: analytical 1D approximation (continuous line) compared to the numerical results of 3D differential (diamonds) and integral (circles) formulations.

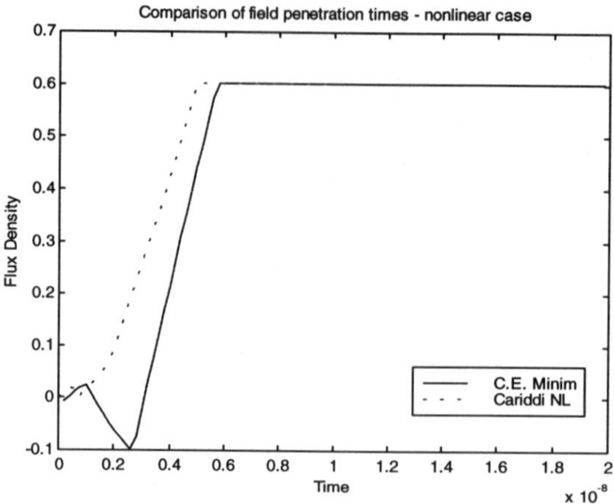

Figure 7. Nonlinear analysis. B_z field (T) vs time (s) at the center of the magnetic core for a step current of 5.64 Ampereturns flowing in the exciting conductor: differential (continuous line) vs differential (dashed line) formulations. The corresponding prediction of the 1D analytical model is simply a 0.65 T step at 5.2 ns.

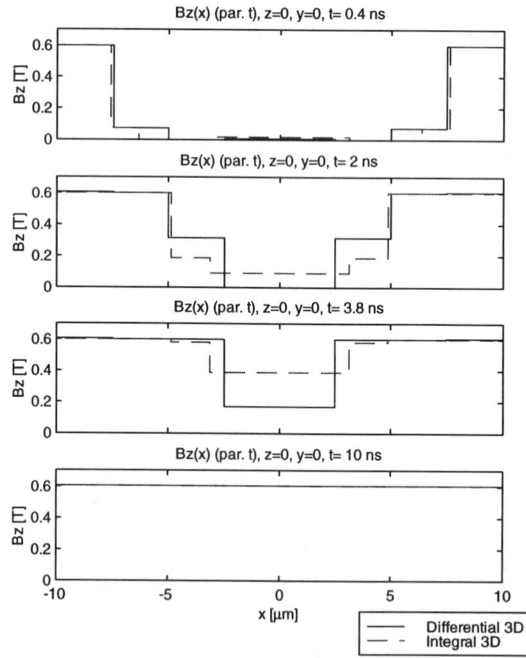

Figure 8. Nonlinear analysis. Bz field at z=0 in the magnetic core of the sensor at various times. A step current of 5.64 Ampereturns flows in the exciting conductor: differential (continuous line) vs differential (dashed line) formulations.

5. Conclusions

We have determined the penetration time of the magnetic flux into the ferromagnetic core of a *FLUXSET* probe.

In the nonlinear case it is about 5 ns, and almost coincides with the penetration time of the magnetic field.

In the linear case, corresponding to small exciting currents, the time scale for the field penetration is about 0.2 μs, but the magnetic flux rise is much faster.

The eddy current effects can then be neglected in the range of frequencies from DC to about 100 MHz. Therefore the magnetostatic model used in [4,8] is acceptable for the characterization of the probe at its terminals.

In any case, for higher frequencies or different cores, Eq. (16) can be used to improve the ingenious circuit schematization introduced in [4].

Finally, the results presented in this paper show that the 3D numerical formulations are able to provide reliable values of global quantities even with coarse meshes for a rather critical problem, characterized by the rapid motion of a sharp discontinuity surface.

Acknowledgements

This work was supported by the INCO-COPERNICUS Project PL 964037 of the European Commission and by the Italian MURST, with the contribution of Regione Calabria and the European Union.

References

[1] S. Daróczi, J. Szöllõsy, G. Vértesy, J. Pávó and K. Miya, "Electromagnetic NDT Material Testing By Magnetic Field Sensor", in *Studies in Applied Electromagnetics and Mechanics 8, Nondestructive Testing of Materials*, Editors: R.Collins, W.D. Dover, J.R.Bowler and K. Miya, IOS Press, Amsterdam (1995), pp.75-86.

[2] P. Ripka, "Review of Fluxgate sensors", *Sensors and Actuators*, A33 (1992), pp. 129-141

[3] R. Albanese, A. Formisano, R. Fresa and G. Rubinacci, "Assessment of Performance for Fluxset Sensor Core", in *Electromagnetic Nondestructive Evaluation (II)*, R. Albanese, G. Rubinacci, T. Takagi and S.S. Udpa (Eds.), IOS Press, Amsterdam (1998), pp. 181-189.

[4] D. Ioan, I. F. Hantila, M. Rebican and C. Costantin, "FLUXSET Sensor Analysis Based on Nonlinear Magnetic Wire Model of the Core", in *Electromagnetic Nondestructive Evaluation (II)*, R. Albanese, G. Rubinacci, T. Takagi and S.S. Udpa (Eds.), IOS Press, Amsterdam (1998), pp. 161-170.

[5] R. Albanese, F. I. Hantila, G. Rubinacci: "A Nonlinear Eddy Current Integral Formulation in Terms of a Two-component Current Density vector potential", *IEEE Transaction on Magnetics*, MAG-32, no.3, May 1996, pp. 784-787.

[6] R. Albanese and G. Rubinacci: "Finite Element Methods for the Solution of 3D Eddy Current Problems", *Advances in Imaging and Electron Physics*, Vol.102, pp. 1-86, 1998.

[7] P.D. Agarwal, "Eddy-Current Losses in Solid and Laminated Iron", AIEE Proc., May 1959, pp. 169-181.

[8] D. Ioan, A. Formisano, A. Gasparics, I. Munteanu, M. Rebican, "High frequency models for the NDT magnetic field sensors", presented at E'NDE '98, Chatou.

Electromagnetic Nondestructive Evaluation (III)
D. Lesselier and A. Razek (Eds.)
IOS Press, 1999

High Frequency Models for the NDT Magnetic Field Sensors

Daniel Ioan*, Alessandro Formisano†, Antal Gasparics‡, Irina Munteanu*, and Mihai Rebican*

1 Introduction

The usual magnetic field sensors encountered in Non–Destructive Testing (NDT) fall in two main categories: sensors based on induction [1]; sensors based on nonlinearity [2], [3]. The first category is suitable for use at high frequencies, over 100 kHz; at low frequency these sensors have low sensitivity and the measurements are affected by noise. The second category is suitable for low frequency, these sensors being able to measure magnetic fields down to DC. The attempt to use FLUXSET– and FLUXGATE–type sensors at high frequencies leads to parasitic effects which limit their frequency domain. However, for industrial application it is desirable to dispose of a sensor which has a frequency domain as extended as possible, in order to allow defects detection at various depths.

Since in the case of the inductive sensors the limitation is of principial nature, the only hope is to increase the maximum frequency at which nonlinear sensors can operate. In order to achieve this goal, it is necessary to understand and to quantitatively characterize the parasitic effects which appear in the functioning of these sensors at high frequencies.

The experimental analysis of nonlinear sensors showed the following types of parasitic field effects which appear at high frequencies and contribute to the depreciation of the sensors metrologic performances:

- eddy currents in the ribbon core [3] [4] [5];
- widening of the hysteresis loop [6];
- capacitive effects between the windings [7].

The paper presents for the first time a complex model for the FLUXGATE/FLUXSET–type sensors, which takes into account all these three effects. The model allows both the analysis of the sensor behavior at high frequencies and the evaluation of the frequency bandwidth for an imposed measurement error.

The FLUXSET sensor we analyze (figure 1) is built of two coils: the driving coil (1) with n_1 turns and the pick-up coil (2) with n_2 turns, winded around a ferromagnetic ribbon core (3) of length l_c. The first coil is fed with a periodical, triangular–shaped AC driving current. The sensor is introduced in an external magnetic field to be measured. For simplicity, at large distance from the sensor, the external field is assumed to be uniform and axially oriented.

The main goal is to compute in an efficient manner the pick-up voltage vs time, when the geometric dimensions, the material characteristics and the time variation of the field sources (driving current and external field) are known.

*"Politehnica" University of Bucharest, Numerical Methods Lab., Spl. Independentei 313, 77206 Romania, Phone/Fax (+401)411 11 90, E-mail:daniel@lmn.pub.ro, http://www2.lmn.pub.ro

†Associazione EURATOM/ENEA/CREATE, Dip. di Ing. Industriale, Università degli Studi di Cassino, Via Di Biasio 43, I-03043 Cassino (FR), Italy

‡Research Institute for Materials Science, H-1525 Budapest, Egry J. u. 19, Hungary

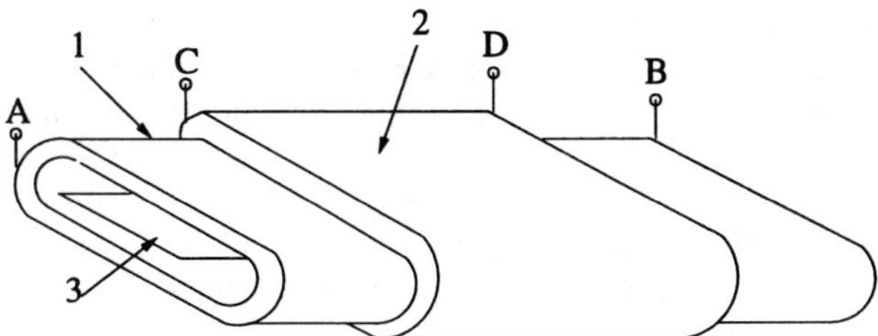

Figure 1: FLUXSET sensor configuration

Due to the extremely large ratio between the geometric dimensions (the length of the core is 500 times larger than its thickness, the insulation between coils is 1000 times smaller than the coil's length), and also due to complicated effects (like hysteresis and displacement currents), the "brute force" approach based on the finite element method (FEM) or other similar discretization does not function in this case.

The goal of our approach is to generate an approximate model of the sensor, described by a system of ordinary differential equations, able to be represented as an electrical circuit with lumped parameters in view of its SPICE simulation. The circuit parameter extraction procedure is based on the numerical solution of the electromagnetic field equations.

The idea of the paper is to analyze each effect independently from others, to extract characteristic parameters and to integrate all subcircuits in a global one, able to model the interactions between effects.

The errors introduced by this approach are finally checked by comparisons with the experimental results.

2 Magnetic Field Computation and Eddy Currents Analysis

In order to compute the magnetic field distribution inside the sensor, taking into account the core's nonlinearity, different methods were used: FEM [8] [9], hybrid numerical–analytical method [4], and integral equations method [3]. Comparing the necessary computing resources (CPU time and memory), accuracy and reliability of the code, we can conclude that the integral equation method seems to be the most suitable for this problem.

Satisfactory numerical results were obtained by discretizing the ribbon core along the axis in a 1D uniform grid of n rectangles, each having as unknown the axial magnetization M_i. The nonlinear system of equations [4]:

$$M_i + \chi_i(M_i) \sum_{j=1}^{n} a_{ij} M_j = b_i, \qquad i = 1, n \tag{1}$$

was successfully solved with a reasonable computing effort, by using an adapted version of the Katzenelson method.

Based on the numerical solution, the pick-up coil's magnetic flux was evaluated as a function of the driving current i and of the external field H, and then approximated with an acceptable error by the following expression $\varphi(i, H) = n_2 g(n_1 i + \alpha H l_c)$, where $g(x) = M_0 x + M_1(|x + x_0| - |x - x_0|)$ and α, M_0, M_1, x_0 are constants which characterize the sensor. For a better accuracy, a piecewise linear approximation of $g(x)$ can be extracted from the solution of the magnetostatic field problem [10].

From a magnetic–circuit point of view, the function g represents the constitutive relation between the one–turn magnetic flux φ/n_2 and the total driving current $n_1 i$ of a nonlinear magnetic reluctance characterizing the ribbon and the surrounding air.

The previous analysis is valid for the static case only. In order to model the eddy currents in the core, the magnetic field diffusion in the ribbon should be analyzed.

The analysis of the eddy currents in the ribbon–shaped core starts from the simplest plate model (figure 2), in which we assume $a \ll l \ll L$ and the core unsaturated with conductivity σ and permeability μ. Under these hypotheses, the magnetic field in the plate is axially oriented: $\mathbf{H} = \mathbf{k}H(x,t)$ and the operational magnetic impedance of the plate is:

$$Z(s) = \frac{\gamma l}{2\mu L}\mathrm{cth}(\gamma a), \tag{2}$$

where $\gamma^2 = s\mu\sigma$, $Z(s) = U_m(s)/\Phi(s)$ with $U_m = \int_0^L H\,\mathrm{dz}$, $\Phi = \int_{-a}^{a}\int_{-l/2}^{l/2} B\,\mathrm{dx}\,\mathrm{dy}$.

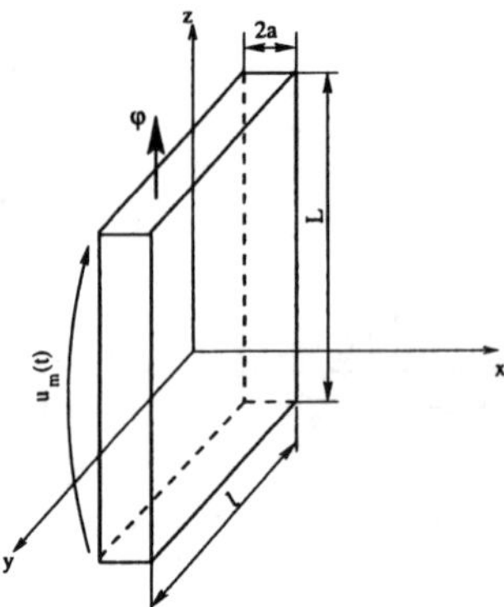

Figure 2: Simple plate model

The equivalent magnetic circuit having impedance (2) contains an infinity of R, L lumped elements. Starting from (2), three canonic schemes can be synthesized: series, parallel and chain [11]. The last one (figure 3) has the advantage that, by truncation, a Padé approximation is obtained and therefore the truncation errors for given order are less than those of other schemes. The elements in the chain scheme have the parameters:

$$R_k = R_0(4k-3), \qquad L_k = L_0/[2(4k-1)], \tag{3}$$

where

$$R_0 = L/(2\mu a l); \qquad L_0 = \sigma a L/l \tag{4}$$

are the dc magnetic reluctance expressed in H^{-1}, and the "dissipative inductance" expressed in $\Omega^{-1} = \mathrm{S}$, respectively; the last one models the dumping effect of eddy currents in the plate. The time constants of this circuit are less than $\tau_k \leq \tau/6$, with $\tau = L_0/r_0 = 2\mu\sigma a^2$, which in our case ($\sigma = 10^5$ S/m, $\mu_r \leq 85,000$) corresponds to $\tau_k \leq 2.135$ μs.

The truncation is not necessarily the best method to obtain a finite approximation of (2). Another method to reduce the order of the approximate system is to use the "transfinite" schemes, which at given order minimize a certain error, e.g. in Hankel norm [5].

The equivalent magnetic circuit technique can be used also to model the nonlinear plate when the skin effect is weak. Using the incremental small variation technique, the scheme in figure 3 remains valid, with parameters given by (3) in which $R_0 = L/(2\mu_d a l)$,

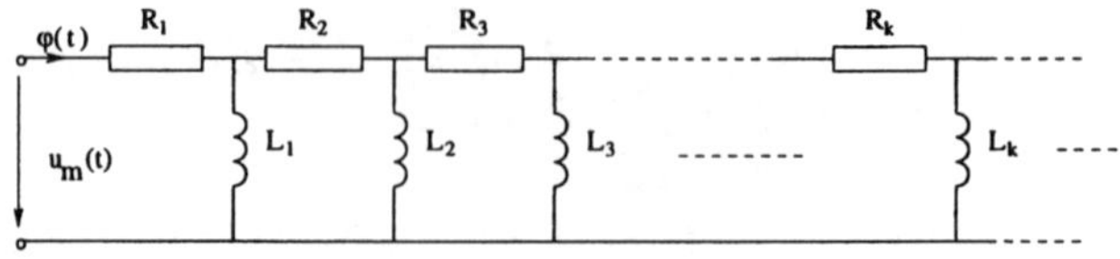

Figure 3: Chain canonic scheme

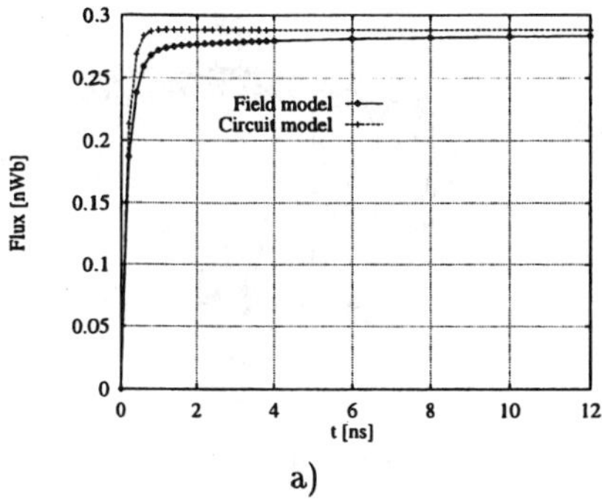

a)

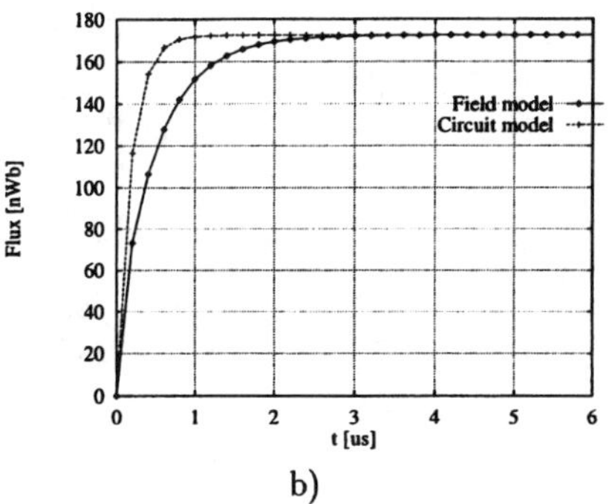

b)

Figure 4: Magnetic flux vs time: a) open magnetic circuit ; b) Closed magnetic circuit

and $\mu_d = \mathrm{d}B/\mathrm{d}H$ is the dynamic permeability of the core. For large variations, the circuit elements in figure 3 are non–linear with parameters (3), in which

$$R_0 = L/(2\mu_s al), \qquad L_0 = \sigma a L\mu_d/(l\mu_s), \tag{5}$$

where $\mu_s = B/H$ is the static permeability when the magnetic field H_0 is $u_m(t)/L$ and $\mu_d = \mathrm{d}B/\mathrm{d}H$ is the dynamic permeability corresponding to the same field H_0 [12]. If the plate is saturated, then $\mu_d = \mu_0$ and the time constants are thousands of times lower, $\tau_k \leq 25$ ps.

The hypotheses used to draw these conclusions are not valid in reality, mainly because the magnetic circuit is an open one and the air reluctance plays an important role in the magnetic field distribution. In order to explain this role, we consider the simple test case of a toroidal coil having as a core the FLUXSET ribbon. The 2D linear magnetodynamic problem, was solved in two cases: a) with core in the half of the torus (open magnetic circuit) and b) with circular core (close magnetic circuit). Considering zero initial conditions and a current step excitation $J_s(t)$, both transient problems were solved by FEM with the MEGA software package [13].

The rise of the magnetic flux is presented in figure 4. Despite the fact that both cases have the same "internal time constant" rising time of the flux is very different in the two cases: $\tau_a \approx 0.148\,\mathrm{ns}$, $\tau_b \approx 0.177\,\mu\mathrm{s}$ the value being first valid FLUXSET sensor too [9]. The explanation is due to the magnetic reluctance of the "air gap".

Indeed, in the very first moment, the internal field in the ribbon is zero and it has the tendency to remain zero, due to the internal diffusion time constant, therefore the external magnetic field "avoids" the ribbon (figure 5a). Assuming a magnetic field distribution in the air similar to that without core, the tangential field strength generates an unexpectedly large flux density on the internal surface of the core, due to the large value of the permeability. This large flux density tends to accelerate the transient process of internal field diffusion (figure 5b). During the transient process, the boundary flux density decreases; meanwhile, the magnetic field penetrates in the ribbon (figure 5c).

In the nonlinear case the magnetic flux is limited by saturation and the dynamic relative permeability is near the air permeability. Therefore, the diffusion process is even more accelerated, and, from the beginning, the magnetic flux along the ribbon has the tendency to be equal to the material saturation flux density B_S multiplied by the ribbon core's cross-section area.

These effects explain why the sensor's time constant has the order of nanoseconds (the "external time constant"), and not of microseconds (the internal field time constant).

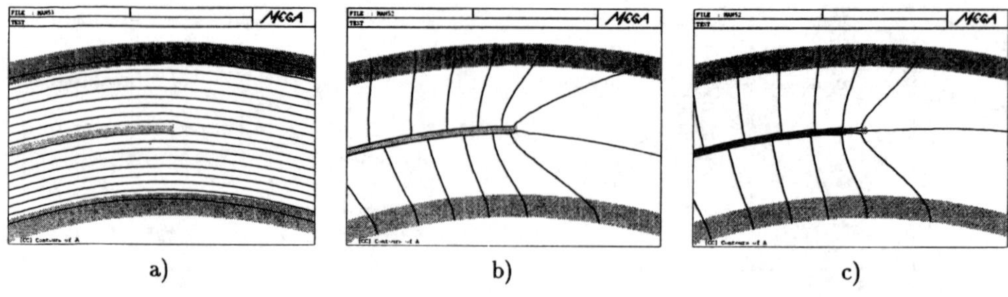

Figure 5: Magnetic field penetration process: a) $t = 0.2$ ps; b) $t = 0.2$ ns; c) $t = 2\mu$s

In order to model this complex phenomenon in a simplified manner, with ordinary differential equations able to be represented as electric circuits, we propose the following technique. The ribbon core and the air of the external domain can be modeled as a global nonlinear reluctance, whose flux–current constitutive relation can easily be extracted from the magnetostatic field solution [10]. An RL circuit which represents in an approximate manner the eddy currents in the ribbon core will be connected in series with this reluctance. In order to have the same static behavior, this RL circuit must have zero dc impedance. Among the different possible equivalent schemes, the chain topology seems to be the most suitable, because it satisfies this condition.

If only one cell is considered, the circuit is a RL series one. The time constant of this circuit is much smaller than the internal time constant, because the global reluctance R also includes the reluctance of the external domain.

The reduced magnetic circuit has the scheme represented in figure 15, with $L_1 = L_0/6 = 20/6$ S, $L_2 = L_0/14$, $L_3 = L_0/22$ and $R_2 = 5R_0 = 5 \times 9.366$ MH^{-1}, $R_3 = 9R_0$. For the frequency band in which FLUXSET sensors usually operate, 1–3 inductances are enough to model the eddy current effects with acceptable accuracy [5]. This is proven by the comparison between transient magnetic flux computed with FEM and those computed with first order circuit approximation (figure 4).

A different perspective on the same argument has been shown in another paper [18].

3 Hysteresis modeling

Starting from the experimental data regarding the driving current $i(t)$ and the pick-up voltage $v(t)$, oscillograms of the flux-current relation for FLUXSET sensor were measured. These exhibit a hysteresis loop which widens as the frequency increases [6].

In order to characterize the magnetic material properties, the analysis will start with the extraction of an one-to-one $B - H$ relation, by solving the following inverse problem.

Let us consider a homogeneous nonlinear ferromagnetic body D_m (with $\mathbf{B}\|\mathbf{H}$ and with a monotonic and bijective $B - H$ relation) surrounded by air, and a non-magnetic coil C with the magnetic flux φ generated by the current i.

The magnetostatic field equations are:

$$\begin{aligned} \operatorname{div}\mathbf{B} = 0 \quad \text{in } \mathbb{R}^3; \qquad \operatorname{rot}\mathbf{H} = \mathbf{J} \quad \text{with } J = 0 \text{ in } \mathbb{R}^3 \backslash C; \\ \mathbf{B} = f(H)\cdot \mathbf{H}/H \quad \text{in } D_m; \qquad \mathbf{B} = \mu_0 \mathbf{H} \quad \text{in } \mathbb{R}^3 \backslash D_m. \end{aligned} \tag{6}$$

Considering the geometrical configuration and the function $g : \mathbb{R} \rightarrow \mathbb{R}$ with $\varphi = g(i)$ known, the problem is to find the function $f : \mathbb{R} \rightarrow \mathbb{R}$, which is the magnetic characteristic $B = f(H)$ of the ferromagnetic material in the domain D_m, assuming that f is not dependent on the point $P \in D_m$.

The following algorithm solves the inverse problem an extracts the $B-H$ characteristic:

- **Generate a piece-wise linear approximation** of the $\varphi - i$ characteristic with an imposed error, by the selection of appropriate m points (φ_k, i_k) from the experimental data.
- Consider a linear magnetostatic field problem with $B = \mu H$ in D_m, and find $\mu = \mu_1$ such that the magnetic flux due to the current i_1 is φ_1, **by solving the nonlinear equation:** $\varphi(i_1) = L(\mu) \cdot i_1 = \varphi_1$.
- Compute the maximum magnetic field in D_m (H_{max}, $B_{max} = \mu_1 H_{max}$) and set **the first breakpoint** on the extracted $B - H$ characteristic: $H_1 = H_{max}$, $B_1 = B_{max}$, which start from the origin: $H_0 = 0$, $B_0 = 0$.
- **Solve the magnetostatic field** problem with a non-linear $B - H$ characteristic (figure 6). **Find** μ_2 such that the magnetic flux due to the current i_2 is: $\varphi(i_2) = \varphi_2$.
- The maximum magnetic field H_{max}, B_{max} due to i_2 represents **the second breakpoint** of the $B - H$ characteristic.
- **Repeat iteratively the last two steps** until the breakpoint number m of the $\varphi - i$ characteristic is reached.

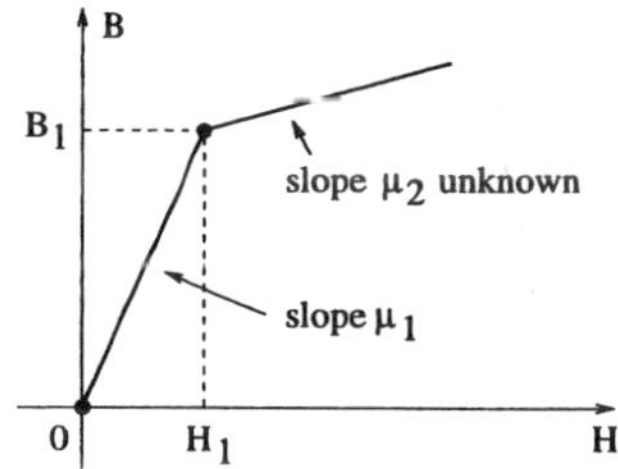

Figure 6: $B - H$ characteristic after the first iteration

The algorithm exploits the monotonicity property of the $B-H$ and $\varphi-i$ relations. It is based on the piece-wise linear approximation of the $\varphi - i$ characteristic and the extracted $B - H$ characteristic with the same number of breakpoints. For reducing the computational effort, the experimental data are filtered and a minimal number of breakpoints is selected, in order to obtain an imposed approximation error. In this manner the computing time is correlated with the error. The detailed form of this algorithm and the analysis of its numerical stability and complexity are presented in [14].

The algorithm was successfully applied to extract the $B - H$ characteristic of the ribbon-shaped core of the *FLUXSET* magnetic field sensor. Starting from 418 experimental data points, this algorithm selected the number of breakpoints $m = 5$ at the imposed error $\varepsilon = 1\%$, and $m = 36$ at $\varepsilon = 0.1\%$.

An important decision to be taken in order to apply the proposed algorithm is concerning the numerical method used to compute the magnetic field. In the case of the FLUXSET Sensor, the nonlinear magnetostatic field problem was solved with the integral equations method [10].

The ribbon was discretized in 15 rectangular elements, the coil was discretized in the axial direction in 10 slides and each oval slide in 24 segments. Starting from the $\varphi - i$ dependence with $m = 10$ breakpoints represented in figure 7, the $B - H$ characteristic represented in figure 8 was extracted.

In the particular case of the 1D FLUXSET model (1) the **B**, **H** and **M** in the magnetic ribbon core are oriented along the same axial direction (Ox). Let us consider the dependence $\varphi = g'(i)$, valid for a decreasing current i, after a strong positive saturation. The **B** − **H** relation has the form:

$$B_x = I_p + f(H_x). \tag{7}$$

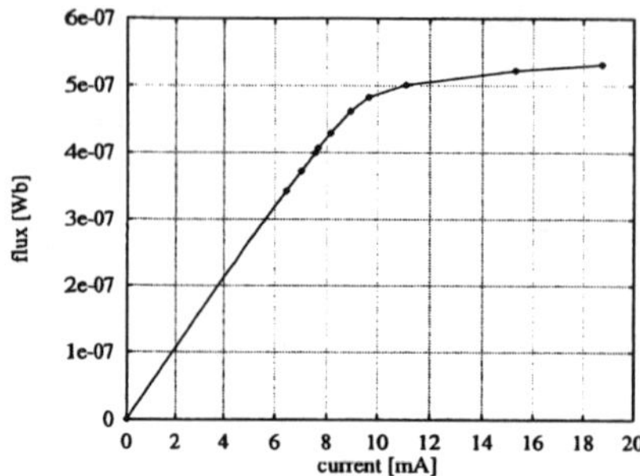

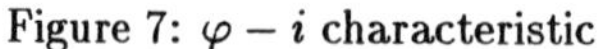
Figure 7: $\varphi - i$ characteristic

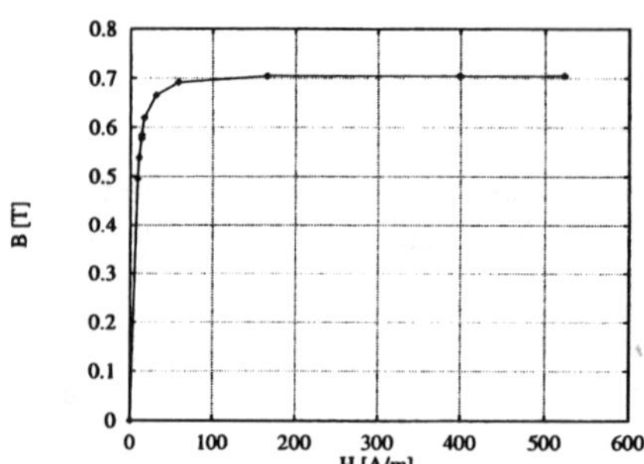

Figure 8: Extracted $B - H$ characteristic

In order to extract the fundamental hysteresis loop of the ribbon, it is enough to determine the constant I_p and the real function f in (7). The algorithm previously presented can be used to extract the function f, starting from the following $\varphi - i$ relation: $\varphi = g(i) = g'(i) - g'(0)$. The value of I_p can be computed solving the magnetic field problem with the ribbon uniformly magnetized.

The extracted $B - H$ relation has, at 100kHz, the remanent flux density $B_r = 0.2245\,\mathrm{T}$ and the coercive field $H_c = 8.0966\,\mathrm{A/m}$. At the frequency 10kHz the values are $B_r = 0.048\,\mathrm{T}$, $H_c = 0.425\,\mathrm{A/m}$.

One of the simplest hysteresis models which allows a lumped equivalent circuit is presented in [15]. It is a dynamic model, completely specified by two monotonically increasing functions: f-the "restoring function" which describes the energy storage and g-the "dissipation function" which describes the energy dissipation, and it is based on:

$$\frac{\mathrm{d}\varphi}{\mathrm{dt}} = g[i(t) - f(\varphi)] \tag{8}$$

or equivalently:

$$i(t) = g^{-1}[v(t)] + f[\varphi(t)] \tag{9}$$

where $v(t) = \mathrm{d}\varphi/\mathrm{d}t$ is the induced voltage. Equation (9) suggests the lumped-circuit model for an iron-core inductor presented in figure 9a. The $v - i_R$ relationship of the resistor is given by $v = g(i_R)$ and the $i_L - \varphi$ relationship of the inductor is given by $i_L = f(\varphi)$.

The relation (9) could have another interpretation: from a magnetic circuit point of view (figure 9b), the lumped magnetic circuit model consists of two elements connected in series. One is a nonlinear magnetic reluctance with $u_1 - \varphi_t$ constitutive relation given by $u_1 = n\, f(n\varphi_t)$ and the other is a nonlinear "dissipative inductor" with the $u_2 - \varphi_t$ constitutive relation given by:

$$\frac{\mathrm{d}\varphi_t}{\mathrm{d}t} = \frac{1}{n} g\left(\frac{u_2}{n}\right), \tag{10}$$

where n is the number of turns, $\varphi_t = \varphi/n$ is the one-turn magnetic flux and $u = u_1 + u_2 = n\,i$ is the total current. The relation (10) can be approximated by a linear one, $u_2 = L_h\,\mathrm{d}\varphi_t/\mathrm{dt}$, where, in the FLUXSET case, $L_h = 70.2\,\mathrm{S}$. This value is 21 times bigger than the inductance representing eddy currents in the ribbon, showing the importance of the hysteresis effect in the FLUXSET behavior.

The simulated hysteresis loop and the experimental results in $\varphi - i$ plane at 100kHz are presented in figure 10. The comparison validates the technique we propose.

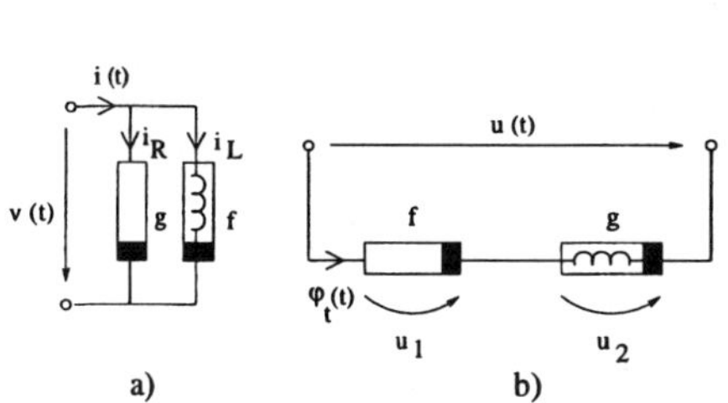

Figure 9: Lumped-circuit models

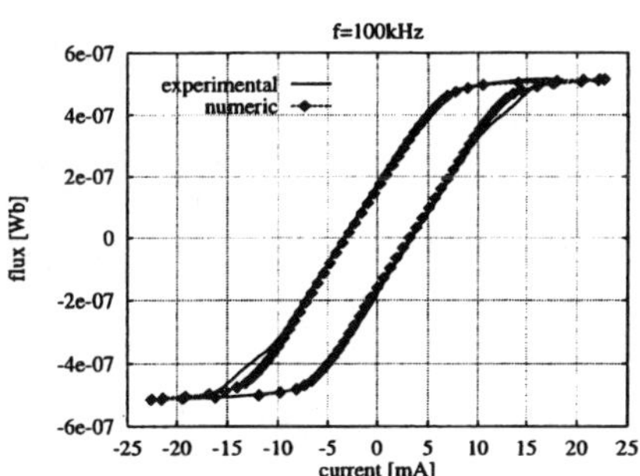

Figure 10: The hysteresis loop at 100kHz

4 Capacitive Effects

Even if the capacitance between the two windings (driving and pick-up) is relatively small, of the order of tens of pF, at high frequencies the capacitive effect becomes important, because it combines with the inductive effect, generating resonances which explain the presence of the spikes in figure 17. In order to model these effects, capacitors need to be added to the sensor circuit model [7].

The capacitance between the driving coil and the pick-up coil depends on the winding technology, therefore it is difficult to be accurately computed. However, an approximate value, suitable for our study, can easily be obtained by solving the electrostatic field problems represented in figure 11. They were solved by FEM with the MEGA.

The total capacitance between the coils is:

$$C = 2\frac{\varepsilon_0 l_t l_1}{d}\Lambda, \tag{11}$$

where l_t is the length of one wire turn, l_1 is the pick-up coil length, d is the diameter of the isolated wire, and Λ is the unit–less 2D geometric permeance of the configurations represented in figure 11. Considering $\varepsilon_r = 4$, the capacitance values corresponding to the two winding technologies are: $C_a = 51.91$ pF, $C_b = 25.03$ pF, while the experimental value is $C_e = 40.2$ pF [6].

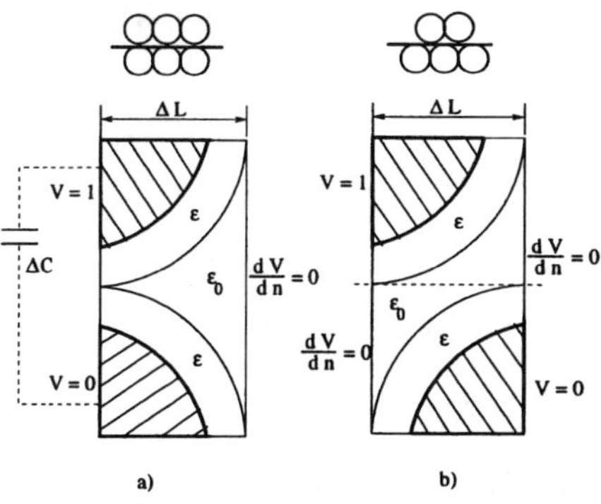

Figure 11: Electrostatic simplified models

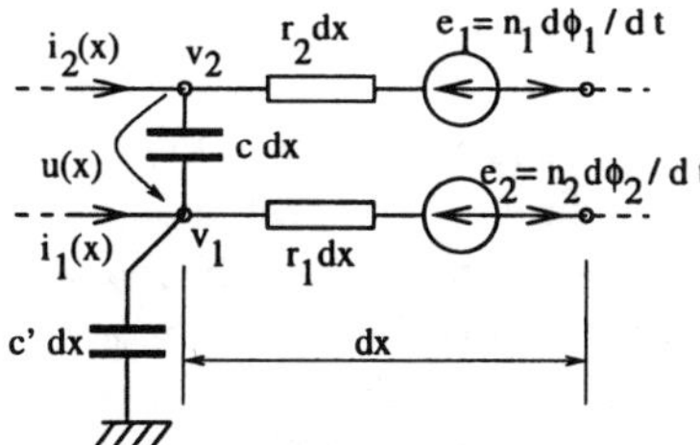

Figure 12: Elementary segment

The capacitive effects in the insulation are mixed with resistive and inductive effects in the windings. The simplest way to model these correlated effects is to use the transmission line approach [7]. Considering an elementary segment of the two coils (figure 12), the equations satisfied in the Laplace domain by the voltages $V_1(x,s)$, $V_2(x,s)$ and currents $I_1(x,s)$, $I_2(x,s)$ along the two coils, $0 \le x \le l$, are:

$$\begin{cases} -\dfrac{dV_1}{dx} = z_1(s)I_1 + n_1 s\varphi_1 \\ -\dfrac{dV_2}{dx} = z_2(s)I_2 + n_2 s\varphi_2 \\ -\dfrac{dI_1}{dx} = sc(V_1 - V_2) + sc'V_1 \\ -\dfrac{dI_2}{dx} = sc(V_2 - V_1) \end{cases} \tag{12}$$

where z_1, z_2 are the wire internal impedances per unit length (equal to the per-unit-length Ohmic resistance, $z_1(s) = r_1$, $z_2(s) = r_2$, if the skin effect inside the wire is neglected), n_1, n_2 are the numbers of turns per coil unit length, φ_1, φ_2 are the one–turn fluxes of each coil, $c = C/l$ is the capacitance between the coils and c' is the capacitance between the driving coil and the ground.

The solution of (12) with current excitation at the input port and short-circuit of the output terminals yields (figure 13a), the input admittance matrix of the transmission line [7], a symmetric 2×2 matrix:

$$\mathbf{Y}_t = \mathbf{Z}^{-1}\left[\gamma l \ c\ \mathrm{th}\gamma l + \mathbf{M}(\mathbf{Z}+\mathbf{M})^{-1}\right]/l, \tag{13}$$

where $\gamma = \sqrt{\mathbf{ZY}}$, $\mathbf{M} = \mathbf{N}s\,l\,Y_m(s)$, $\mathbf{N} = \begin{bmatrix} n_1^2 & n_1n_2 \\ n_1n_2 & n_2^2 \end{bmatrix}$, $\mathbf{Z} = diag(\ z_1(s), z_2(s)\ )$, $\mathbf{Y} = sc\begin{bmatrix} 1+c'/c & -1 \\ -1 & 1 \end{bmatrix}$, and $Y_m(s)$ is the operational magnetic admittance of the core ($Y_m(s) = 1/R_m$, the reciprocal value of the magnetic reluctance, if the eddy currents in the ribbon are neglected). $\mathbf{Y}_t$ can be separated in two terms:

$$\mathbf{Y}_t = \mathbf{Y}_0 + \mathbf{Y}_p, \tag{14}$$

in which

$$\mathbf{Y}_0 = \mathbf{Z}^{-1}\left(\mathbf{I} + \mathbf{M}(\mathbf{Z}+\mathbf{M})^{-1}\right)/l \tag{15}$$

represents the particular value of $\mathbf{Y}_t$ in the case $\gamma = 0$ (corresponding to neglecting the transversal capacitive effects), while

$$\mathbf{Y}_p(s) = \mathbf{Z}^{-1}\left(\gamma l\ \mathrm{cth}\gamma l - \mathbf{I}\right)/l, \tag{16}$$

represents the admittance of a tri–polar element (figure 13 b) which **concentrates at the beginning of the line** the parasitic capacitive effects, in reality distributed along the line.

Since c'/c is neglectable, the tri-pole characterized by $\mathbf{Y}_p$ can be reduced to the two–terminal infinite RC circuit represented in figure 14.

In order to be effective in modeling, this circuit shall be reduced to an approximate finite circuit, with the same topology as the initial one (figure 14), but with parameters $\widetilde{C}_k$, $\widetilde{R}_k$ and a finite number q of cells.

The reduction algorithms that gave the most accurate results were the Lanczos algorithm [16], and the technique based on compensated truncation (CT) of balanced realizations [17]. The values of the parameters $\widetilde{C}_k$, $\widetilde{R}_k$ in the two reduced schemes and the corresponding approximation errors are given in [7].

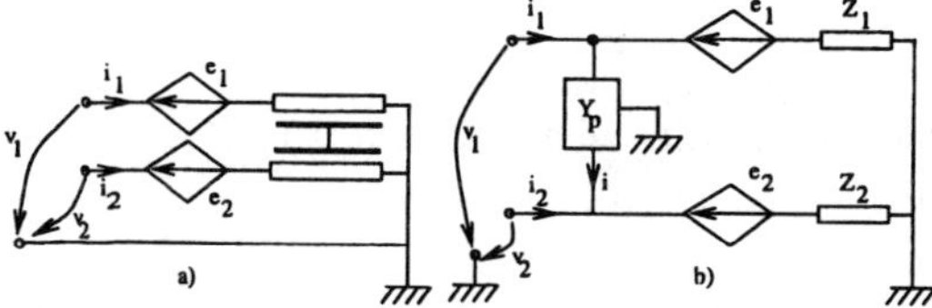

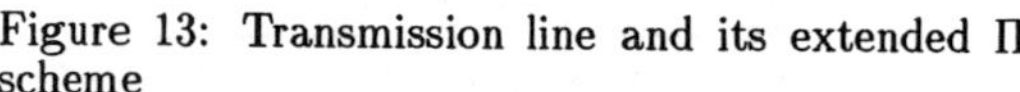

Figure 13: Transmission line and its extended Π scheme

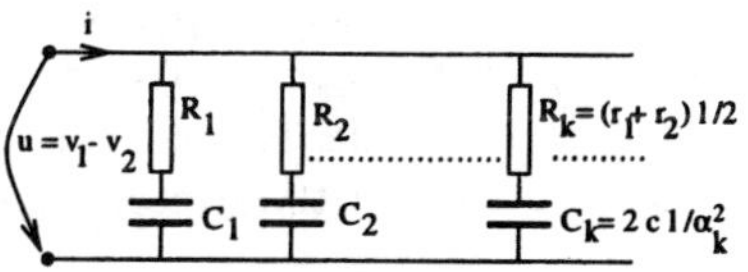

Figure 14: Equivalent parasitic circuit

5 The Nonlinear SPICE Model

The approximate nonlinear circuit of the entire sensor (figure 15) contains three subcircuits: one for the driving coil, one for the pick-up coil and the third one for the magnetic circuit.

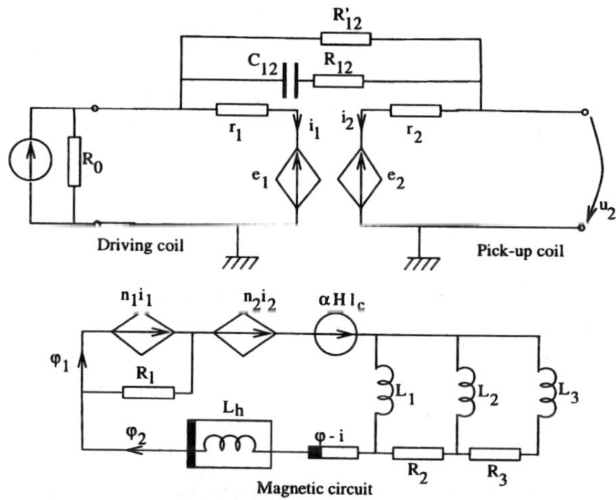

Figure 15: The SPICE model of the FLUXSET sensor

The three subcircuits are interconnected by means of controlled sources. The **driving coil** circuit contains an induced voltage source $e_1 = n_1 \mathrm{d}\varphi_1/\mathrm{d}t$, controlled by the derivative of the magnetic flux φ_1 of the magnetic circuit.

The **pick-up coil** subcircuit contains the induced voltage source $e_2 = n_2 \mathrm{d}\varphi_2/\mathrm{d}t$, controlled by the derivative of the flux φ_2. The magnetic circuit is driven by the linear controlled "voltage" sources with parameters $n_1 i_1$ and $n_2 i_2$, equal to the total currents in the two coils.

Besides the ohmic resistances r_1 and r_2 of the coils and the internal resistance R_0 of the driving source, the corresponding circuits are interconnected by the reduced circuit (C_{12}, R_{12}), which models the capacitive effects.

The **magnetic circuit** contains a nonlinear resistance, actually the magnetic reluctance with the characteristics $\varphi - i$ extracted from the magnetic field numerical solution.

A linear reluctance R_l represents the leakage flux path. Both eddy currents and hysteresis effects in the ribbon are modeled by some "dumping inductances" (L_h, L_1, L_2 and L_3) [5] and the external field H by a "voltage" source $\alpha H l_c$ [10], both placed in the magnetic circuit.

The equivalent circuit represented in figure 15 was simulated using the SPICE package at different frequencies. The experimental and simulation results for 10 kHz (without hysteresis) are presented in figure 16. Figure 17 presents the results obtained at 100 kHz, taking into account the hysteresis effect.

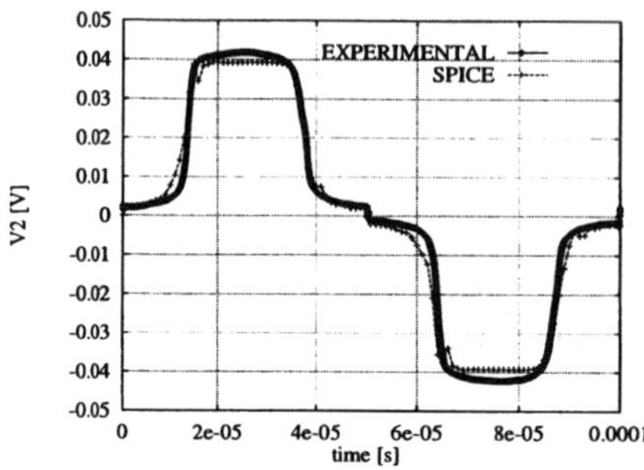

Figure 16: Pick-up coil voltage vs. time at 10 kHz

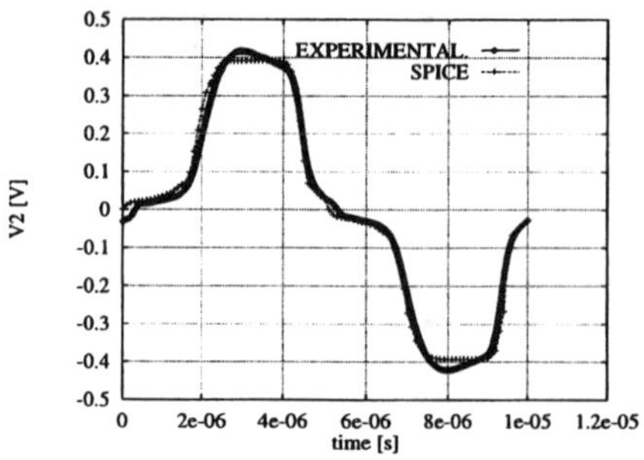

Figure 17: Pick-up coil voltage vs. time at 100 kHz

6 Conclusions

The following effects of an NDT magnetic field sensor were modeled: internal impedance of the driving source, capacitive effects between coils, ohmic resistance of coils, voltage induced in coils, magnetic flux leakage between coils, nonlinear magnetic characteristic of the core including hysteresis, 3D non–uniformity of the magnetic field and eddy currents in the core. The paper presented an effective parameter extraction procedure for each effect mentioned above.

The main original contribution of the paper consists in the method to model simultaneously eddy currents, hysteresis and parasitic capacitive effects which become important at high frequency in all electromagnetic devices. A SPICE nonlinear equivalent circuit of the FLUXSET sensor incorporating all these effects is proposed. The lumped circuit parameters are extracted from the field equations solution. Eddy currents are modeled by an RL chain magnetic scheme and hysteresis, which in this case prevails, is modeled by an electric inductance computed with the Chua–Stromsmoe method. For the capacitive effects, instead of the classical Π approximation of a distributed RC line, based on two transverse capacitances $C/2$ connected at the two terminals, a more sophisticated approximation is proposed in the paper.

Using SPICE simulation of the extracted circuit, good agreement with the experimental data was obtained. The proposed method is able to determine the frequency bandwidth of a newly designed FLUXSET sensor.

Acknowledgments

This work was supported in part by the European Commission, in the frame of the INCO-COPERNICUS MANODET project (Contract no. ERBIC15CT969703).

The authors thank all the partners of the MANODET–project team for their support and interesting comments, David Rodger for the donation of the MEGA software licenses, and Raffaele Fresa for his help in performing the analysis of the field penetration.

References

[1] Zhenmao Chen, Kenzo Miya, and Masaaki Kurokawa. Reconstruction of crack shapes using a newly developed ECT probe. In *Electromagnetic Nondestructive Evaluation*, volume II, pages 225–232. R. Albanese et al. (Eds), IOS Press, 1998.

[2] A. Gasparics, Cs. S. Daróczi, G. Vértesy, and J. Pávó. Improvement of ECT probes based on FLUXSET type magnetic field sensor. In *Electromagnetic Nondestructive Evaluation*, volume II, pages 146–151. R. Albanese et al. (Eds), IOS Press, 1998.

[3] R. Albanese, A. Formisano, G. Rubinacci, and R. Fresa. Assessment of performance for FLUXSET sensor core. In *Electromagnetic Nondestructive Evaluation*, volume II, pages 180–187. R. Albanese et al. (Eds), IOS Press, 1998.

[4] Daniel Ioan, Ioan Florea Hănţilă, Mihai Rebican, and Cristian Constantin. FLUXSET sensor analysis based on nonlinear magnetic wire model of the core. In *Electromagnetic Nondestructive Evaluation*, volume II, pages 160–169. R. Albanese et al. (Eds), IOS Press, 1998.

[5] Daniel Ioan, Irina Munteanu, and Cristian-George Constantin. The best approximation of the field effects in electric circuits coupled problems. *IEEE Trans. Magn.*, 34(5), September 1998. in press.

[6] ECT – NDT. WWW page `http://Alag3.atki. kfki. hu/ect.`

[7] Daniel Ioan, Irina Munteanu, and Corneliu Popeea. Models for the capacitive effects in a magnetic field sensor. In *Proc. of 8th IGTE Symposium*, 1998.

[8] Kurt Preis, Oszkár Biro, Kurt R. Richter, József Pávó, Antal Gasparics, and Igor Ticar. Numerical simulation and design of a Fluxset sensor by finite element method. In *Conf. Record of the 11th Conf. on the Comp. of Electromagn. Fields COMPUMAG/RIO*, pages 99–100, Rio de Janeiro, Brazil, November 1997.

[9] Daniel Ioan, Mihai Rebican, Gabriela Ciuprina, and Paul Leonard. 3D FEM model of a FLUXSET sensor. In *Electromagnetic Nondestructive Evaluation*, volume II, pages 152–159. R. Albanese et al. (Eds), IOS Press, 1998.

[10] Daniel Ioan, Mihai Rebican, and Mihai Iordache. Approximate SPICE models of nonlinear magnetic circuits based on field solution. In *Electromagnetic Nondestructive Evaluation*, volume II, pages 120–128. R. Albanese et al. (Eds), IOS Press, 1998.

[11] C. I. Mocanu. Canonic PSC schemes of the bar and circular cylinder, with transient skin effect (in Romanian). *St. cerc. energ. electr.*, 21(4):807–829, 1971.

[12] D.C. Ioan. Equivalent circuits of solid iron core for transient problems. In *COMPUMAG 78 - Conference of the Computation of Magnetic Field*, page 10.5, Grenoble, France, September 1978.

[13] P.J. Leonard and D. Rodger. A finite element scheme for transient 3D eddy currents. *IEEE Trans. Magn.*, 24(1):90–93, August 1982.

[14] D.C. Ioan, M. Rebican, and A. Gasparics. B-H characteristic extraction using devices with non-uniform magnetic field. In *Proc.of 8th International IGTE Symposium*, Graz, Austria, September 1998.

[15] L.O. Chua and K.A. Stromsmoe. Lumped-circuit models for nonlinear inductors exhibiting hysteresis loops. *IEEE Trans. Circ. Theory*, CT-17, November 1970.

[16] Gene H. Golub and Charles F. Van Loan. *Matrix computations*. The Johns Hopkins University Press, Baltimore, 3 edition, 1996.

[17] B. C. Moore. Principal component analysis in linear systems controlability, observability and model reduction. *IEEE Trans. AC*, 25(2):17–32, 1991.

[18] R. Albanese, M. Federico, A. Formisano, R. Fresa, and G. Rubinacci. Eddy current effects in the core of a magnetic sensor. In *Electromagnetic Nondestructive Evaluation – ENDE'98*, Paris, France, September 1998.

Electromagnetic Nondestructive Evaluation (III)
D. Lesselier and A. Razek (Eds.)
IOS Press, 1999

Eddy Current Probe Based on Local Field Excitation

Pavel NOVOTNÝ, Maxim MOROZOV

Institute of Chemical Technology, Technická 5, 166 28 Prague 6, Czech Republic

Gábor VÉRTESY

Research Institute for Technical Physics and Materials Science

H-1525 Budapest, PO. Box 49, Hungary

Abstract. This paper deals with a way of the resolution power in the process of the non-destructive material evaluation. A novel eddy current probe, based on a magnetic circuit generating a local excitation field was experimentally tested. Responses, which were obtained by the lock-in system on artificial cracks in stainless steel and Inconel samples, are presented and the results are discussed.

1. Introduction

A number of novel eddy current (EC) probes have been developed and tested recently. The probes in a mutual conjunction with the whole measuring system are usually characterized by special benchmark steps qualifying their abilities [1]. These experiments should make it possible to compare various types of probes for applications.

One of the final steps of the non-destructive evaluation is related to the solution of inverse problems. A theoretical approach to this solution is derived from Maxwell's equations and finished by minimizing an error function [2]. An experimental solution of the inverse problems is based on the correlation of the probe output voltage (response) and various types of artificial defects. A main attention of the both approaches is devoted to determination of geometrical shape of artificial cracks in the tested specimen.

An advanced non-destructive testing should provide more detailed information about material structures and defects, including a distinction of multiple cracks [3,4]. As far as cracks are concerned, their harmful effect is mainly caused by the stress singularities occurring along the cracks. A resulting instability is principally related to the dimensions of the crack front. One way of how to reach a better evaluation of material defects is increasing of the resolution power of EC probes. From this point of view the analogy to a near field optics [5] indicates: the smaller excited material volume is, the better resolution power can be achieved. This paper deals with an optimization of the resolution power of EC set-up with respect to dimensions and material parameters of the inspected samples.

2. Experimental set-up and samples

A typical experimental set-up for EC testing consists of generators, probe and assessing equipment (LOCK-IN amplifier, PC, ...). There are a number of parameters characterizing this system , e.g. sensitivity, standard response to noise ratio, electrical power consumption, dimensions, inspection velocity, and so on. Considering the EC investigation as a method of "magnetic" optics, then a parameter of magnetic resolution power (RP) can be introduced. This quantity is a measure of the system ability to detect and to resolve local changes of material characteristics including defects. Higher RP makes it possible to visualize smaller defects and to distinguish multiple ones. A powerful advantage of the EC magnetic optics - to see through the metals - is limited by the skin effect.

In our case the EC probe was proposed to generate the exciting electromagnetic field in a focused form due to a suitable magnetic circuit. The probe consists of a double-ring ferrite core with two identical gaps (0.5 mm x 2 mm) having a distance of about 7 mm. Exciting field was energized by a joint winding. The flux density was estimated to have a scale of 0.5 mm in the direction of scanning and a depth of the penetration to a material less than 1 mm depending on the skin effect. Two pick-up coils in the differential connection were experimentally compensated to balance a difference of ferrite rings. The probe (matched for flat samples) was carried by a holder and shifted by a scanning mechanism along straight lines. A schematic drawing of the probe arrangement is shown in Fig. 1.

A two channel lock-in amplifier was one of the important parts of the experimental system. The change of the voltage of the first channel - U_X - and the second channel -U_Y -, are sometimes called real and imaginary parts of the output signal, respectively. Both response components U_X and U_Y (or absolute value U_R and phase angle θ) vs. position, x , were step by step registered and recorded to the PC memory. Typical steps were ranging from 0.25 to 1 mm. Then the experimental data were processed by the EXCEL software and the results were printed. Lock-in amplifier was able to operate in the audio bandwidth (20 Hz - 20 kHz).

Samples of various materials were prepared in our laboratory for the testing procedures. Most of them were in the form of flat sheets with artificial defects (cracks and

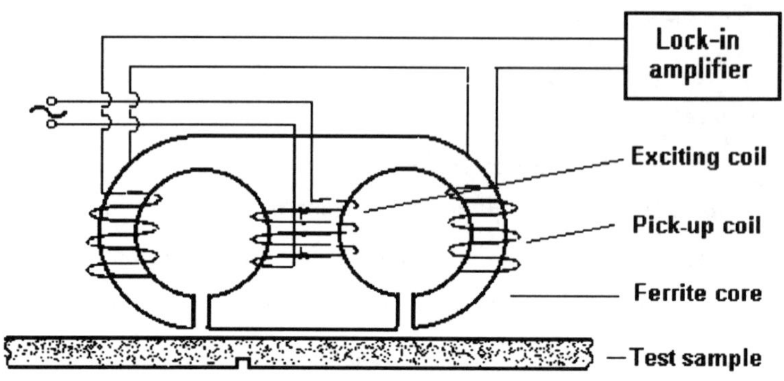

Fig.1. Schematic drawing of the EC probe

holes). Results obtained with magnetic specimens were presented elsewhere [6]. This paper is concentrated on non-magnetic stainless steel and Inconel samples. Inspection frequency was 20 kHz.

3. Results and discussion

In Fig. 2 responses U_X and U_Y of the probe on a crack in the 2 mm thick stainless steel sample vs. position are shown. The artificial 100 % crack (length x depth x width =5 mm x 2mm x 0.5 mm) was located in the middle of the flat sample (90 mm x 90 mm) and scanned across. The course of dependencies shows more distinct double peaks (or local minimum) on the U_Y component plot. The nature of this result may be related to the profile of the exciting field around the probe gap.

To verify this assumption, a distribution of the exciting fields was made visible by suitable garnet films (Fig. 3a,b) and measured on the base of knowledge of a change of the domain structure in the large scale fields. In Fig. 4 (curve **a**) a dependence of the normal flux density component -B_n- vs. position generated along the axis of the EC probe is shown. This field excited eddy currents within the closer surface of the investigated samples. The distribution of an exciting field (the same component and again in the air) at a distance of 1 mm is also shown in Fig. 4 (curve **b**). This field would excite the material of the samples on the distance 1 mm under the closer surface if the skin effect was negligible.

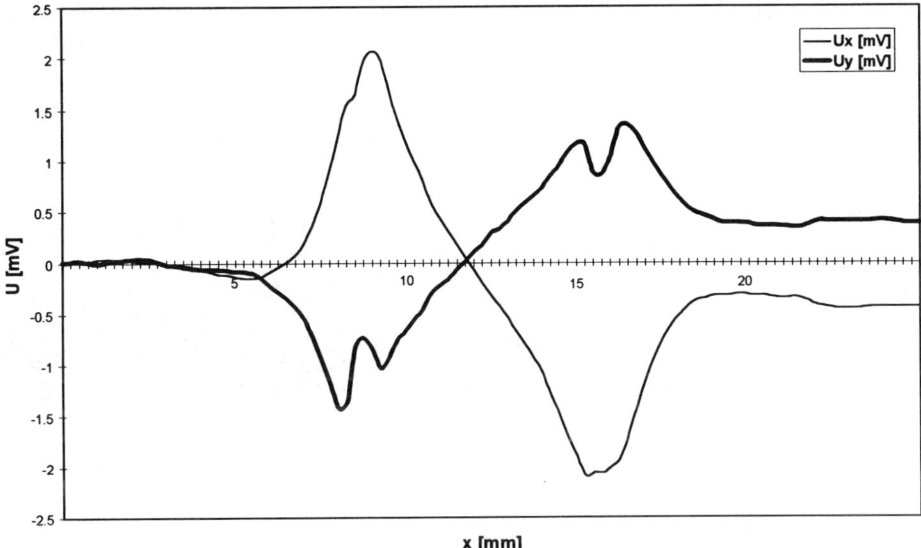

Fig.2. Responses U_X and U_Y of the EC probe on the crack in the stainless steel sample vs. position x

An in-plane component of the exciting field -B_i- was also measured. The dependence of B_i vs. position x has "nearly" rectangular shape in the case of the closest distance. The value of B_i corresponds to the maximum of the component B_n. The width of this course is approx. 0.6 mm. More detailed description of the measurements of local fields by magneto-optical films will be published elsewhere [7] .

In Fig.5 the defect responses (U_R) vs. the probe position (x) for a sample Inconel are shown. An artificial 20% deep crack (9.3 mm x 0.2 mm x 0.25 mm) was situated in the center of an Inconel-600 sheet (80 mm x 80 mm x 1.25 mm). Scanning was carried out across the central part of the crack from internal (ID - internal defect) and opposite (OD - opposite defect) sides of the sample, respectively.

The results which are shown in Fig. 5, can be summarized as follows:

a) The experimental system was able to detect the ID 20% and OD 20% in the 1.25 mm thick Inconel sheet.

b) There is a potential to identify defects as small as corresponding to the signal obtained during the scan over the defect-free sample (10 times).

c) The output pulse width value (PW 50 = 1.6 mm) determines that the resolution power of the present probe is 1.4 mm for the Inconel sample.

d) The ratio of both of the responses (5 times) corresponds to a decrease of the flux density inside material (see Fig.4).

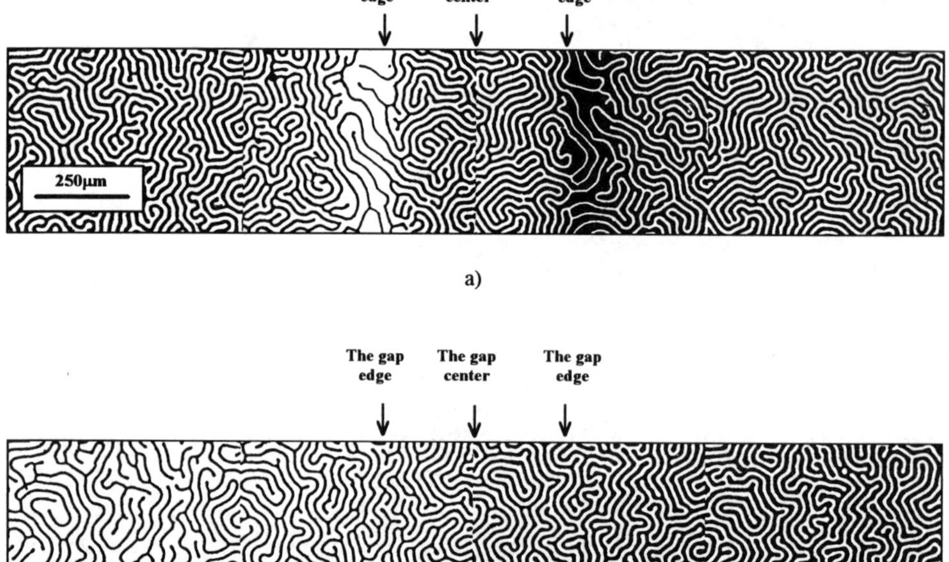

Fig. 3. Magneto-optical images of the gap of the EC probe (the domain structure is changed by a normal component of the flux density):

a) 0.12 mm above surface (closest distance); exciting DC current I= 0.25 A;

b) 1.05 mm above surface; exciting DC current I= 1 A

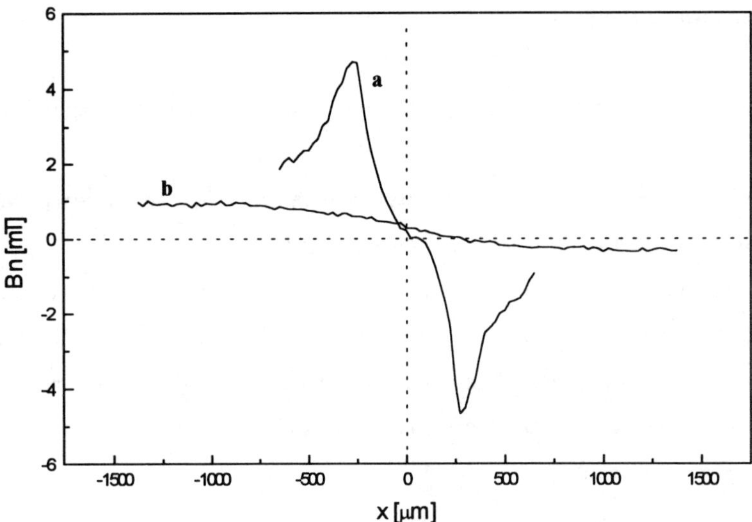

Fig. 4. Dependencies of the normal component of the flux density B_n vs. axis position x from the centre of the gap for DC current I=0.25 A (line **a** corresponds to the closest distance (0.12 mm) and line **b** corresponds to the distance of about 1mm above the gap)

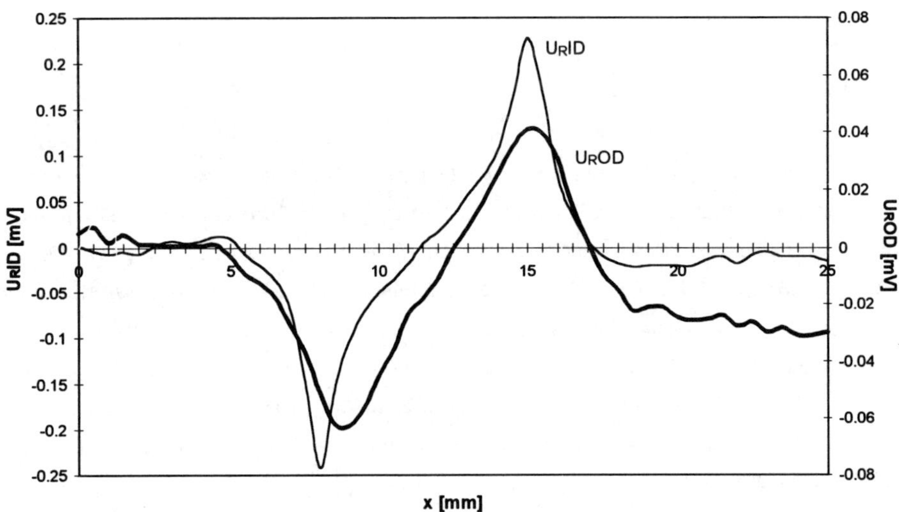

Fig.5. Dependencies of the ID 20% and OD 20% responses vs. the probe position x (Inconel sample)

4. Conclusion and outlook

A) The basic responses of the novel EC probe on artificial cracks in nonmagnetic materials were measured and evaluated.

B) The results of investigation indicate us that improvement of the resolution power of the probe demands a more local distribution of the exciting field.

C) The resolution power has been proposed to belong to the parameters characterizing EC probes (with respect to dimensions and parameters of the inspected material structure).

D) Advanced electromagnetic nondestructive evaluation (E-NDE) of materials and industrial components will demand more complex investigation ranging from metallography to ultrasonic measurements.

Acknowledgements

This work was partially supported by the EU INCO-COPERNICUS Project ERBIC-15-CT-960703 and by Hungarian Scientific Research Fund through the project T-023559.

References

[1] J.Pávó, A.Gasparics, C.S.Daróczi and G.Vértesy, "Proposal for Benchmark Problem Qualifying Some Aspects of the Performance of ECT Probes", in "Electromagnetic nondestructive evaluation (II)", R.Albanese, G.Rubinacci, T.Takagi and S.S.Udpa (Eds), IOS Press, Amsterdam, 1998, pp. 337-342.

[2] M.Yan, M.Afzal, S.S.Udpa, S.Mandayam, Y.Sun, L.Udpa and P.Sacks, "Iterative Algorithms for Electromagnetic NDE Signal Inversion", in "Electromagnetic nondestructive evaluation (II)", R.Albanese, G.Rubinacci, T.Takagi and S.S.Udpa (Eds), IOS Press, Amsterdam, 1998, pp.287-296.

[3] W.Cheng, Z.Chen, K.Miya and Y.Yoshida, "Numerical Evaluation of ECT Signal from Multiple Cracks", in "Electromagnetic nondestructive evaluation (II)", R.Albanese, G.Rubinacci, T.Takagi and S.S.Udpa (Eds), IOS Press, Amsterdam, 1998, pp.92-100.

[4] R.Albanese, F.C.Morabito, A.Formisano, R.Martone, "Multiple Defect Analysis via Neuro-Fuzzy Approaches", in "Electromagnetic nondestructive evaluation (II)", R.Albanese, G.Rubinacci, T.Takagi and S.S.Udpa (Eds), IOS Press, Amsterdam, 1998, pp. 101-106.

[5] D. W. Pohl, "Light on the Nanoworld", Europhysics News, 26 (1995), 75.

[6] P. Novotný, "High resolution eddy current detector", Conference"Defectoscopy-97", Prague, November 1997.

[7] P. Novotný and M. Morozov, "Measurement of local magnetic fields using magneto-optical films" IEEE trans. on Magnetics, in press.

Electromagnetic Nondestructive Evaluation (III)
D. Lesselier and A. Razek (Eds.)
IOS Press, 1999

Multi-Detector Eddy Current Probe for the Non-Destructive Evaluation of Steam Generator Tubes, Designed for an Imaging Approach

P.Y. Joubert*, D. Miller*, D. Placko*, E. Savin**
**L.E.Si.R., E.N.S de Cachan, 61 Avenue du Pdt. Wilson, 94 235 Cachan Cedex, France*
***Framatome, Centre Technique, B.P. 13 - 71 380 Saint-Marcel, France*

Abstract. The imaging approach of eddy current Non-Destructive Evaluation (NDE) should be able in the near future to compute a direct diagnosis of the part under control. Therefore this approach raises great interest in the NDE community. However, in the context of steam generator tubes NDE, usual eddy current probes do not answer the specific requirements of imaging techniques. In this paper, we present a structure and a prototype of an "imaging" probe. We proceed to the first evaluation of the performances of the probe with tube samples containing calibrated notches. In addition, we present a detection method based on continuous wavelet transform which increases the probe performances and allows efficient small notch detection.

1. Towards Eddy Current Non-Destructive Evaluation using Imaging Techniques

As is generally known, a Non-Destructive Evaluation (NDE) process can be described by a four-step algorithm, as depicted in Figure 1. First, the *measurement device* (probe) used for the control provides the signals. The second step consists of the *analysis* of the signals, which extracts pertinent data, thanks to appropriate signal processing techniques. In the third step, the data are supplemented with the *a priori* knowledge of the physical phenomenon implemented by the probe, in order to provide an *interpretation* of the signals. According to this interpretation, the last step performs a *diagnosis* of the part under control.

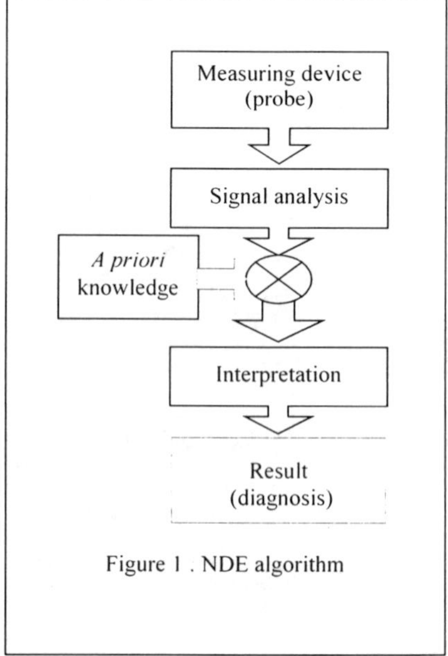

Figure 1 . NDE algorithm

In the context of Steam Generator Tube NDE using eddy current techniques, the second and third steps are usually performed by a human operator, who compares the signatures provided by the probe during the control with the signatures of well-known defects. This method known as signature analysis is mainly based on the operator's experience. To perform the diagnosis, the operator may be helped by expert systems or neural networks. But, in both cases, the efficiency of the sustems is directly related to the richness of the database used for training.

In fact, for both, the *a priori* information added to the collected data is *quantified*, because the database used is made of a quantified number of well-known situations. This points out the problem of the system's robustness when confronted with untypical situations during the control.

To avoid this major inconvenience, the use of imaging techniques should provide a solution. In the future, these promising methods will perform a direct reconstruction of the image of the conductivity at every point of the tube wall. This imaging technique replaces the interpretation step with a reconstruction based on the data provided by the probe and on a *generalized a priori* knowledge of all the interactions between the target and the probe. therefore, this promising diagnosis tool raises great interest in the NDE community.

The implementation of this imaging technique requires the setting up of an iterative process in which the following threefold problem is to be solved : the *instrumentation problem*, the *forward problem*, and the *inverse problem*. The resolution of the *instrumentation problem* consists of the elaboration of a measurement device simultaneously able to excite the inspected material, and to pick up exploitable data. The *forward problem* deals with the elaboration of a mathematical model linking the conductivity in the inspected material to the magnetic field induced at the surface of the target and sampled by the measurement device. The *inverse problem* consists of the actual reconstruction of the conductivity thanks to the data provided by the measurement device, and based on the forward model.

Recent progress made in the forward and inverse problem resolution techniques [1] give hopes for an implementation, in near future, of this NDE imaging approach. However, only a judicious choice of the probe structure could lead to the success of this method. Indeed, the solving of the instrumentation problem requires the design of a probe that answers the specific needs of the imaging process, without sacrificing the detection performances required for the control. In this paper we will present the structure and prototype of a probe designed for imaging purposes.

Besides, in this imaging approach, the signal analysis cannot be skipped, for two main reasons. In the first place, the iterative reconstruction process requires time-consuming numerical calculations. Hence it is yet not possible to reconstruct the conductivity at every point of the tube wall for a whole steam generator, which is equivalent to a 60 km long tube. However, imaging techniques can be used in addition to classical methods to facilitate the diagnosis of suspicious regions which have been previously detected by an appropriate signal analysis.

Secondly, it is hardly possible in practice to take into account all the perturbation sources in the forward problem resolution. These perturbations, namely the bad centering of the probe, the irregularity of the tube wall, and the pilgrim noise, have great magnitude. The signal analysis is therefore needed to reduce the effects of these perturbations on the signal.

In section 3 of this paper we will present a signal analysis method based on a time / frequency analysis technique, which allows efficient detection and which can be extended to a denoising process.

2. Structure and prototype of "imaging" probe

The ideal imaging probe structure should lead to a "simple" solving of the forward problem, that is to say, simple enough to be exploitable in the solving of the inverse problem, or even simple enough to be computable at all. Furthermore, it should provide an adequate sampling of the magnetic field induced at the surface of the tube wall, in order to allow a correct image reconstruction to be performed when solving the inverse problem.

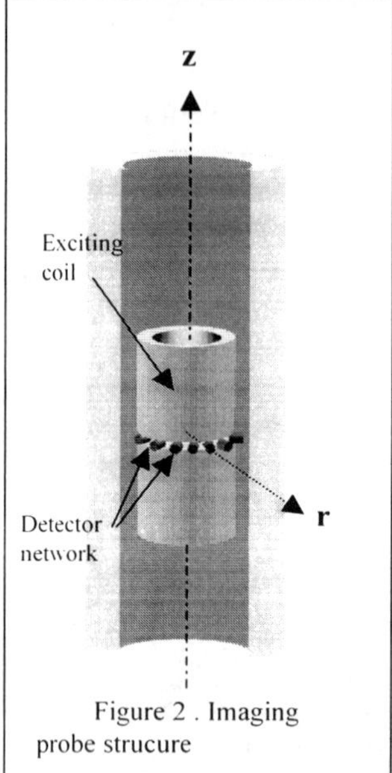

Figure 2 . Imaging probe strucure

Probes currently used for the NDE of steam generator tubes, namely axial probes and rotating probes, do not answer all these requirements. Indeed, axial probes do allow an easy calculation of the forward model, thanks to the axisymmetry of their geometry, but do not allow circumferential resolution. On the other hand, rotating probes enable a good spatial resolution in the measurement of the field, but the geometry of these probes does not permit an easy modelling of the implemented physical phenomenon. Furthermore they have a rather slow inspection speed, which is not acceptable for the control of the whole set of tubes.

It is therefore necessary to turn to another probe structure. We have chosen to design a probe with separate function as depicted in Figure 2, using a single exciting device (see # *2.1. Exciting device*) and a detector network (see # *2.2. Detector network*). This probe is designed for 22.22 mm outer diameter tubes with 1.27 mm wall thickness.

2.1. Exciting device

The solving of the forward problem is based on the computation of the eddy current density J_0 when there are no flaws, then of the perturbation induced by the presence of a flaw. The magnetic field can then be computed at any point in space. The calculation is easier if the exciting device is a simple axial coil (Figure 2), as the computation of J_0 becomes a 2D problem. 3D computations are only needed for perturbations and for the induced magnetic field.

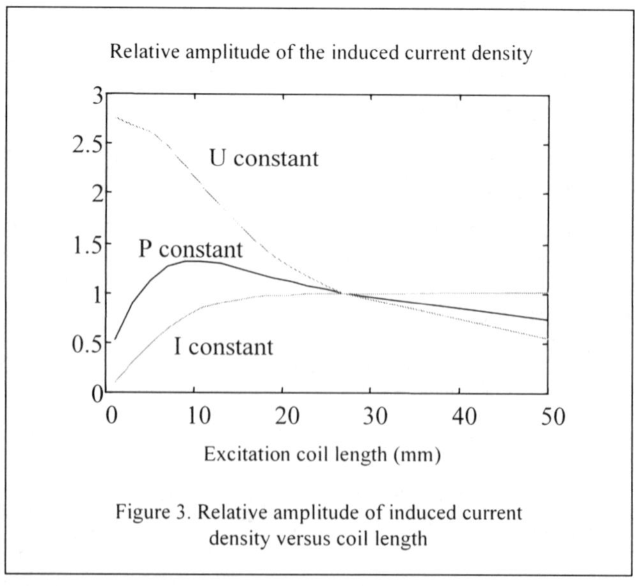

Figure 3. Relative amplitude of induced current density versus coil length

Furthermore, the model calculation is even easier to make if the coil is "long" enough to be considered as "infinite" in comparison with the area of interest. Finally, the actual length of the exciting coil is determined according to the amplitude of the eddy current density induced in the wall of the tube. Indeed, 2D axisymmetric finite element computations show that for a given exciting power, the amplitude of J_0 is maximized (that is to say, the sensitivity of the probe is maximized [2]) for a 10 mm long coil at 240 kHz standard frequency (Figure 3).

Moreover, the sensitivity to flaws being proportional to the magnitude of J_0, we choose to use an exciting coil large enough to induce currents through the whole thickness of the tube wall. The diameter is only limited by mechanical considerations.

2.2. Detector network

For symmetry reasons, the radial component of the magnetic field is null in the median plane of the exciting coil in the absence of flaw (Figure 2). Therefore, it seems judicious to measure this component, because the radial detectors behave like differential detectors. This allows the use of large gain amplifiers.

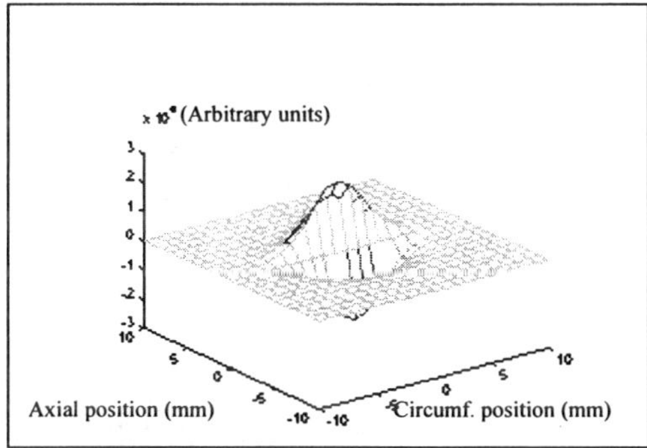

Figure 4 . Radial magnetic field component for a "punctual" flaw

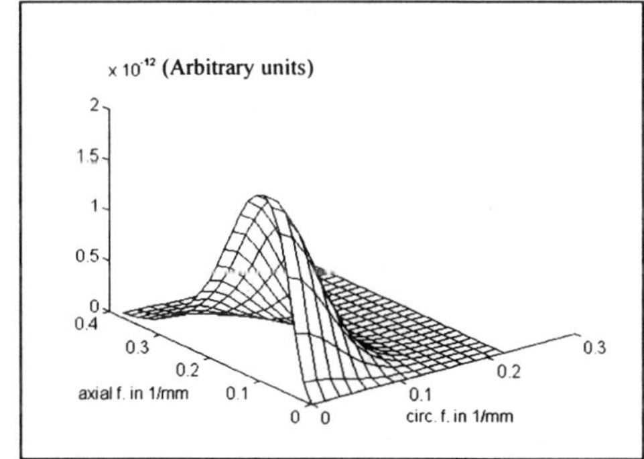

Figure 5 . Spectral density of the radial magnetic field component due to a "punctual" flaw

The sampling step of the magnetic field in both axial and circumferential directions is determined in relation to the 2D impulse response of the magnetic field to a "punctual" flaw. In fact no actual *punctual* flaw is considered, but we calculate the field created by the smallest flaw of interest, which is a 100μm x 100μm, 10% deep outer notch, and we assume that this field is the impulse response of the probe (Figure 4).

From this impulse response, we can compute the spectral density corresponding to the "punctual" flaw (Figure 5). A $0.4\ mm^{-1}$ bandwidth in the axial direction and a $0.1\ mm^{-1}$ bandwidth in the circumferential direction allows us to keep more than 99.9% of the total energy. This determines the minimum sampling step. With a 2.5 oversampling ratio, this leads to an axial step of 0.5 mm and a circumferential step of 2 mm, that is to say 32 sensors on a 9 mm radius, and a 1 kHz spatial sampling frequency if the probe speed is 50 cm/s. Considering the different existing technologies (Hall effect, magneto-resistance), only the use of coils seems to be realistic nowadays.

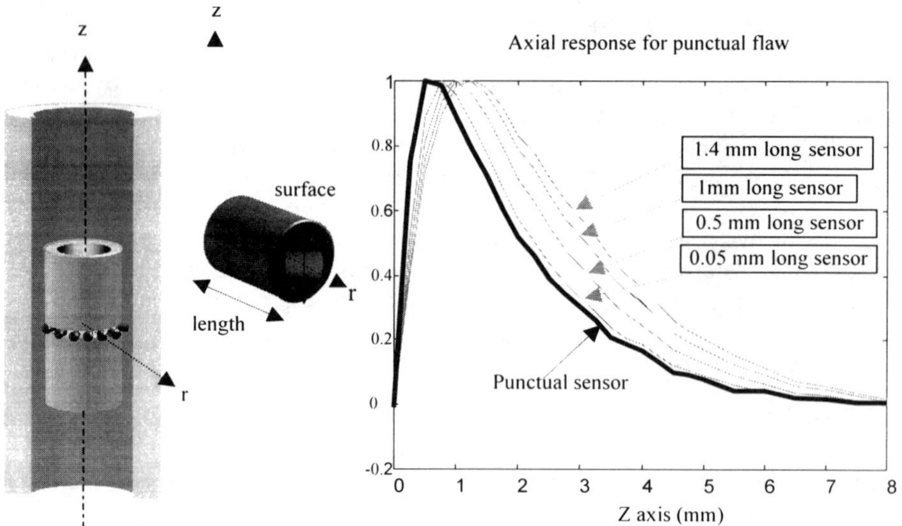

Figure 6 . Influence of the coil length along the radial axis

Moreover, the size of the detectors has a great influence on the quality of the measurement. In particular, the best measurement is achieved when taken directly at the surface of the inner wall of the tube. To be more specific, we established by 2D finite element computations, that the magnetic field response becomes narrow-banded when the distance to the inner wall increases. That is to say, the shorter the radial coil, the more pertinent the measured information (Figure 6). These results suggest the use of flat coil detectors.

Besides, the sensor measures a magnetic flux and integrates the magnetic field across its section. This integration operates a spatial low-pass filtering. To avoid a critical deconvolution, the error due to this integration must be kept negligible. More precisely, since the inverse problem solving computes a quadratic error with every point of a local area around a flaw, we shall limit the sensor surface so that the quadratic error induced by the integration lets us separate two nearby flaws and remains negligible in comparison with other noises or errors. An inevitable noise is the electronic noise due to the coil resistance, which we can estimate from the geometrical and physical properties of the sensor.

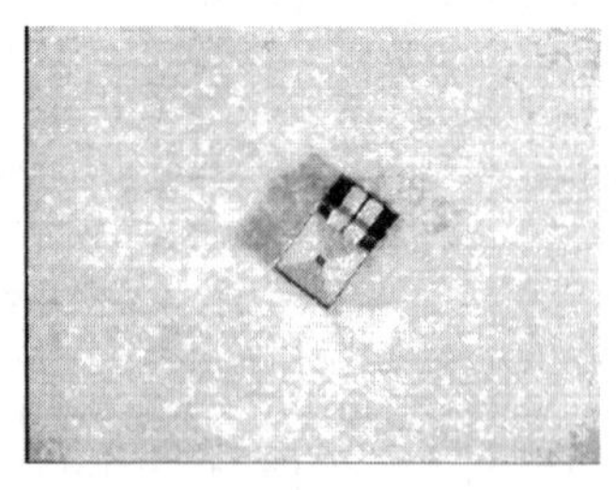

Figure 7 . Flat coil on silicon substrate (2 layers with 21 turns each)

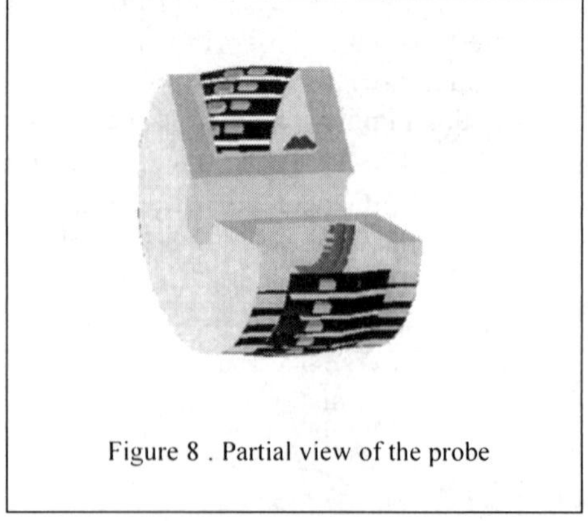

Figure 8 . Partial view of the probe

We calculated the error due to integration effect for several flat coil configurations and reached the following conclusion : in order to be able to separate 2 "punctual" notches that are only 1 mm apart from each other, it is necessary to limit the surface of the detector to 0.5 mm x 0.5mm. Such detectors have been built by the LETI (Grenoble, France) on a silicon substrate (Figure 7).

2.3. "Imaging" probe prototypes and first evaluation

Two imaging probe prototypes have been designed using the structure presented above. The first one is depicted in Figure 8. Thirty-two flat detectors have been used with a gain 100 pre-amplifier added next to each detector, to compensate for the loss of signal amplitude caused by the small size of the detectors. Both coils and amplifiers have been mounted on flexible ribbons around the axial exciting coil. A prototype is still under fabrication, and no performance evaluation is available for the moment.

However, we have also built a previous prototype designed with only 16 detectors. These detectors are conventional bobbin coils with no pre-amplification (as depicted in Figure 2). With this configuration we cannot hope for any imaging reconstruction, however this "easy to build" probe has been developed to validate the structure and give a first evaluation of detection performances.

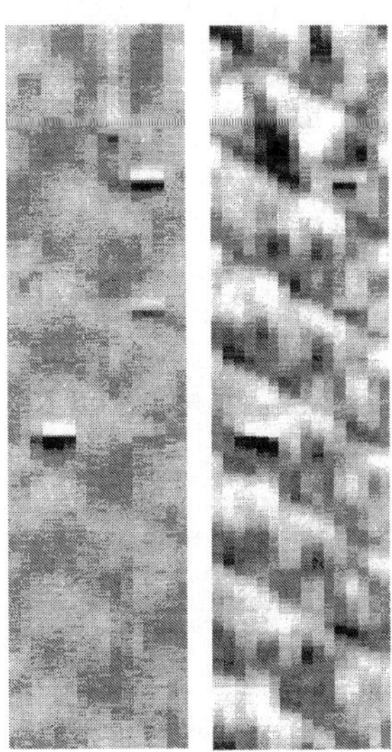

Figure 9 . Complex field images provided by the 16 detector probe. Acquisitions have been performed at 240 kHz standard frequency on 2 tube samples containing circumferential notches. Tube 1 (left) : 4 inner 100μm wide, 15 mm long notches of 60%, 40% 20% and 10% depth (up to down). Tube 2 (right) : 5 outer 100μm wide 15 mm long notches of 100% (left of the acquisitions) and 60%, 40% 20% and 10% depth (up to down).

With this probe we inspected some tube samples containing calibrated notches, internal or external, 100μm wide and 15mm long, axially or circumferentially oriented, and of different depth. Our attention has been specially focused on circumferential notches, because they are particularly difficult to detect with a probe using axial exciting coil. For example, measurement signals at 240 kHz standard frequency are shown Figure 9.

In these acquisitions, the presence, orientation and position of the major notches are all visible. However, the smallest notches (10% outer and inner notches) are masked by surface noises. Indeed surface noises due to the misalignment and the bad centering of the probe, the evolution of the tube wall thickness, the pilgrim noise and so on, are clearly visible because of the high sensitivity of the probe. This is the price paid for using a long exciting coil. These results demonstrate the necessity of prior signal analysis in order to perform efficient detection.

3. Signal analysis using Continuous Wavelet Transform

The Fourier transform is widely used for signal analysis purposes and is satisfactory when applied to signals whose stationary features are of particular interest. However, it turns out to be inefficient when dealing with defect detection, where it is the non-stationary characteristics of the signal which needs to be highlighted. The main reason for this is that in the Fourier analysis, the *time* parameter is discarded. Therefore, it seems natural to consider time-frequency techniques in order to perform an efficient defect detection through the signal provided by the sensors.

Of course, it is well known that because of the duality of the *time* and *frequency* parameters, there is no universal time-frequency technique. Indeed the two parameters cannot be simultaneously known with arbitrarily high resolution. This is expressed by the Heisenberg - Gabor inequality :

$$\Delta t \cdot \Delta\omega \geq \frac{1}{2} \qquad (1)$$

where Δt and $\Delta\omega$ respectively denote a measure of the time resolution and of the frequency resolution of a signal. However, several techniques have been developed depending on the characteristics of the signal to be analyzed, and lead to interpretable time-frequency representations. Among these techniques, the Continuous Wavelet Transform is particularly well suited to the eddy current signal coming from the tube inspection [2], and provides efficient detection results.

3.1. Basic principle

Like the Fourier Transform (FT), the Continuous Wavelet Transform (CWT) is expressed by the means of an inner product between the signal to analyze $s(t)$ and a set of analyzing functions :

$$CWT(s(t)) = \int_{-\infty}^{+\infty} s(t).\psi^{*}_{a,b}dt \qquad (2)$$

But the major difference lies in the fact that for the CWT the analyzing functions $\psi_{a,b}(t)$, called wavelets if they satisfy to the admissibility condition [3], are located both in time and frequency.

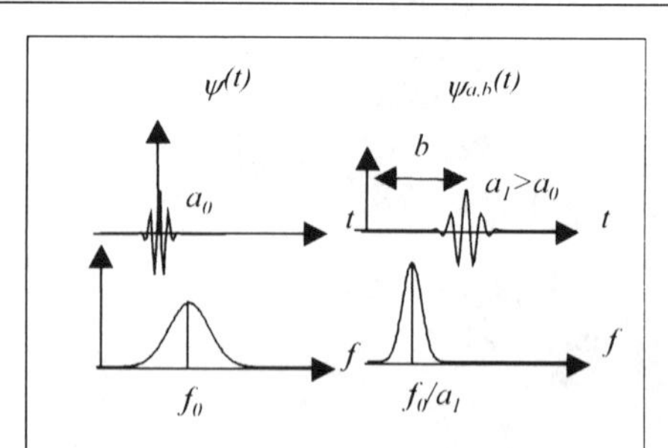

Figure 10 . mother wavelet $\psi(t)$ (left) and wavelet $\psi_{a,b}(t)$ (right) built out of the mother wavelet by time shift b, and dilatation a. Both functions are represented in the time-domain and the frequency-domain.

The whole set of function $\psi_{a,b}(t)$ is constructed out of a basic (or "mother") wavelet $\psi(t)$ by the means of time shifting and dilatation :

$$\psi_{a,b}(t) = \frac{1}{\sqrt{a}}\psi\left(\frac{t-b}{a}\right) \quad (3)$$

where b denotes the time shift, a the dilatation coefficient and the term $a^{-1/2}$ stands for normalization purposes. From the successive dilatation of the mother function arises the notion of "scale analysis", in which the signal $s(t)$ is successively analyzed by a set of functions of same shape, but with different scales. Indeed the CWT performs a "time-scale" rather than a "time-frequency" analysis.

However, it is easily shown that if the mother wavelet is located in the frequency domain "around" f_0 (Figure 10), then the wavelet $\psi_{a,b}(t)$ is located around f_0/a. That is to say, by the means of the formal identification $f \equiv f_0/a$ it is possible to interpret a time-scale representation as a time-frequency representation [3].

3.2. Time and frequency resolution

A measure of the time and frequency resolution of the mother wavelet is given respectively by :

$$\Delta t^2 = \frac{1}{E_\psi}\int (t-t_0)^2 |\psi(t)|^2 dt \quad \text{and} \quad \Delta\omega^2 = \frac{1}{E_\psi}\int (\omega-\omega_0)^2 |\Psi(\omega)|^2 d\omega \quad (5)$$

where E denotes the energy of the wavelet, and t_0 and ω_0 the center of gravity of the wavelet in the time domain and the frequency domain respectively. Let us introduce the time and frequency resolution of the $\psi_{a,b}(t)$ wavelets, respectively $\Delta t_{a,b}$ and $\Delta\omega_{a,b}$. It is easily shown that :

$$\Delta t_{a,b} = a.\Delta t$$
$$\Delta\omega_{a,b} = \frac{\Delta\omega}{a} \quad (6)$$

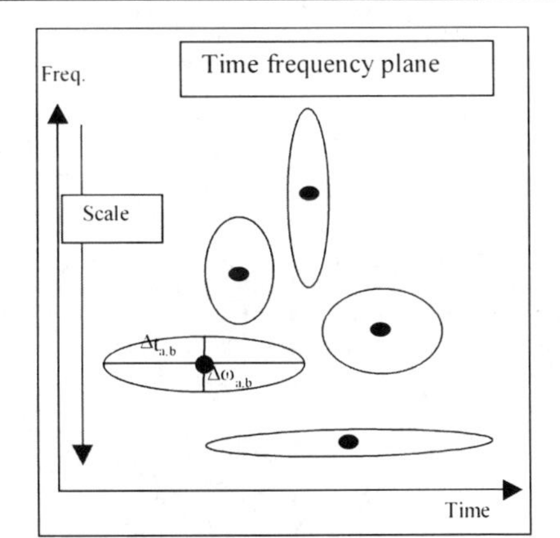

Figure 11 . Representation of the time-frequency resolution of $\psi_{a,b}(t)$ wavelets in the time frequency plane, with the help of the Heisenberg ellipses. The axes of the ellipses are sized with respect to the RMS value of the time resolution and frequency resolution.

From the two equations given in (6), we can observe that the time-frequency compromise expressed by (1) evolves within the scale changes performed by the CWT. To be more specific, given a mother wavelet with its own time and frequency properties, the small values of scale coefficient a (high frequencies) lead to a high time resolution (and a poor frequency resolution). Correspondingly, high values of the scale coefficient (low frequencies) lead to a high frequency resolution (and a poor time resolution), (see Figure 11).

The evolution of the time-frequency compromise in the time-frequency plane is the main characteristic of the Wavelet Transform, and is particularly well suited for the analysis of the eddy current signal. Indeed defect signatures are signals with large frequency bandwidth and have to be located with a good time resolution. On the other hand, noise due to the roughness of the tube wall, and especially the pilgrim noise, is a slowly variable signal (low frequencies), which is best analyzed with good frequency resolution.

3.3. Time frequency representation

The choice of the mother wavelet $\psi(t)$ is a factor of major importance for the quality of the representation. We choose the Morlet wavelet expressed by :

$$\psi_{morlet} = e^{\frac{-\sigma^2 t^2}{2}} e^{j\omega_0 t} \quad (7)$$

because, as its Gaussian modulus provides the best time / frequency compromise, inequality (1) becomes the equality :

$$\Delta t \cdot \Delta\omega = \frac{1}{2} \quad (8)$$

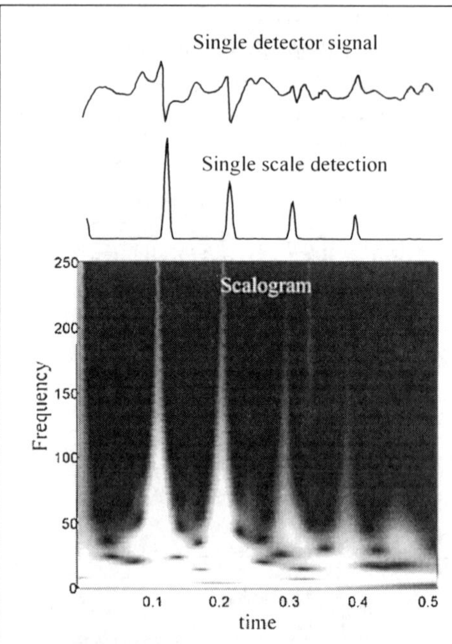

Figure 12 . Scalogram computed for single detector acquisition on tube 1 (see Figure 9). The time and frequency axes have been scaled according to the speed of the probe (50 cm/sec).

3.4. Scalogram

The computation of the CWT leads to complex coefficients. Therefore the total information provided by the transform needs a double representation (amplitude and phase). However, as the representation of the CWT phase is generally quite difficult to interpret, we shall focus on the amplitude of the CWT. Furthermore, it is known that the square amplitude of the transform, $|CWT(s(t))|^2$, corresponds to a distribution of the energy of $s(t)$ in the time-frequency plane [3]. This property enhances the interpretability of the analysis. Indeed, each pattern formed in the representation can be understood as a part of the signal total energy. This representation is called scalogram.

3.5. Application to notch detection

Scalograms were calculated with Morlet wavelet for all tube samples, and gave efficient detection results, even for the smallest notches. A scalogram calculated for a single detector signal and for inner circumferential notch detection, is shown as an example, in Figure 12. Easily interpretable patterns are observed : the first one is a horizontal pattern, located around 15Hz, and is due to the pilgrim noise. Four regularly spaced vertical patterns are also observable, and show the detection efficiency. They are due to the presence of the 4 inner circumferential notches of successive 60%, 40% 20% and 10% depth. The selection of one particularly pertinent scale, corresponding to the bandwidth located around 100Hz for example, leads to a fully informative detection signal where the presence of every notch is detected (Figure 12).

The same process applied to the signals coming from the 16 detectors leads to a 2D detection image. For the 2 examples shown Figure 13 the detection image computed achieves an efficient notch detection even for the 10% deep notches.

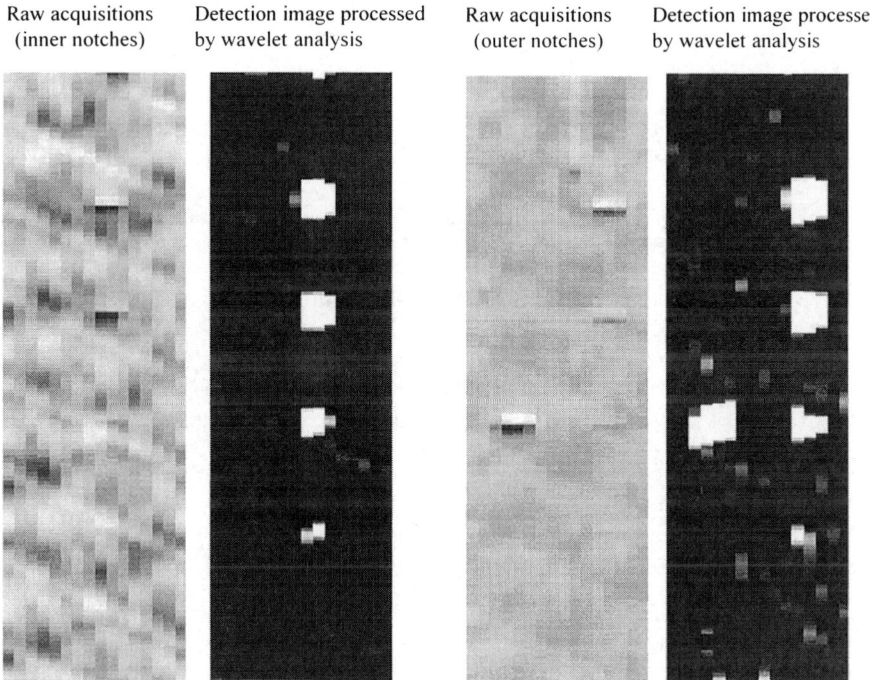

Figure 13 . Raw acquisitions of the magnetic field and corresponding detection images processed by wavelet analysis. Left : tube 1 containing 4 inner 100µm wide, 15 mm long notches of 60%, 40% 20% and 10% depth (up to down). Right : tube 2 containing 5 outer 100µm wide 15 mm long notches of 100% (left hand side of the image) and 60%, 40% 20% and 10% depth (up to down, right hand side of map).

Similar results have been obtained for axially-oriented notches. In fact, those are perpendicular to the current circulation, and therefore induce greater perturbation than the circumferential notches. That is to say, better detection results are expected in the axial case. However, as the detectors are radially oriented, only the extremities of the axial notches are detected by the probe. Therefore, the signal processing described above is also needed to detect the smallest axial notches such as 10% inner or outer notches.

4. Conclusion

In this paper we have presented a new eddy current multi-detector probe specially designed for an imaging purpose. The first evaluation of the probe structure, performed by an "easy to build" earlier probe, shows good performances. We have also presented a signal analysis based on Continuous Wavelet Transform which allows efficient detection even for the smallest notch considered. Furthermore this signal processing technique using wavelet theory can be extended to "denoising" process [4][5]. The prototype of a more advanced probe, with 32 flat detectors and pre-amplifiers, is still under construction. With the developments of these techniques and probes, we hope to perform a direct diagnosis of steam generator tubes by means of eddy current NDE using the imaging approach.

References

[1] D. Prémel, N. Madaoui, O. Venard, E. Savin, An Eddy Current Imaging System for the Reconstruction of Three-Dimensional Flaws. 7th ECNDT'98, Copenhagen, 26-29 Mai 1998, ISBN 87-986898-0-0, pp. 2584-2591.

[2] P.Y. Joubert, D. Miller, D. Placko, Optimisation d'une sonde différentielle à courants de Foucault et analyse des signaux par transformée en ondelettes continue, pour la détection de défauts circonférentiels dans les tubes de générateurs de vapeur. Journées SEE, Toulouse, November 1997.

[3] A. Cohen, J. Kovacevic, Wavelets the Mathematical Background. Proceedings of the IEEE, vol. 84 n°4 April 1996, pp. 514-522.

[4] G. Chen, Y. Yoshida, K. Miya, M. Uesaka, Application of Eddy Current Testing Inspection to the First Wall of Fusion Reactor with Wavelet Analysis. Fusion Engineering and Design, vol. 29, 1995, pp. 309-316.

[5] P.Y. Joubert, N. Madaoui, Dispositif expérimental, modèle associé et traitement par ondelettes des signaux capteur pour la reconstruction tomographique par courants de Foucault. In : Instrumentation, Interdisciplinarité et Innovation. ISBN 2-86601-730-7, Editions HERMES, Paris, 1998, pp. 135-142.

Electromagnetic Nondestructive Evaluation (III)
D. Lesselier and A. Razek (Eds.)
IOS Press, 1999

Differential Eddy Current Transducer for Non-destructive Testing of Pipes

Ryszard SIKORA, Mieczysław KOMOROWSKI, Stanisław GRATKOWSKI
Technical University of Szczecin, Al. Piastów 19, 70-310 Szczecin, Poland

Ermanno CARDELLI
Institute of Energetics, University of Perugia, via G. Duranti 1\a-4, 06125 Perugia, Italy

Masato ENOKIZONO, Tomasz CHADY
Faculty of Engineering, Oita University, 700 Dannoharu Oita, 870-1192, Japan

Abstract. This paper describes a new differential eddy current transducer for testing pipes made of conducting material. The transducer was developed and constructed at the Technical University of Szczecin, Poland, in co-operation with the University of Perugia, Italy, and Oita University, Japan. The results of experiments clearly show that the differential eddy current probe is very effective for testing pipes.

1. Introduction

The eddy current method of non-destructive testing uses an alternating current excitation to induce secondary currents and fields in the specimen under test. Flaws within the specimen affect the induced currents, causing measurable changes in the impedance of the exciting coil or in the voltage induced in a secondary coil, if separate coils are used for excitation and detection. Unfortunately, in practically important cases the changes are usually very small. In this paper we describe a new differential eddy current transducer for testing pipes, developed and constructed at the Technical University of Szczecin, Poland in cooperation with the University of Perugia, Italy and Oita University, Japan.

2. Differential eddy current transducer

The cross-section of the newly developed differential eddy current probe is shown in Fig. 1. The probe is placed inside the pipe under test (6), having the simulated defect (7). The probe is coaxial with the pipe thanks to the support wheels (4). The tube (5) enables electrical contact with the probe and serves as a drive rod. The transducer comprises three coils mounted on the ferrite core (1). The exciting coils (2) are connected differentially. These coils create opposite directed magnetic fluxes Φ_1 and Φ_2, which penetrate the pipe wall. The fluxes Φ_1 and Φ_2 induce the circumferential eddy currents J_{w1} and J_{w2}. The signal coil (3) is placed between exciting coils. If there is no influence of a defect, the transducer is in equilibrium. The defect causes changes in the eddy current J_{w1} or J_{w2} and results in non-zero output signal from the coil (3). The signal depends on a flaw type and distance between the flaw and the probe.

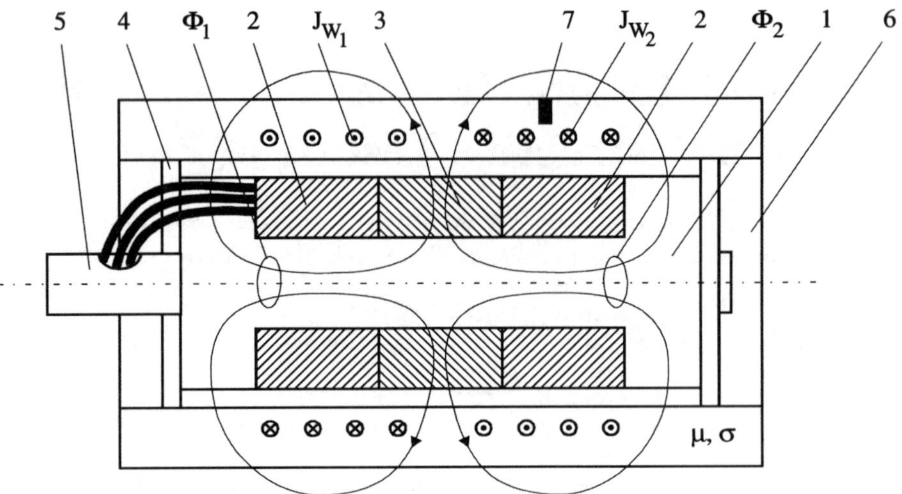

Fig.1. Sectional view of the differential eddy current probe

3. Simplified theoretical analysis of the transducer

A simplified electric model of the probe and its equivalent magnetic circuit are shown in Fig. 2 and Fig. 2a, respectively. Theoretically, the transducer is in equilibrium if there is no flaw in the neighbourhood of the probe or if the symmetric flaw is placed exactly above the centre of the probe. The adjustment can be made by using the potentiometer *P* (see Fig. 2).

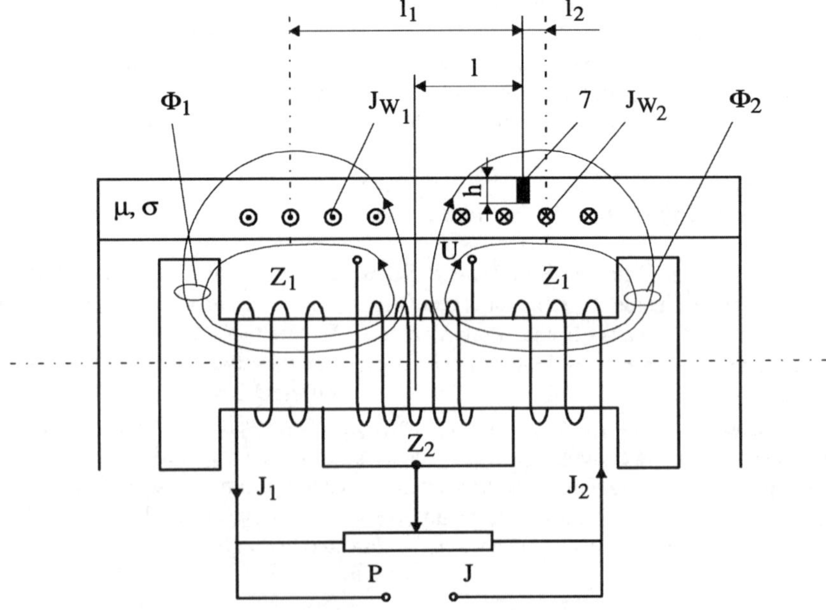

Fig. 2. Simplified electric model of the probe

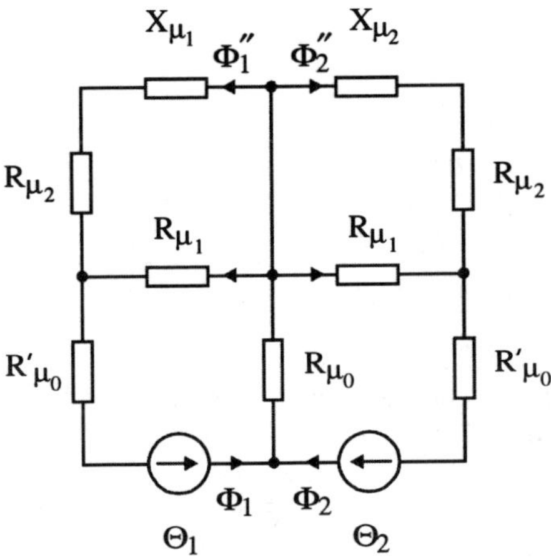

Fig. 2a. Equivalent magnetic circuit

In order to limit the mathematical efforts and yet gain physical insight regarding the problem, the following simplifying assumptions are made in this study. First, saturation effects are neglected. Second, the magnetic circuit is symmetric unless a flaw is considered. Third, a forced current in the exciting coils is assumed. And last, a flaw influences only the values of $x_{\mu 1}$ and $x_{\mu 2}$ in the equivalent magnetic circuit shown in Fig. 2a.

Owing to the simplifying assumptions, with reference to Fig. 2a, the magnetic fluxes can be written as:

$$\Phi_1 = \frac{\Theta_1}{R_{\mu 0} + R_{\mu 0}^{'} + \dfrac{R_{\mu 1}(R_{\mu 2} + jx_{\mu 1})}{R_{\mu 1} + R_{\mu 2} + jx_1}}, \quad \Phi_2 = \frac{\Theta_2}{R_{\mu 0} + R_{\mu 0}^{'} + \dfrac{R_{\mu 1}(R_{\mu 2} + jx_{\mu 2})}{R_{\mu 1} + R_{\mu 2} + jx_{\mu 2}}} \tag{1}$$

where: $\Theta_1 = I_1 z_1$, $\Theta_2 = I_1$ – magnetomotive forces,

$R_{\mu 0}^{'}$, $R_{\mu 0}$ reluctances of the core and air,

$R_{\mu 1}$ – reluctance shunting the path of flux penetrating the pipe,

$x_{\mu 1}$, $x_{\mu 2}$ –magnetic reactances caused by eddy current,

$R_{\mu 2} + jx_{\mu 1}$, $R_{\mu 2} + jx_{\mu 2}$ – impedances of the eddy current circuits.

The output voltage from the transducer is given by:

$$U = -\, j\omega z_2(\Phi_2 - \Phi_1), \tag{2}$$

where: z_2 – number of turns of the signal coil,

ω – angular frequency.

Combining (1) with (2) and assuming additionally: $R_{\mu 2} \langle\langle x_{\mu 1}$, $R_{\mu 2} \langle\langle x_{\mu 2}$, $\Theta_1 = \Theta_2 = \Theta$, yields:

$$U = \omega z_2 \Theta \frac{R_{\mu 1}^{'}(x_{\mu 1} - x_{\mu 2})}{R_{\mu}^2 R_{\mu 1}^2 - x_{\mu 1} x_{\mu 2}(R_{\mu} + R_{\mu 1})^2 + j(x_{\mu 1} + x_{\mu 2})(R_{\mu 1} + R_{\mu 2}) R_{\mu} R_{\mu 1}}, \tag{3}$$

where:

$$R_{\mu} = R_{\mu 0} + R_{\mu 0}^{'}.$$

Substituting $R_{\mu}=kR_{\mu 1}$, one can obtain:

$$U = \omega z_2 \Theta \frac{x_{\mu 1} - x_{\mu 2}}{R_{\mu}^2 - x_{\mu 1} x_{\mu 2} (k+1)^2 + j(x_{\mu 1} + x_{\mu 2}) R_{\mu} (k+1)}. \quad (4)$$

The magnetic reactances $x_{\mu 1}$ and $x_{\mu 2}$ depend on a position of the probe in relation to a flaw and on the geometric dimensions of the flaw.
Then:

$$x_{\mu 1} = x_{\mu 0} k_1, \quad x_{\mu 2} = x_{\mu 0} k_2, \quad (5)$$

where:

$x_{\mu 0}$ – reactance of the magnetic circuit when the probe is far from a flaw (the transducer is in equilibrium),
$k_1 = f(g, l_1)$, $k_2 = f(g, l_2)$ – coefficients of linkage with a flaw, dependent on geometry (g) of the flaw and on the distances (l_1, l_2) from the flaw (see Fig. 2).

In what follows it is assumed that the flaw related effect on eddy current is dependent on the ratio

$$p = \frac{S_c}{S_e} \quad (6)$$

where: p – relative flaw area,
S_c – cross-sectional flaw area,
S_e – cross-sectional eddy-current area.

Another factor having an influence on the eddy current is the distance between the flaw and the centres of exciting coils. This may be expressed as

$$l_1 = l_0 + l \quad ; \quad l_2 = l_0 - l$$

$$x_1 = \frac{l_0 + l}{l_0} = 1 + x \quad ; \quad x_2 = \frac{l_0 - l}{l_0} = 1 - x \quad (7)$$

where: l_0 – distance between the transducer and the centres of exciting coils,
x_1, x_2 – relative position of exciting coils,
x – relative position of the flaw.

In view of Eqs. (6) and (7), the coupling factors may be expressed as

$$k_1 = -\frac{p}{1+x} \quad ; \quad k = -\frac{p}{1-x} \quad (8)$$

Substituting Eq. (8) into Eq. (5) gives

$$x_{\mu 1} = x_{\mu 0}(1 - \frac{p}{1+x}) \quad ; \quad x_{\mu 2} = x_{\mu 0}(1 - \frac{p}{1-x}) \tag{9}$$

Once the flaw is situated along the transducer axis, the transducer output becomes zero, hence, on the strength of Eq. (4) we get

$$R_{\mu}^2 - x_{\mu 1}x_{\mu 2}(k+1)^2 = 0 \tag{10}$$

whence, after entering Eq. (9) for x = 0, the reluctance for the state of equilibrium is obtained:

$$R_{\mu} = x_{\mu 0}(1-p)(k+1) \tag{11}$$

Substituting Eqs. (9) and (11) into Eq. (4) and rearranging gives us the output magnitude

$$|U| = A\frac{2px}{\sqrt{x^4\left[1+(1-p)^2\right]^2 - 8x^2(1-p)^3 + 4(1-p)^4}} \tag{12}$$

where: $A = \frac{z_2 l_e \Theta}{z_e \sigma S_e}[mV]$ – factor dependent on properties of the transducer and the material under test,

z_e, l_e, S_e, σ – turns of winding, length, area and conductivity respectively of the eddy current circuit.

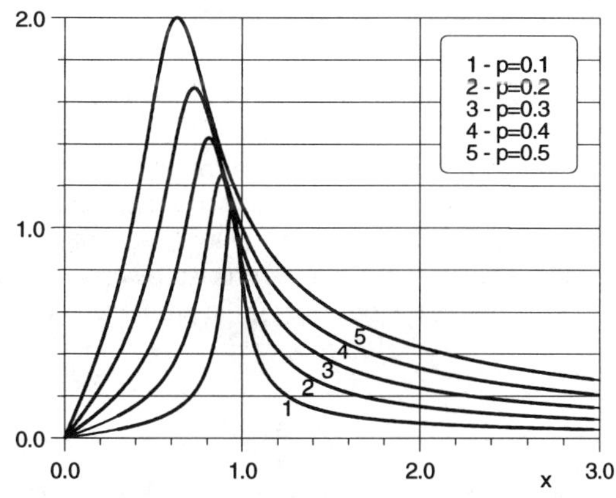

Fig. 3. Theoretically predicted transducer characteristics according to Eq. (12).

Equation (12) provides a basis for determining the transducer theoretical characteristics shown in Fig. 3. It may be noted to what extent the distance between the flaw and the transducer axis affects the output and the position of the characteristic maximum. The same applies to the flaw size (p).

The results delivered by Eq. (12) show good agreement with those of measurements made for the flaws 4, 5 and 6 (see Fig. 6). The curves obtained are almost overlapping in both cases. For the remaining flaws the results obtained are in disagreement. Efforts are underway to improve the formula (12) to cover a wider class of flaws.

4. Measuring system

Schematic representation of a measuring system for the investigation is shown in Fig. 4.

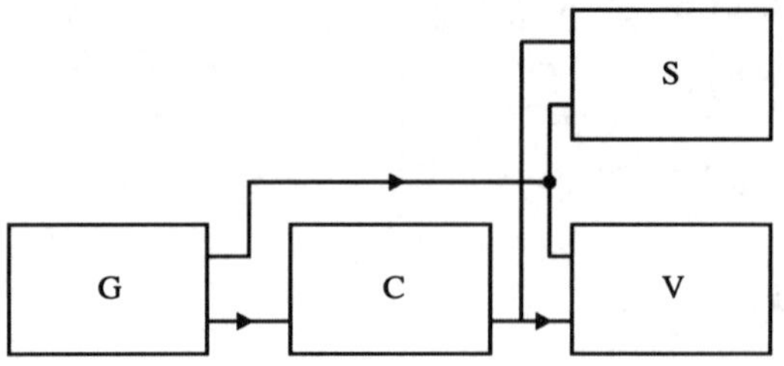

Fig. 4. Schematic representation of the measuring system

The measuring system consists of: generator (G), converter (C), oscilloscope (S) and digital lock-in nano-voltmeter (V), which provides in-phase and quadrature information. A positioning system moves the probe inside the pipe under test. The positioning system is shown in Fig. 5.

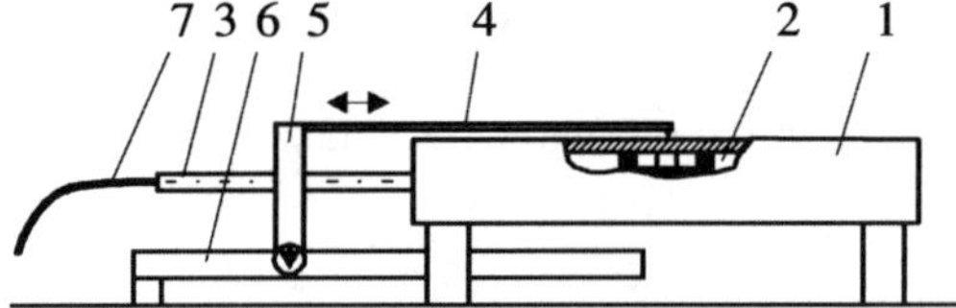

Fig. 5. The positioning system: 1-pipe under test; 2-probe; 3-tube with cables (7); 4, 5, 6-indicators

The measuring system is connected with the positioning system and with the probe (2) by the cables (7), which are placed in the tube (3). The tube serves also as a drive rod. The probe moves in the pipe under test (1) together with the indicators (4) and (5). The indicator (5) is equipped with a magnifying glass, which enables precise reading of the probe position from the ruler (6). The indicator (4) shows the position of the probe, which is unseen during testing.

5. Experiments

The specimen was a pipe made of brass (relative permeability $\mu = 1$ and electrical conductivity $\sigma = 12.5 \cdot 10^6$ S/m) with the internal and outer diameter equal to 30 mm and 40 mm, respectively. The length of the pipe was equal to 1100 mm. The specimen with 7 different kinds of flaws is shown in Fig. 6.

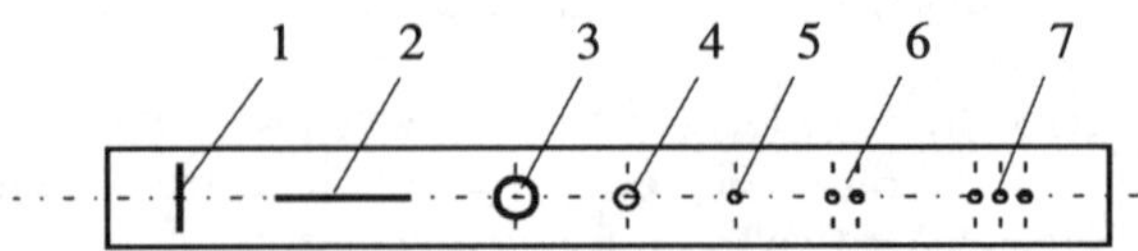

Fig. 6. The pipe with 7 different types of flaws (not in scale): 1, 2– long surface cracks (0.5 mm in width); 3, 4, 5– holes; 6, 7– sets of holes

The parameters of the probe were as follows: outer diameter – 28 mm, length – 42 mm, $z_1 = 160$ turns, $z_2 = 2500$ turns. Experiments were made first for 1-kHz excitation frequency and for the exciting current amplitude equal to 60 mA. For this frequency the equivalent depth of penetration of the electromagnetic wave is approximately equal to 4.5 mm and is comparable with the thickness of the pipe under test. Next, the experiments were repeated at different frequencies and amplitudes of the exciting current, only for the flaw marked by 3 in Fig. 6. Results of the experiments are shown in graphic form as plots of the probe output voltage and its components versus position of the probe in relation to flaws and as patterns on the complex plane. The measurements were made starting from a position when a flaw is over the probe and then moving the probe by 40 mm inside the pipe.

5.1. Circumferentially and axially oriented cracks

Dimensions of the flaw marked by 1 in Fig. 6 (circumferentially oriented crack) are shown in Fig. 7. Figs. 7a and 7b show the results of the experiments.

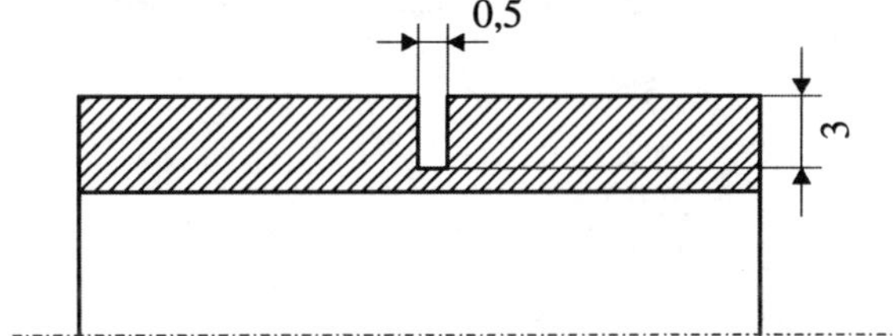

Fig. 7. Circumferentially oriented crack

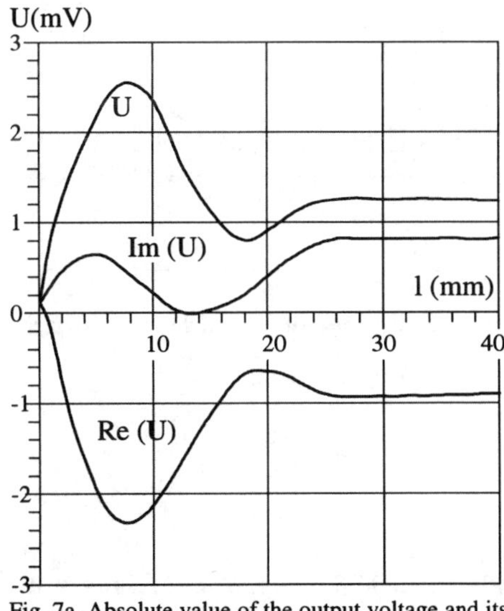

Fig. 7a. Absolute value of the output voltage and its components (real and imaginary part)

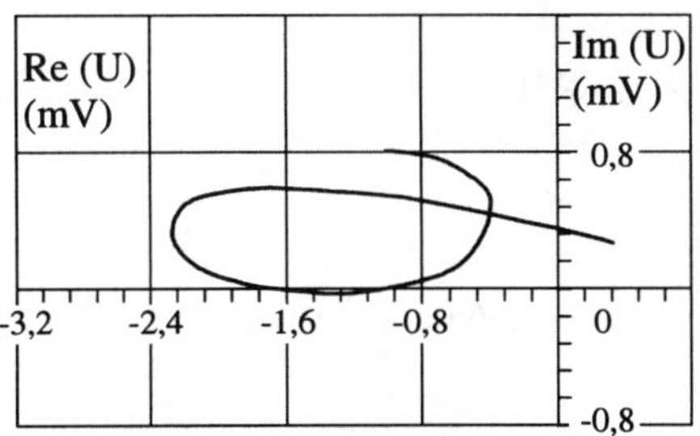

Fig. 7b. Real and imaginary part of the output voltage on the complex plane

Fig. 8 shows dimensions of the flaw marked by 2 in Fig. 6. The results of the experiments are shown in Figs. 8a and 8b.

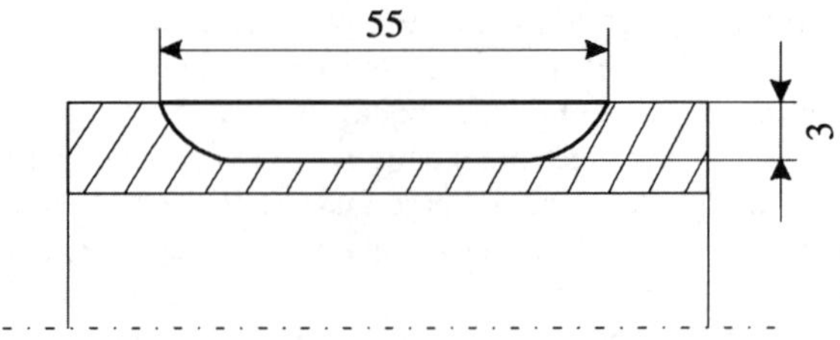

Fig. 8. Axially oriented crack

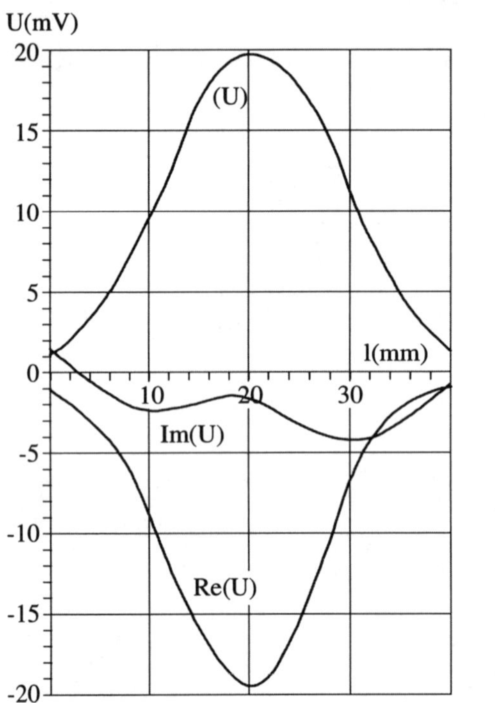

Fig. 8a. Absolute value of the output voltage and its components (real and imaginary part)

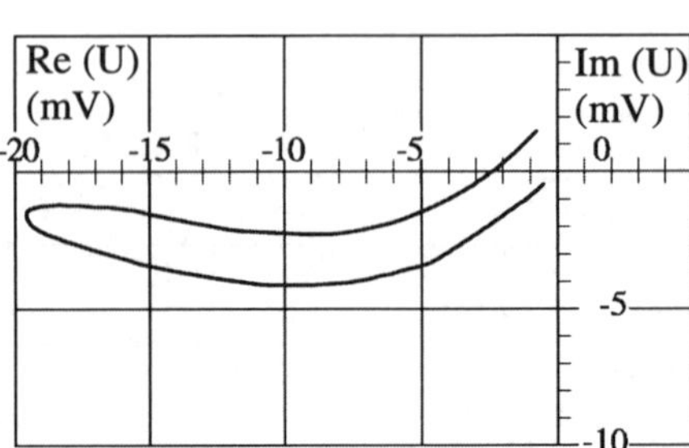

Fig. 8b. Real and imaginary part of the output voltage on the complex plane

5.2. *Single holes*

Figs. 9 shows in general way the flaws marked by 3, 4 and 5 in Fig. 6. The results of the experiments for the three flaws are shown in Figs. 9a and 9b, 10a and 10b, 11a and 11b, respectively.

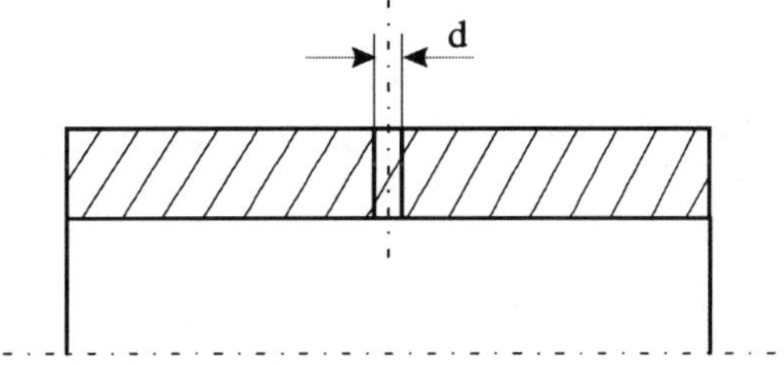

Fig. 9. Flaw as a single hole: $d_1 = 4$ mm, $d_2 = 2$ mm $d_3 = 1$ mm

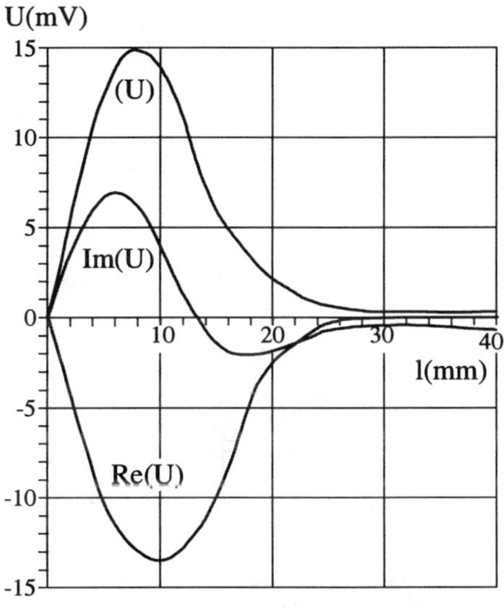

Fig. 9a. Absolute value of the output voltage and its components (real and imaginary part), $d_1 = 4$ mm

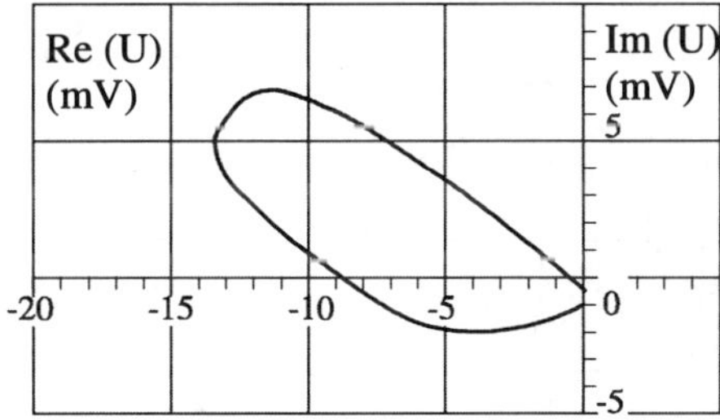

Fig. 9b. Real and imaginary part of the output voltage on the complex plane, $d_1 = 4$ mm

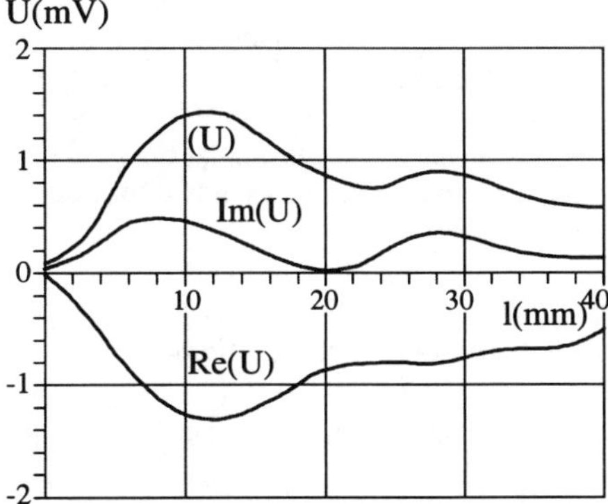

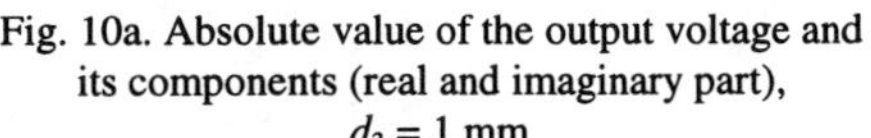

Fig. 10a. Absolute value of the output voltage and its components (real and imaginary part), $d_3 = 1$ mm

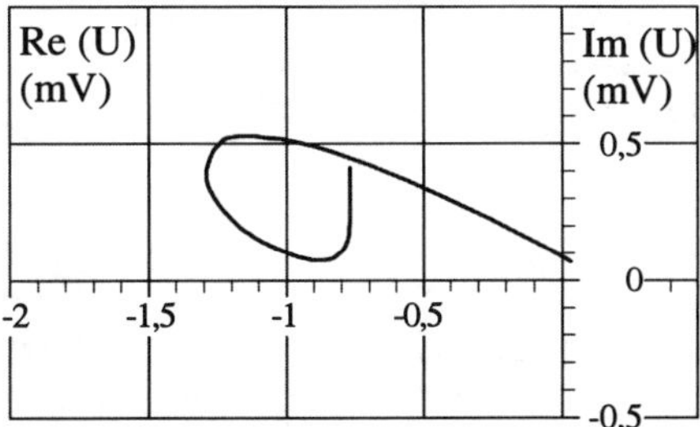

Fig. 10b. Real and imaginary part of the output voltage on the complex plane, $d_3 = 1$ mm

5.3. Sets of holes

Fig. 11 shows the flaws marked by 6 in Fig. 6. The results of the experiments for the set of two holes are shown in Figs. 11a and 11b.

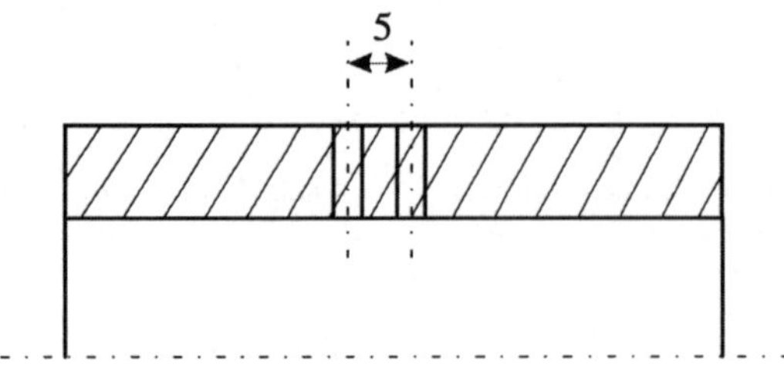

Fig. 11. Set of two holes of diameters equal to 1 mm

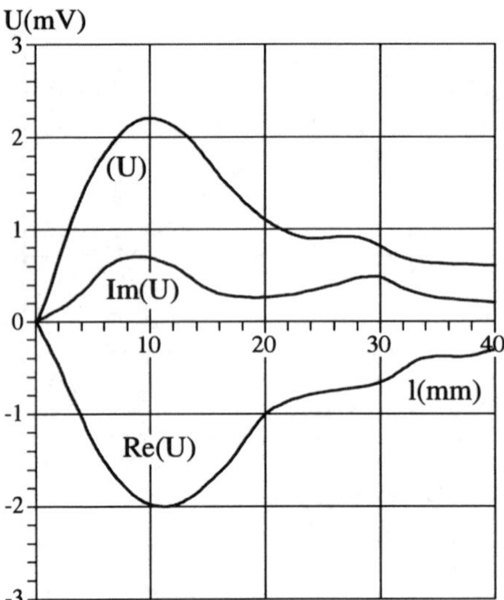

Fig. 11a. Absolute value of the output voltage and its components (real and imaginary part) for the set of two holes of diameters equal to 1 mm

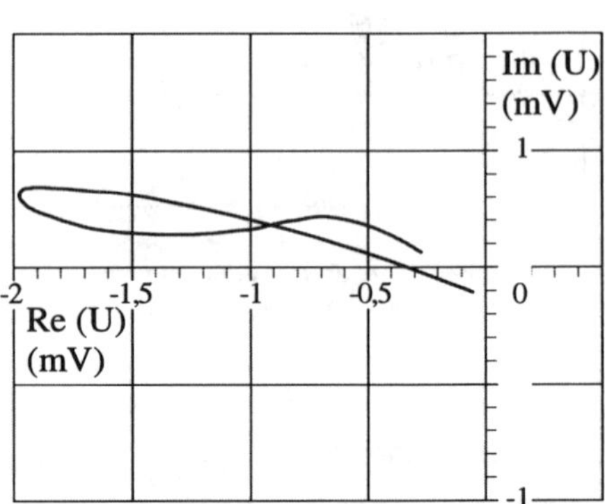

Fig. 11b. Real and imaginary part of the output voltage on the complex plane for the set of two holes of diameters equal to 1 mm

5.4. Influence of frequency and amplitude of the exciting current on output signal

The experiments were repeated at different frequencies and amplitudes of the exciting current for a single hole of diameter equal to 4 mm. The results are shown in Figs. 12 – 15.

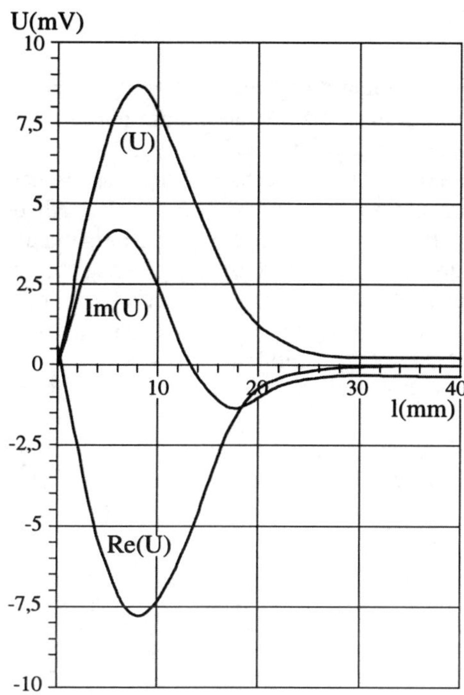

Fig. 12. Absolute value of the output voltage and its components (real and imaginary part) for the single hole of diameter equal to 4 mm, f = 1 kHz and I = 30 mA

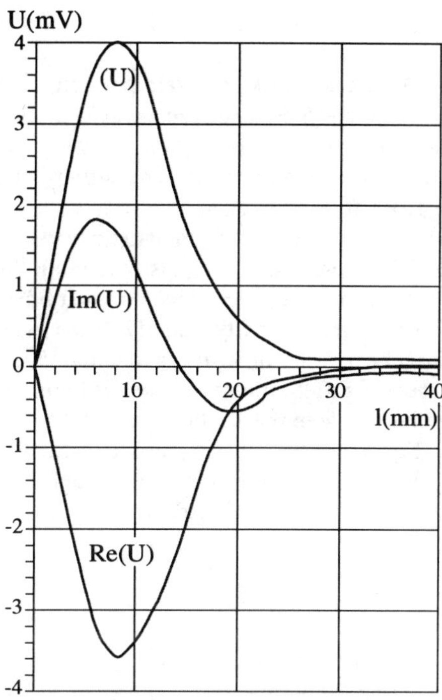

Fig. 13. Absolute value of the output voltage and its components (real and imaginary part) for the single hole of diameter equal to 4 mm, f = 1 kHz and I = 10 mA

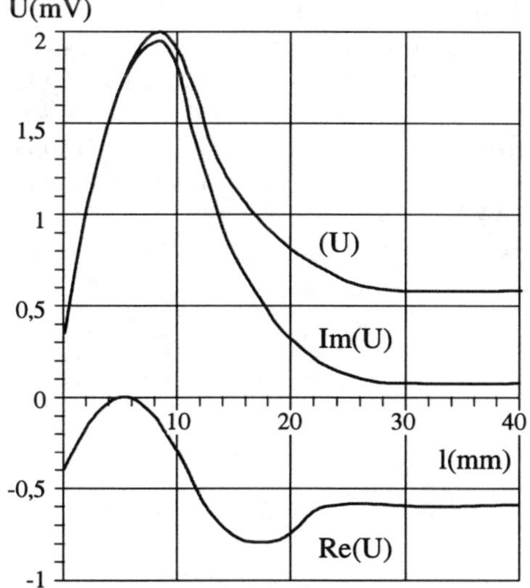

Fig. 14. Absolute value of the output voltage and its components (real and imaginary part) for the single hole of diameter equal to 4 mm, f = 500 Hz and I = 10 mA

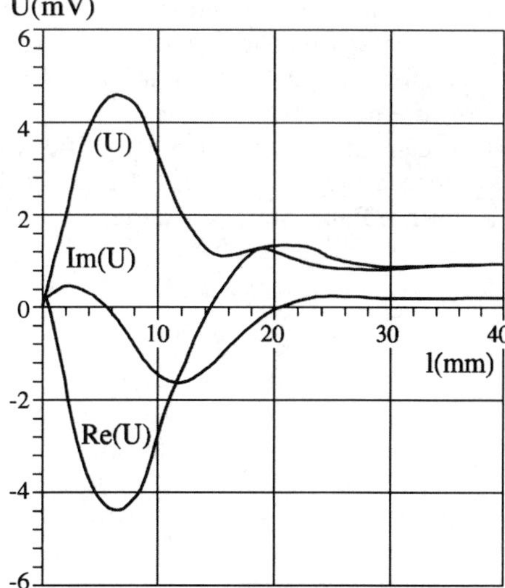

Fig. 15. Absolute value of the output voltage and its components (real and imaginary part) for the single hole of diameter equal to 4 mm, f = 2 kHz and I = 10 mA

6. Conclusions

One of the most important problems in eddy current non-destructive testing is the design of the probe. The results of investigations reported in Figs. 7 ÷ 21 (and those not reported here due to space limitation) make it possible to evaluate the efficiency of the probe presented in the paper. The sensitivity of detection of surface cracks and holes is similar, because for the excitation frequencies used in the experiments, the equivalent depth of penetration of the electromagnetic wave is comparable with the thickness of the pipe under test. The sensitivity depends mainly on the section of a flaw, which perturbs eddy current streamlines. The dependence is non-linear. The probe can easily detect axially oriented cracks. The sensitivity linearly depends on the amplitude of the exciting current, and in a non-linear fashion on the excitation frequency. Due to its simplicity and sensitivity, the proposed differential type probe is highly recommended. Detailed theoretical analysis of the probe is now in preparation.

Equation (12) derived from theoretical analysis is sufficiently accurate for flaws 4, 5 and 6 (Fig. 6). However, the results yielded in case of the remaining flaws seem to be highly unsatisfactory. A modified formula free from the above limitation is being currently tested.

7. Acknowledgement

The authors acknowledge the financial support provided by the Polish Scientific Research Committee under the Grant 8 T 10A 058 12

References

[1] USA Standard ASTM E 243–71 T: Electromagnetic Testing of Seamles Copper Alloys Heat Exchanger and Condenser Tubes.
[2] B. S. 3889: Part 2B: 1966: Eddy–Current Testing of Non–Ferrous Tubes.
[3] S. Fukui et al., Progress of Verification Test Program of Eddy Current Inspection Technology for Steam Generator Tube of Nuclear Power in Japan. In: Electromagnetic Nondestructive Evaluation by T. Takagi, J.R.Bowler, Y.Yoshida (Eds.), IOS Press, Amsterdam, 1997, pp. 1-8.
[4] B. Giovanni, Fabrication of Artificial Reference Defects for Non–Destructive Testing. Proceedings of the 12th World Conference on Non–Destructive Testing. Elsevier Science Publishers, Amsterdam, Vol. 1, pp 548–553, 1989.

Electromagnetic Nondestructive Evaluation (III)
D. Lesselier and A. Razek (Eds.)
IOS Press, 1999

New Eddy Current Probe for NDE of Steam Generator Tubes

Masaaki KUROKAWA, Reizou MIYAUCHI, and Koji ENAMI
Takasago R&D Center, Mitsubishi Heavy Industries, LTD.
Shinhama 2-1-1, Arai-cho, Takasago, Hyogo, 676-8686, Japan

Mitsuyoshi MATSUMOTO
Kobe Shipyard & Machinery Works, Mitsubishi Heavy Industries, LTD.
1-1, Wadasaki-cho, 1-chome, Hyogo-ku, Kobe, 652-8585, Japan

Abstract. A new eddy current probe for the steam generator tube inspection was produced and examined. In order to inspect the full length of steam generator tubes with the fast scanning speed and the high detectability, the high density multi-segment probe with transmitter and receiver segments was adopted. Especially for the receiver segment, 4 coil configuration was applied. To reduce the number of electric wires and also to prevent from the degradation of electrical SN ratio, electronic circuits were installed in the probe. The preliminary function test of this probe showed the high performance of overall function and the potential detectability to detect the small defects.

1. Introduction

The eddy current testing is applied for the steam generator tube inspection. In Japan, there are more than 20 PWR plants, and each plant has 2~4 steam generators with about 3000 tubes for each generator. The full length of each tube is inspected almost annually. So, a lot of inspections are performed every year.

For the eddy current testing, various kinds of eddy current probes such as the bobbin probe, the multi-segment probe and the rotating probe were developed and applied. As for the bobbin probe, it can inspect with a high scanning speed, but the detectability between axial and circumferential defects are different. In case of the multi-segment probe, it can perform high speed inspection, but it usually has the low sensitivity zones between segments. The rotating probe has a good detectability, but it cannot inspect with a high scanning speed. Therefore, we produced and examined the new eddy current probe which can perform the fast inspection with the high detectability.

To realize the fast inspection speed and the durability, the non-contact multi-segment type was adopted. As for the detectability , various coil alignments were investigated and the optimum type was selected. As for the low sensitivity zones, high density multi-segments were applied to reduce them. Thin film coils were adopted to obtain uniform detectability. And electronic circuits were installed in the probe to increase the SN ratio and to reduce the number of electric wires.

2. Coil Alignment

As for the coil alignments, we investigated the induction method (self and mutual),coil types (pancake and 4 coil). Consequently, we selected the mutual induction probe with 4 coil receiver [1]. Furthermore, we investigated driving coils to increase the detectability.

Fig.1 shows 3 coil alignments which we investigated by the numerical analysis. For the driving coils, we investigated the angular coil and the pancake coil. The angular coil is also a pancake coil, but it is installed in the probe such as perpendicular to the pickup coil plane. The pancake coil probe was analysed as a contact probe, but the other probes were analysed as non-contact. Fig.2 is the tube expanded region which was used as the analytical model. The absolute values of impedance changes by the defect and the tube transition zone were calculated, and the SN ratio was estimated. The analysed results are shown in Fig.3. The detectability of the angular coil drive-4 coil pickup probe was best among 3 types of probes.

Probe type	Coil alignment	Specification
Angular coil drive 4 coil pickup (non-contact)	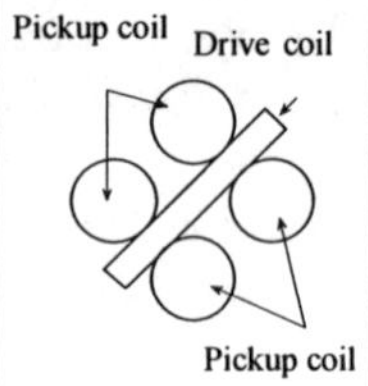 Pickup coil Drive coil Pickup coil	○ Drive coil · Pancake coil · Perpendicular to the pickup coil plane ○ Pickup coil · 4 pancake coils
Pancake drive 4 coil pickup (non-contact)	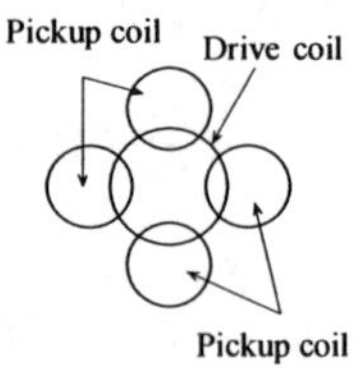 Pickup coil Drive coil Pickup coil	○ Drive coil · Pancake coil ○ Pickup coil · 4 pancake coils
Pancake coil (contact)	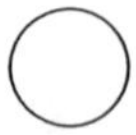 Drive and pickup	○ Self-induction · Pancake coil

Fig.1 Specifications of the coil segments (Analytical model)

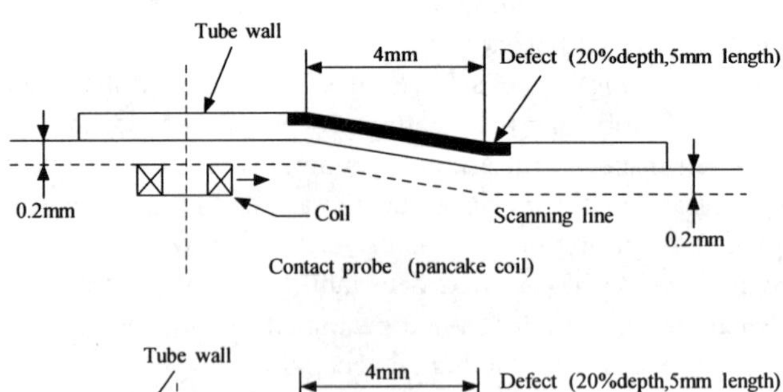

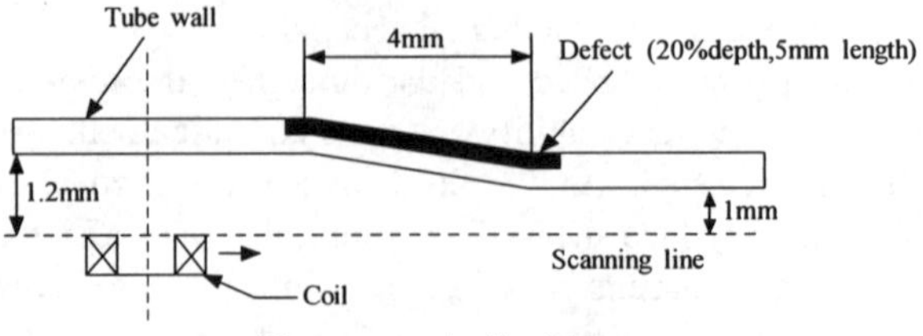

Fig.2 Analytical models of the tube expanded region

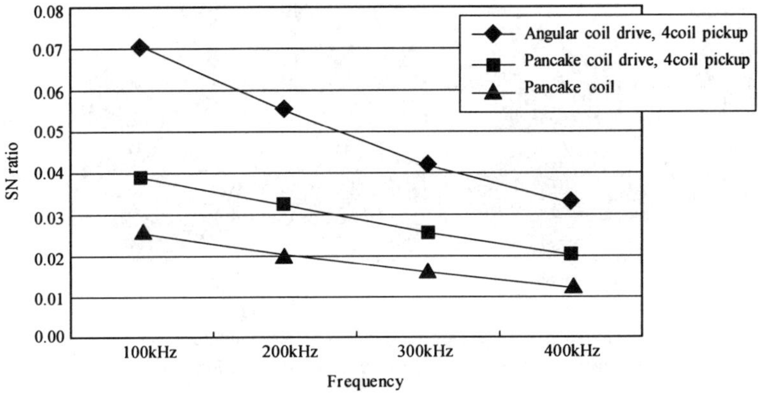

Fig.3 Analytical results

In order to confirm this analytical result, the experimental work was also conducted. One coil segment of each probe was made, put on a rotating mechanism, and examined as a rotating probe using a test piece of the tube expanded region. Fig.4 shows an example of the experimental results. These results also showed the good performance of the angular drive-4 coil pickup probe.

For the next step, we selected the number of coil segments for the angular drive-4 coil pickup probe which is closely related to the low sensitivity zones of the probe. Fig.5 showed the experimental data obtained by the rotating probe with a non-contact angular drive-4 coil pickup segment.

Circumferential sampling points are 24 and 64 in Fig.5. In case of 24 sampling points, data are interpolated into 64 points. All signals are processed as to reduce noise signals. This data showed that the difference between 24 and 64 points was small, and the 20% defects signals were apparent in both cases. So we selected the number of segments to be 24.

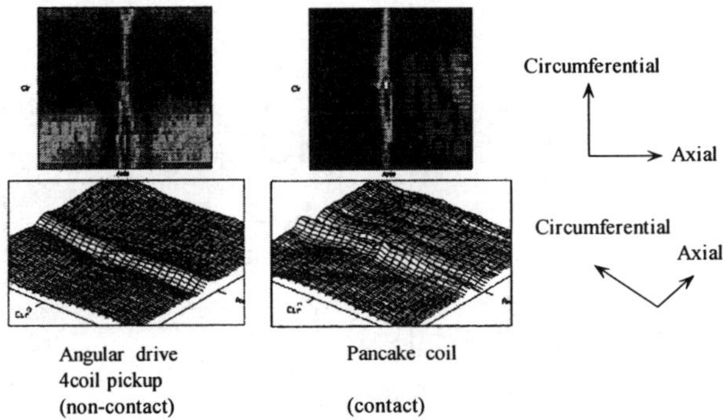

Fig.4 Comparison of detected signals
(Tube expanded region. OA20%)

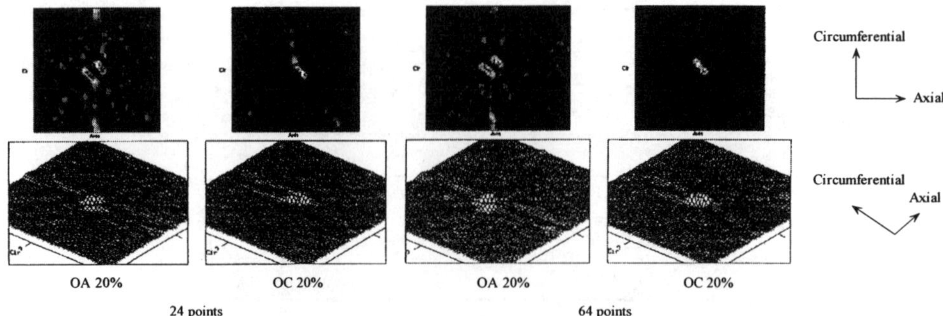

Fig.5 Effect of circumferential sampling point number
(Tube expanded region. Angular drive-4coil pickup. Filtering results)

3. Electronic Circuits and the New ECT Probe

In order to increase SN ratio and to reduce electric wires, electronic circuits are installed in the probe. Fig.6 shows the block diagram of the electronic circuits. Based on the addressing signals from the flaw detector, receiving signals from pickup coils are selected by the multiplexing circuits. Pre-amplifier circuits are also involved in the probe.

Fig.7 shows the new ECT probe. This probe mainly consists of the coil segment block and the electronic circuit blocks. There are 24 segments in the coil segment block. Each segment has an angular drive coil and 4 pickup coils. Thin film coils are adopted for the pickup coils in order to obtain uniform detectabillity.

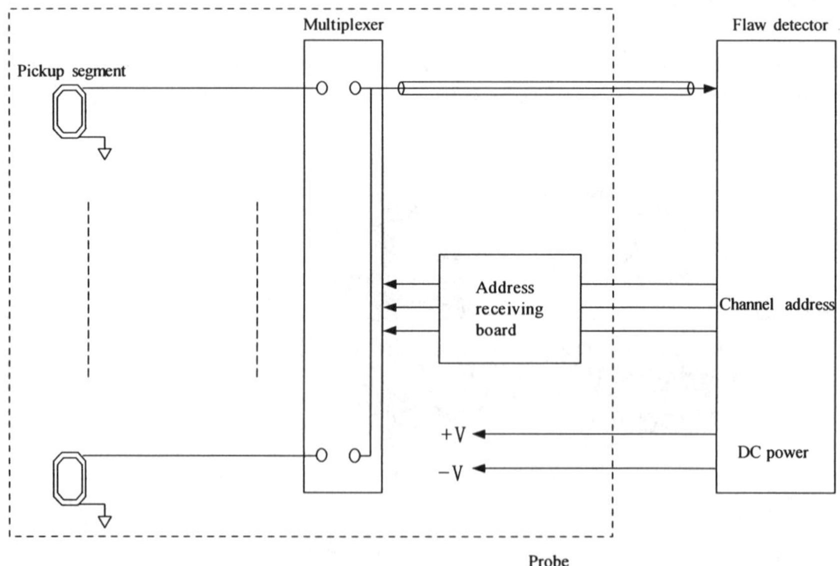

Fig.6 Configuration of electronic circuits

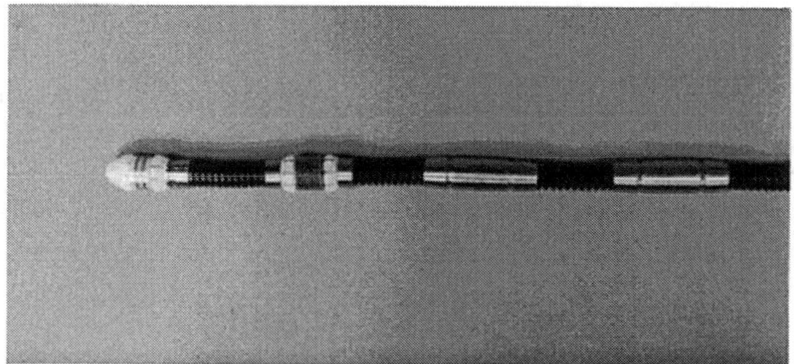

Fig.7 Main part of the new eddy current probe

4. Functional Tests of the New ECT Probe

Preliminary functional tests were performed using the new ECT probe. One is the circumferential sensitivity of the probe and the other is detectability.

As for the circumferential sensitivity, axial EDM notches were used to measure it. Fig.8 is the test result. The horizontal axis of the graph means the circumferential distance between a segment and the notch. Sensitivity characteristics obtained by the adjacent segments were also included in these graphs. It was confirmed from these data that there are not low sensitivity zones between segments as expected from the previous experiment.

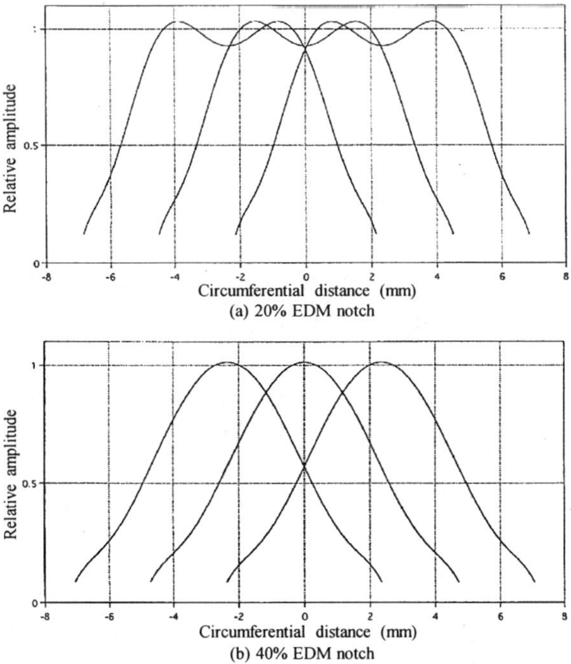

Fig.8 Circumferential sensitivity obtained by OD axial notch

Following the circumferential sensitivity test, the preliminary detectability test was conducted. Fig.9 shows the test piece used for this test. Fig.10 is the chart data, and Fig.11 is C scan representations for 20% EDM notches. There are original signals, filtering results and multi-frequency processing results in this figure. These data showed that the new ECT probe has enough potential to detect small defects.

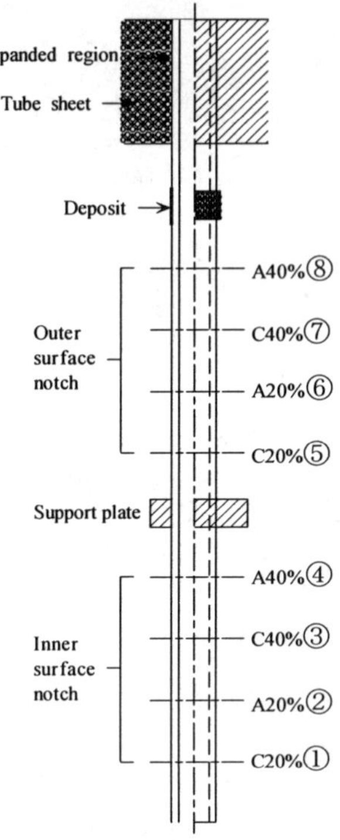

Fig.9 Test piece

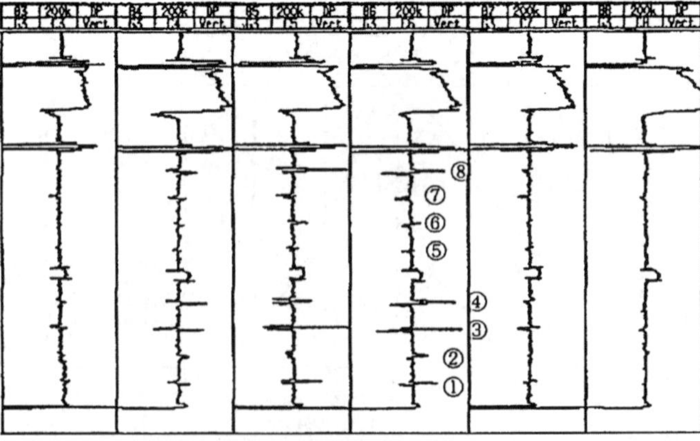

Fig.10 Experimental result by test piece

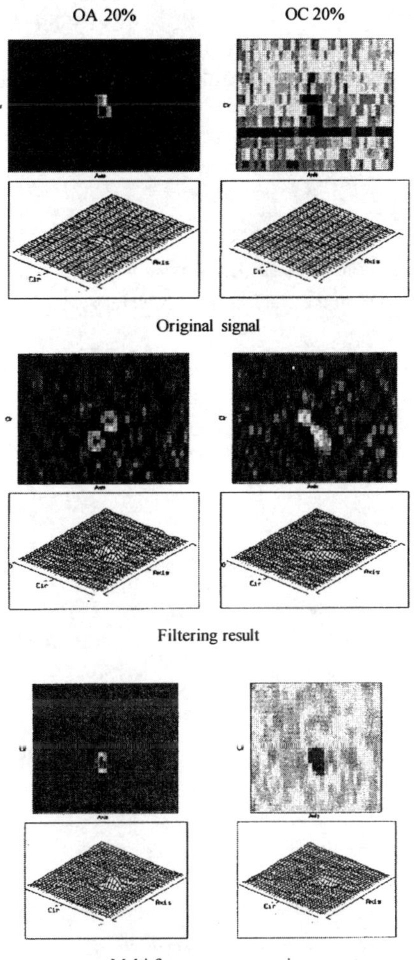

Fig.11 Defect signals obtained by OD 20% notches

5. Conclusions

In order to realize the new ECT probe with the fast scanning speed and the high detectability, the angular drive-4 coil pickup non-contact segment was selected based on the analytical and experimental results. The probe is composed of 24 coil segments and electronic circuits. The thin film coils were applied for pickup coils to obtain uniform sensitivity between coils. Preliminary function tests showed that there are no circumferential low sensitivity zones and this probe has enough potential to detect small defects.

Acknowledgement

The verification test of this probe is funded by MITI and managed by Japan Power Engineering and Inspection Corporation.

Reference

[1] M.Kurokawa, T.Kamimura, K.Enami and A.Kurokawa, "Development of New Eddy Current Testing Probe", 14th Int. Conf. on NDE in the Nuclear and Pressure Vessel Industries, pp.113-119, September, 1996

Defect Evaluation with Emphasis on Stochastic Methods

Electromagnetic Nondestructive Evaluation (III)
D. Lesselier and A. Razek (Eds.)
IOS Press, 1999

Direct and Inverse Calculations for Cracks of Different Shapes

Raffaele ALBANESE
Associazione EURATOM/ENEA/CREATE, DIMET, Università degli Studi di Reggio Calabria, Via Graziella, Loc. Feo di Vito, I-89128 Reggio Calabria, Italy

Guglielmo RUBINACCI, Fabio VILLONE
Associazione EURATOM/ENEA/CREATE, Dip. di Ing. Industriale, Università degli Studi di Cassino, Via Di Biasio 43, I-03043 Cassino (FR), Italy

Abstract. An integral formulation in terms of a two-component electric vector potential (based on a tree-cotree decomposition of the mesh and edge-element basis functions) is used for the solution of the JSAEM Problem 6, which consists in the reconstruction of the shapes of various cracks from the impedance change of a pancake coil. The cracks are assumed to be infinitely thin. Superposition and reciprocity are employed in order to get more accurate impedance calculations. Woodbury's algorithm is used to speed up the solution process: the solution of each direct problem requires the inversion of a linear system whose order is less than the number of degrees of freedom related to the crack region. The inverse problem is formulated as the search of the mesh facets belonging to the crack. Due to the binary nature of the unknown, a genetic-like inversion algorithm is used for the determination of the crack shape, minimizing the normalized root mean square error between the predictions and the measurements of the impedance variation.

1. Introduction

One of the greatest challenges of electromagnetic nondestructive evaluation is the reconstruction of the shapes of cracks using external measurements. In order to validate the existing procedures several benchmark problems have been proposed in the last years [1,2].

In this paper, we focus our attention on the JSAEM Problem 6 - crack of different shapes [2]. A pancake coil is moved over a cracked metallic plate, and the impedance change with respect to the unperturbed situation is recorded at two different frequencies for various crack shapes and locations.

In order to solve this problem, an efficient integral formulation in terms of the electric vector potential is used for the direct analysis [3-5]. The crack is assumed to be infinitely thin, and superposition is used in order to impose that the normal component of the current density is zero on the crack. In addition, reciprocity is employed in order to get more accurate impedance calculations.

The features of a non-conventional gauge condition (two-component gauge condition imposed by means of a tree-cotree decomposition of the mesh) and of the edge-elements basis functions are fully exploited [3-4]. In this way, a solution of the direct problem can be obtained, which is very efficient in terms both of the number of unknowns required and of the computational time needed [5-6]. We are able to exactly enforce that the current flowing through a mesh facet is the opposite of the unperturbed one, hence forcing the total current through the crack to be zero. An analytical method [7] is used in order to obtain the unperturbed solution, although this is not a limitation of the method, since a numerical solution can be easily used as well.

The inverse problem is formulated as the search of the mesh facets belonging to the crack. Due to the binary nature of the unknown, a genetic-like inversion algorithm is used for the determination of the crack shape, minimizing the normalized root mean square error between the predictions and the measurements of the impedance variation.

2. Direct and inverse analysis methods

The unknown current density is expanded as:

$$\mathbf{J} = \Sigma_k I_k \nabla\times\mathbf{N}_k \tag{1}$$

where the $\mathbf{N}_k$'s are edge element based shape functions for a current density vector potential $\mathbf{T}$ whose uniqueness is guaranteed by the gauge condition:

$$\mathbf{T}\cdot\mathbf{w} = 0 \tag{2}$$

where the direction of $\mathbf{w}$ is identified by any tree of the graph made by nodes and edges of the mesh [3-4]. Therefore the shape functions considered in Eq. (1) are only those corresponding to the cotree edges.

The condition

$$\mathbf{J}\cdot\mathbf{n} = 0 \tag{3}$$

at the boundary of the conducting domain, for simply connected regions and a suitable choice of the tree, is usually enforced by imposing the tangential component of $\mathbf{T}$ to be zero at the boundary.

The resulting linear system of equation is obtained by applying Galerkin's method to Ohm's law [3-4]:

$$\mathbf{Z}\,\mathbf{I} = \mathbf{V} \tag{4}$$

with:

$$\mathbf{Z} = \mathbf{R} + j\omega\,\mathbf{L} \tag{5}$$

$$L_{ij} = \frac{\mu_0}{4\pi}\int_{V_c}\int_{V_c}\frac{\nabla\times\mathbf{N}_i(\mathbf{x})\cdot\nabla\times\mathbf{N}_j(\mathbf{x}')}{|\mathbf{x}-\mathbf{x}'|}dVdV' \tag{6}$$

$$R_{ij} = \int_{V_c}\nabla\times\mathbf{N}_i(\mathbf{x})\cdot\eta\nabla\times\mathbf{N}_j(\mathbf{x})dV \tag{7}$$

$$V_i = -j\omega\int_{V_c}\nabla\times\mathbf{N}_i(\mathbf{x})\cdot\mathbf{A}_s(\mathbf{x})dV \tag{8}$$

where V_c is the conducting domain, η the electric resistivity, μ_0 the magnetic permeability of the vacuum, j the imaginary unit, ω the angular frequency, $\mathbf{A}_s$ the magnetic vector potential due to the exciting current density $\mathbf{J}_s$.

2.1 Solution of the direct problem

Using superposition, the forward problem is reformulated as the determination of the modified eddy current pattern due to the presence of the defect. The current density is given by:

$$\mathbf{J} = \mathbf{J}_0 + \delta\mathbf{J} \tag{9}$$

where the unperturbed current density $\mathbf{J}_0 = \Sigma\, I_{0k}\, \mathbf{J}_{0k}$ is obtained by means of Eq. (4) or alternative methods.

We also expand $\delta\mathbf{J}$, the perturbation due to the crack, in terms of solenoidal shape functions, which not necessarily coincide with those used for the flawless plate:

$$\delta\mathbf{J} = \Sigma_{k=1,n}\, \delta I_k\, \mathbf{J}_k \qquad (10)$$

A thin crack can be described as a surface Σ_d, discretized via a set of finite element faces characterized by the constraint:

$$\mathbf{J} \cdot \mathbf{n} = 0 \qquad (11)$$

where $\mathbf{n}$ is the normal unit vector on the face.

The flux of $\mathbf{J}=\nabla\times\mathbf{T}$ across any elementary face is given by the circulation of $\mathbf{T}$ along the edges identifying the face. As the values of the unknowns I_k's represent the line integrals of $\mathbf{T}$ along the active edges, the net current crossing an elementary face is given by the algebraic sum of the unknowns associated with the active edges of the face:

$$\mathbf{G} = \mathbf{P}\,\mathbf{I} \qquad (12)$$

where $\mathbf{G}$ is the set of net currents crossing the set T formed by m candidate independent crack facets (a subset of them is Σ_d, which is unknown in the inverse problem), and $\mathbf{P}$ is a (m,n) sub-matrix of the edge-facet incidence matrix with coefficients 0, +1 or -1.

We then make a change of variables [5]:

$$\delta\mathbf{I} = \mathbf{K}\,\delta\mathbf{X} + \mathbf{S}\,\delta\mathbf{G} = \mathbf{M}\,\delta\mathbf{I}' \qquad (13)$$

where $\mathbf{K}$ is a (n,n-m) matrix given by an orthonormal basis for the null space of $\mathbf{P}$, $\mathbf{S}$ is the pseudo-inverse of $\mathbf{P}$, $\delta\mathbf{X}$ is an auxiliary variable, $\mathbf{M} = [\mathbf{K}, \mathbf{S}]$, and $\delta\mathbf{I}' = [\delta\mathbf{X}\,,\, \delta\mathbf{G}]^T$. In this way we get a new representation of $\delta\mathbf{J}$:

$$\delta\mathbf{J} = \Sigma_{i=1,n}\, \Sigma_{k=1,n}\, \delta I'_i\, M_{ki}\, \mathbf{J}_{0k} = \Sigma_{i=1,n}\, \delta I'_i\, \mathbf{J}'_i \qquad (14)$$

with the set of new non-local shape functions $\{\mathbf{J}'\}$ obtained by the old ones by:

$$\{\mathbf{J}'\} = \mathbf{M}^T\, \{\mathbf{J}\} \qquad (15)$$

where the first (n–m) shape functions verify condition (11) on all the m candidate crack facets, whereas the remaining m are characterized by a unit current flux across a single candidate crack facet.

Galerkin's procedure in terms of the new variables yields:

$$\mathbf{M}^T\mathbf{Z}\mathbf{M}\, \delta\mathbf{I}' = \delta\mathbf{V}' = \mathbf{M}^T\, \delta\mathbf{V} = \mathbf{0} \qquad (16)$$

where the right hand side is zero because the exciting current must not be considered in the perturbed problem: the solution is therefore trivial if we do not take the crack into account, by imposing:

$$\delta\mathbf{G} = -\mathbf{G}_0 = -\,\mathbf{P}\,\delta\mathbf{I}_0 \qquad (17)$$

at the crack facets, which is somehow similar to imposing essential boundary conditions for finite element problems.

2.2 Solution of the inverse problem

The inverse problem consists in finding the subset $B \subseteq T$ such that the solution obtained imposing $\delta\mathbf{G}_B = -\mathbf{G}_{0B}$ gives the best fit to the experimental data, characterized by an error function χ. Genetic algorithms can be used due to the binary nature of the problem.

Therefore, several function evaluations are needed and therefore a fast method for solving the direct problem must be available.

Let $F=T-B$ be the set of facets belonging to the tentative set T but not to the crack B. The corresponding current fluxes $\delta\mathbf{G}_F$ are not known *a priori*, and hence the relative equations must be imposed, together with the equations corresponding to $\delta\mathbf{X}$, which are unknown, as in the direct problem:

$$\mathbf{K}^T\mathbf{Z}\mathbf{K}\,\delta X + \mathbf{K}^T\mathbf{Z}\mathbf{S}_F\,\delta G_F + \mathbf{K}^T\mathbf{Z}\mathbf{S}_B\,\delta G_B = 0 \tag{18}$$

$$\mathbf{S}_F{}^T\mathbf{Z}\mathbf{K}\,\delta X + \mathbf{S}_F{}^T\mathbf{Z}\mathbf{S}_F\,\delta G_F + \mathbf{S}_F{}^T\mathbf{Z}\mathbf{S}_B\,\delta G_B = 0 \tag{19}$$

$$\delta G_B = -G_{0B} \tag{20}$$

where the subscripts B and F characterize fluxes and submatrices relative to the facets belonging to B and F respectively.

Obtaining $\delta\mathbf{X}$ from (18) and substituting in (19) we obtain

$$\delta X = -(\mathbf{K}^T\mathbf{Z}\mathbf{K})^{-1}(\mathbf{K}^T\mathbf{Z}\mathbf{S}_F\,\delta G_F + \mathbf{K}^T\mathbf{Z}\mathbf{S}_B\,\delta G_B) \tag{21}$$

$$\mathbf{A}\,\delta G_F = \mathbf{b} \tag{22}$$

where:

$$\mathbf{A} = \mathbf{S}_F{}^T\mathbf{Z}\,\mathbf{K}\,(\mathbf{K}^T\mathbf{Z}\mathbf{K})^{-1}\,\mathbf{K}^T\mathbf{Z}\mathbf{S}_F - \mathbf{S}_F{}^T\mathbf{Z}\mathbf{S}_F \tag{23}$$

$$\mathbf{b} = [\mathbf{S}_F{}^T\mathbf{Z}\mathbf{K}\,(\mathbf{K}^T\mathbf{Z}\mathbf{K})^{-1}\,\mathbf{K}^T\mathbf{Z}\mathbf{S}_B - \mathbf{S}_F{}^T\mathbf{Z}\mathbf{S}_B]\,G_{0B} \tag{24}$$

Hence, we can invert $\mathbf{K}^T\mathbf{Z}\mathbf{K}$ once for all, and only the $\mathbf{A}$ matrix must be inverted at each step in order to find $\delta\mathbf{G}_F$, and only products are needed for $\delta\mathbf{X}$. Once $\delta\mathbf{X}$ and $\delta\mathbf{G}$ are known, it is possible to get the original unknowns $\delta\mathbf{I}$ from (13) and in turn any other possible output variable (field measurements, impedance variation, etc.). Notice that the dimension of the $\mathbf{A}$ matrix is much smaller than $\mathbf{Z}$ if we roughly know in which part of the specimen the crack is located. A further speed-up is obtained by applying Woodbury's algorithm, which is very efficient whenever we modify a few equations in a linear system [8].

The binary nature of the unknown and the effectiveness of the direct solver immediately call for a genetic-like inversion algorithm. Starting from an initial random population of binary strings, each corresponding to a tentative crack configuration, the individuals are evolved using the standard rules of genetic algorithms: crossover and mutation [9]. The population is classified on the basis of the normalized root mean square error between the predictions and the measurements of the impedance variation.

3. Analysis of JSAEM Problem 6

In the JSAEM Problem 6 [2] a pancake coil (internal radius 0.6 mm, external radius 1.6 mm, height 0.8 mm, lift off 0.5 mm) is moved over a cracked metallic plate of dimensions 140 mm × 140 mm × 1.25 mm, having a resistivity of $10^{-6}\Omega$m. The impedance change with respect to the unperturbed situation is recorded at two different frequencies (150 kHz and 300 kHz). The coil movement is along the crack major dimension. Four cracks of different shapes are presented: elliptical, rectangular, stepwise and sloping, both on the inner side (superficial crack) and on the outer side (buried crack) of the plate.

The unperturbed eddy current pattern $\mathbf{J}_0$ is obtained by the analytical solution reported in [7] for an infinite plate. The perturbation $\delta\mathbf{J}$ due to the crack is calculated using a finite element mesh of a small region around the crack itself, neglecting the effects of the perturbation far from the crack. For the 150 kHz excitation we considered a 25 mm × 1.25 mm × 16 mm region, meshed with 840 elements, 1200 nodes, 3202 edges, 1477 degrees of freedom, exploiting symmetry condition at the plane z=0. For the 300 kHz excitation we used a mesh referring to a smaller region (11 mm instead of 16 mm along the z direction).

As the crack thickness (0.2 mm) is much smaller than the other dimensions and the skin depth, (0.92 mm) we employed the zero-thickness crack approximation. The cracks were then schematized as a set of mesh faces (see Figure 1).

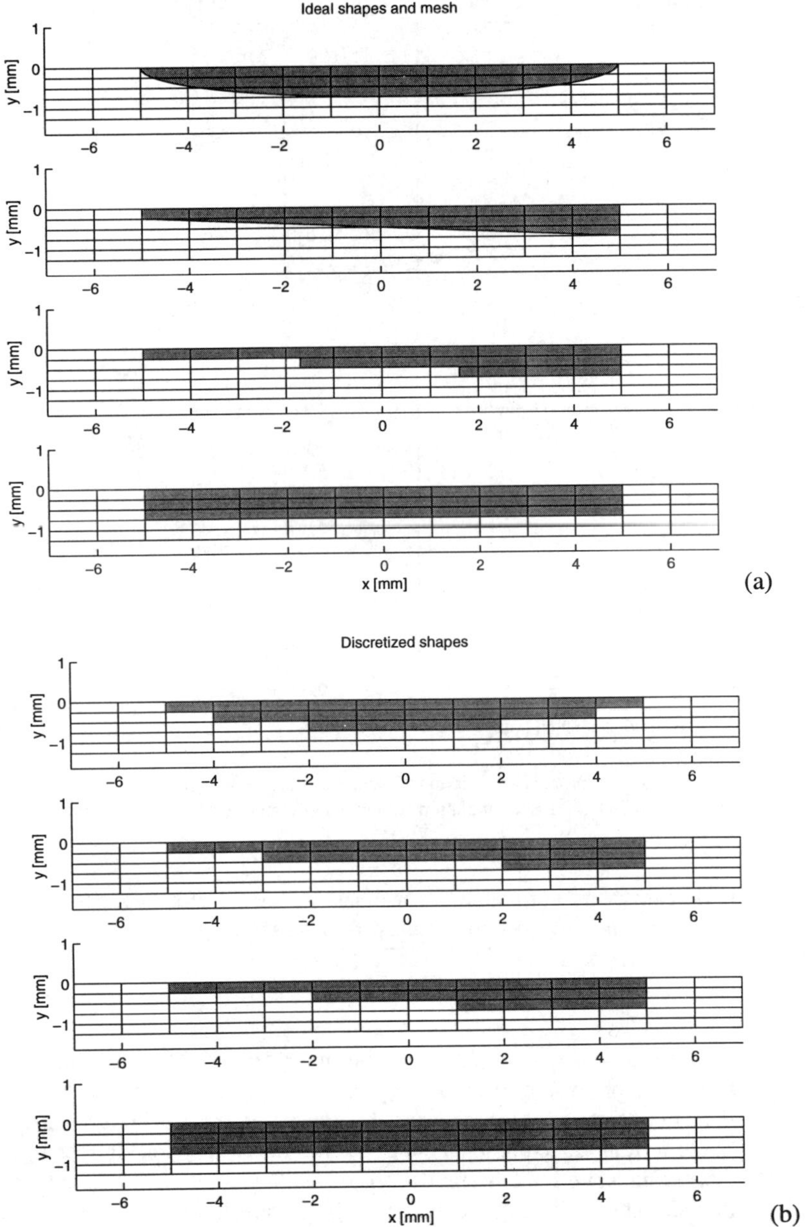

Figure 1. JSAEM problem 6: (a) nominal defect shapes; (b) approximated defect shapes considered in the direct analysis

Figure 2 shows the impedance variations predicted in the cases of inner elliptical defect (150 kHz) and outer stepwise crack (300 kHz). In both cases the results are compared to the experimental measurements [10]. The offset in the experimental data was assumed to be such that ΔZ=0 at x=10 mm.

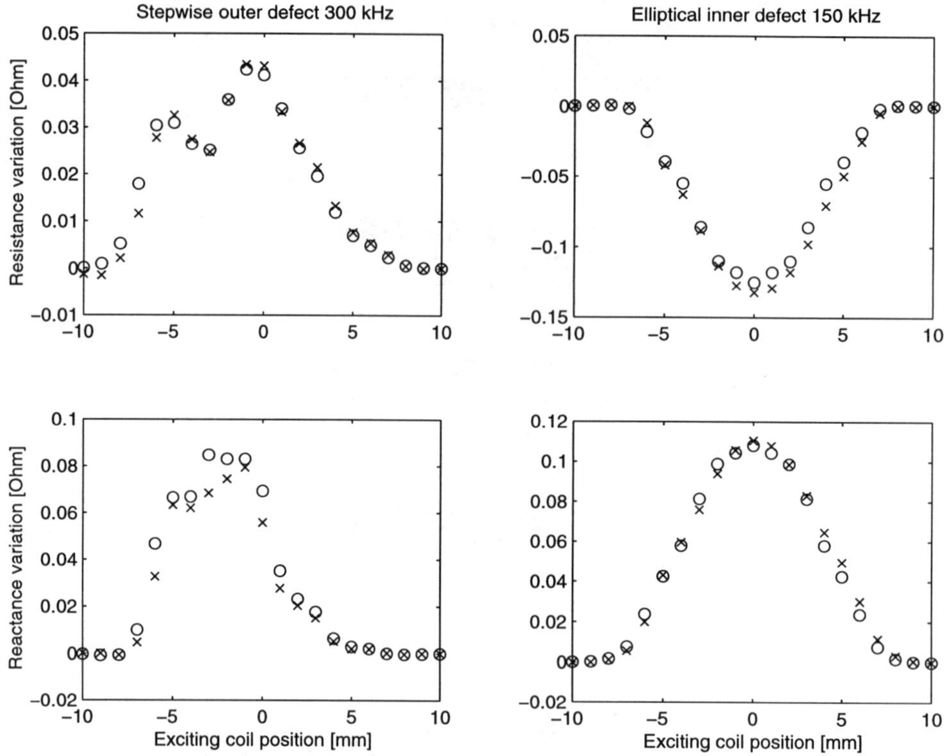

Figure 2. Direct problem. Impedance variations for the outer stepwise defect at 300 kHz and the inner elliptical crack at 150 kHz: simulation results (o) vs experimental measurements (x).

Figure 3 reports the normalized root mean square difference χ between numerical results n_k and experimental measurements m_k for all defects:

$$\chi = \frac{\sqrt{\sum_k (n_k - m_k)^2}}{\sqrt{\sum_k m_k^2}} \tag{25}$$

As expected, the results are better for the inner defects. The normalized results for the outer defects are more distant from the experimental values at 150 kHz, but it should be noted that the level of the signal is smaller at the lower frequency. The discretization of the crack area is able to reproduce the ideal shape of the defect only for the rectangular crack. For the other cases the solid models and the corresponding solutions to the direct problems might be improved either adapting the shape of the hexahedral elements or increasing the number of elements. The latter choice would increase the computational effort but is more practical when considering the inverse problem in which the defect shape is unknown.

To solve the inverse problem more efficiently, a regularizing assumption has been made on the crack shape. In fact, it has been assumed that the only thing that must be determined is the crack depth at each position, either starting from the inner (superficial crack) or the outer (buried crack) plate surface.

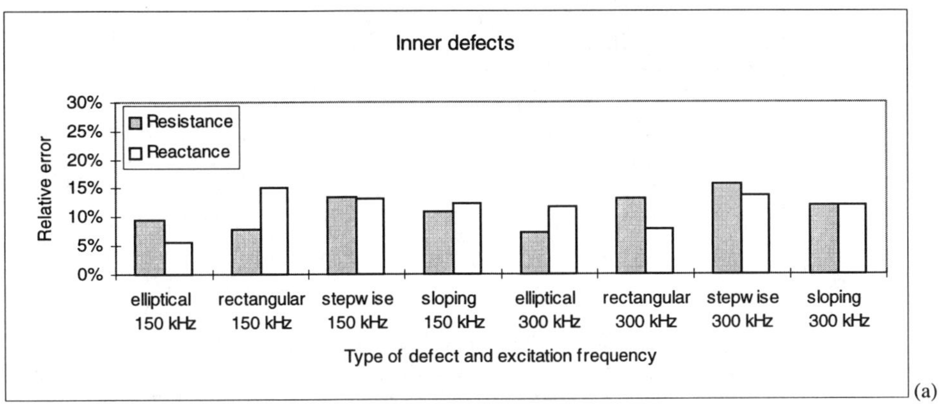

(a)

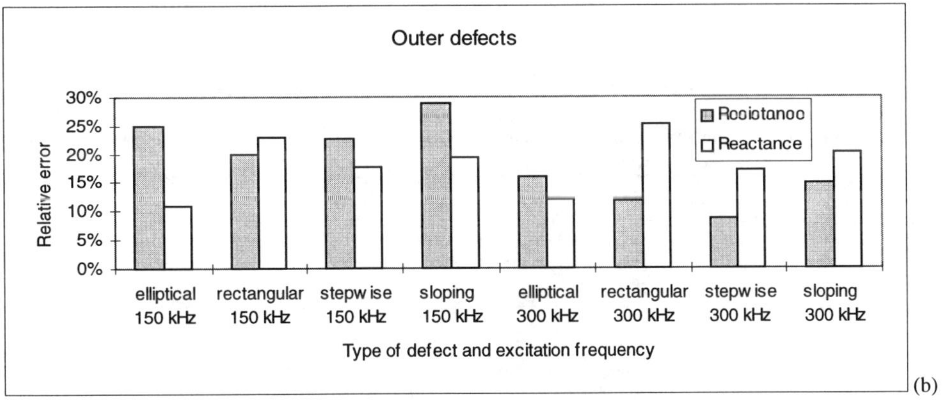

(b)

Figure 3. Direct problem. Normalized root mean square difference between numerical results and experimental measurements: (a) inner defects; (b) outer defects.

Figure 4 shows the defect shapes identified by minimizing the sum of the normalized root mean square difference between numerical results and experimental measurements for the impedance variations. In all cases the value of the cost function achieved by the genetic algorithm was smaller than that corresponding to the nominal shape considered in the direct analysis. Table 1 reports the values of the cost function in the cases of inner elliptical defect (150 kHz) and outer rectangular crack (300 kHz). Figure 5 shows the numerical values of the impedance variations for the reconstructed shapes in comparison to the experimental measurements.

The number of function evaluations was around 2500 (respectively 3000) for the inner elliptical (respectively outer rectangular) crack and the corresponding calculation time was around 250 s (respectively 400 s) on a Personal Computer with a clock of 200 MHz.

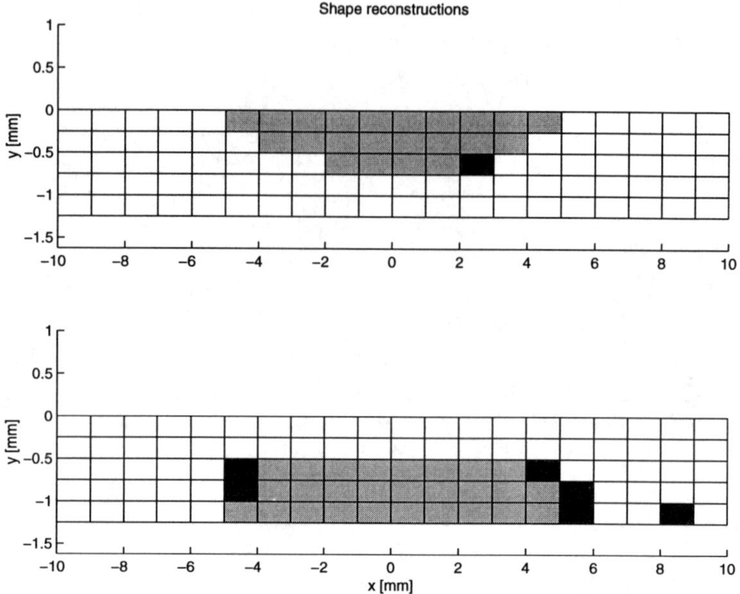

Figure 4. Inverse problem. Defect shapes identified by minimizing the sum of the normalized root mean square difference between numerical results and experimental measurements for the impedance variations: inner elliptical crack at 150 kHz (upper graph) and outer rectangular defect at 300 kHz (lower graph). The colors indicate whether the mesh faces belong to the crack surface or not:

	reconstructed shape	direct problem schematization
white	no	no
light gray	yes	yes
dark gray	no	yes
black	yes	no

Table 1. Values of the cost function (minimizing the sum of the normalized root mean square difference between numerical results and experimental measurements for the impedance variations ΔZ.

	Inner elliptical defect at 150 kHz		Outer rectangular defect at 300 kHz	
	Nominal configuration (direct problem)	Identified configuration (inverse problem)	Nominal configuration (direct problem)	Identified configuration (inverse problem)
Normalized RMS difference of ΔR	9.56%	5.74%	11.91%	7.32%
Normalized RMS difference of ΔX	5.65%	6.42%	25.30%	16.08%
Normalized RMS difference of ΔZ	15.21%	12.16%	37.21%	23.40%

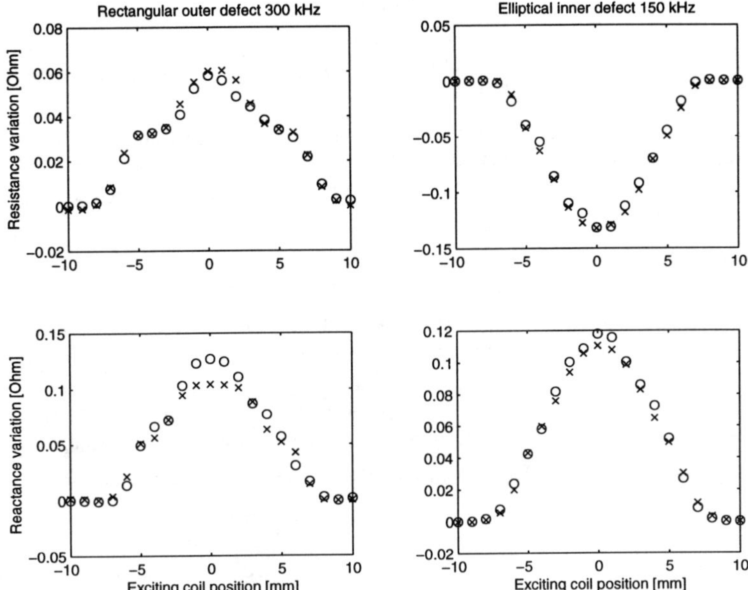

Figure 5. Inverse problem. Impedance variations for the outer rectangular defect at 300 kHz and the inner elliptical crack at 150 kHz: simulation results corresponding to the reconstructed shapes shown in Figure 4 (o) vs experimental measurements (x).

4. Conclusions

We have applied a crack shape identification procedure based on the integral formulation in terms of a two-component electric vector potential for the analysis of JSAEM problem 6, minimizing the sum of the normalized root mean square difference between numerical results and experimental measurements for the impedance variations.

For the direct problem we found a reasonable agreement with a rather coarse mesh in all cases even if for the sloping crack the solid model, mainly devoted to the solution of the inverse problem, was not well suited.

In both inverse problems examined, the minimization procedure based on a genetic algorithm achieved values of the cost function smaller than those obtained in the solution to the direct problem, and the reconstructed crack shapes were fairly close to the nominal ones.

Acknowledgements

This work was supported by the INCO-COPERNICUS Project PL 964037 of the European Commission and by the Italian MURST, with the contribution of Regione Calabria and the European Union.

References

[1] S. K. Burke, "A Benchmark Problem for Computation of ΔZ in Eddy-Current Non-Destructive Evaluation (NDE)", *J. Nondestr. Eval.*, Vol. 7, 1988, pp.35-41

[2] T. Takagi, M. Uesaka, K. Miya, "Electromagnetic NDE Research Activities in JSAEM", *Electromagnetic Nondestructive Evaluation*, T. Takagi et al. (Eds.), IOS press, pp. 9-16 (1997)

[3] R. Albanese, G. Rubinacci, "An Integral Formulation for 3-D Eddy-Current Computation Using Edge Elements", *IEE Proc.*, Pt. A, **135**, 1988, pp. 457-462

[4] R. Albanese, G. Rubinacci, "Finite element methods for the solution of 3D eddy current problems", *Advances in Imaging and Electron Physics*, Vol. 102, pp. 1-86 (1998)

[5] R. Albanese, G. Rubinacci, A. Tamburrino, F. Villone, "Reconstruction of Cracks with Integral Methods", *Electromagnetic Nondestructive Evaluation (II)*, R. Albanese et al. (Eds.), IOS press, pp. 107-113 (1998)

[6] R. Albanese, G. Rubinacci, F. Villone, "Analysis of TEAM Workshop Problem 15 Using an Integral Formulation", *Proceedings of TEAM Workshop*, Rio de Janeiro (Brazil), Nov. 1997, pp. 7-9

[7] C.V. Dodd., W. E. Deeds, "Analytical Solutions to Eddy-Current Probe-Coil Problems", *J. Appl. Phys.*, Vol. 39, No. 6, 1968, pp. 2829-2838

[8] A.S. Householder, *The theory of matrix in numerical analysis*, Dover (1975)

[9] J.A. Scales, M. L. Smith, T.L. Fischer, "Global Optimization Methods for Multimodal Inverse Problems", J. Comp. Phys., Vol. 103, pp. 258-268 (1992)

[10] Huang Haoyu, Tohoku University, Sendai, Japan, personal communication, 1998

Electromagnetic Nondestructive Evaluation (III)
D. Lesselier and A. Razek (Eds.)
IOS Press, 1999

Statistical Inverse Mapping in ECT Based on a Shifting Aperture Approach

Radu C. POPA and Kenzo MIYA
Nuclear Engineering Research Laboratory, The University of Tokyo, Tokai-mura, Naka-gun, Ibaraki 319-1106, Japan

Abstract. The inversion procedure presented in this paper is based on statistically regressing the inverse map between the spaces of ECT scan data, and of crack parameters. The study reveals the benefits provided by the various regularization techniques proposed for treating the detrimental conditioning of the problem. A suitable combination between a principal component transformation step, and a special neural network training algorithm ensures the incremental resolution of the map regression. The novel introduction of a 'shifting aperture' mapping and a data fusion approach both avoided the usual anomalous-region-based inversion techniques, and significantly enhanced the method, allowing the dynamical reconstruction of complex crack shapes even in the presence of high noise levels. Reconstructions from synthetic, and from benchmark data will be presented.

1. Introduction

The definite industrial acceptance of the so far developed ECT inversion methodologies has been hampered both by their failing to provide an optimal tradeoff between speed and reliability of defect shape estimations, and by the yet unsolved problem of natural crack reconstruction. The first mentioned issue has been spurring the emulation of two classes of solutions, denoted as model-based, and model-free approaches.

The model-based inversions consider the underlying physical process as part of the solution procedure, and have basically followed two directions, *i.e.*, the solution of an uncoupled or coupled pair of Fredholm equations (*e.g.*, [1-3]), or the iterative optimization of defect shape in directions obtained from sensitivity computations (*e.g.*, [4-8]).

Problems related to complexity, or low speed encountered in the model-based inversion approaches have suggested that their model-free counterparts could represent a good candidate to be used in hybrid or standalone schemes. The tradeoff between accuracy, on the one hand, and speed and robustness on the other hand, seems to be better achieved based on these latter approaches. The optimality of this tradeoff derives from the fact that the model-free methods employ *rules* and/or *maps*, determined from the whole past experience and the longer and better this accumulated knowledge, the greater the accuracy and robustness achieved. Moreover, relying on a pre-stored data bank which can contain both synthetic, and measurement information, a model-free inversion possesses an increased independence to the employed numerical model, thus making more approachable the estimation of real cracks.

2. Mapping-based inversion

The inverse *mapping* methods aim to discover the signal-to-shape relationship (as a hypersurface) from a set of non-uniformly distributed examples. The procedure is equivalent to a multidimensional interpolation, as the approximation of the corresponding crack shape is expected (at the output), based on a fresh signal (at the input).

The most successful regression methods for extracting these maps are based on artificial neural network (NN) applications [9]. However, the NN-related research area is at a rather incipient stage, thus lacking of satisfactory foundation for ensuring non-empirical application. It has been noticed that NN learning algorithms often ignore much of the information contained in the data and therefore the preprocessing of the training data may improve the conditioning of the problem. Basically, this is a means of both smoothing the map, and reducing the dimensionality of the problem through feature extraction, and has been used, *e.g.*, in [10-11]. Another class of methods for dealing with the non-convexity of the mapping procedure is represented by the resolution-based regularization techniques. The underlying assumption of these techniques is that the 'good' minimum from the ill-behaved, 'bumpy' hypersurface of the objective function can be approached with greater probability of success by ensuring various levels of detail awareness. In this context, one of the options is that based on a multi-level-resolution representation of the input-output relationship. Wavelet functions are currently used in this purpose[12,13].

In the present study, a different multiresolution approach was introduced, in fact a progressive, or incremental resolution of the map approximation. This is realized by the combination of two modules[14]: the statistical analysis and transformation of the input data – by Principal Component Analysis (PCA)[15] –, and the NN with an incremental-resolution learning[16]. The first module discovers the principal directions (features) of data variances and rotates the original coordinate axes along these directions. The transformed sets of input data are *uncorrelated* and *ordered* according to their relevance in the total knowledge. The training algorithm of the NN permits an incremental creation of new nodes, until the map becomes accurate. This association allows thc parameters (weights) of the NN to comply with increasingly detailed features of the map. With the appearance of each new node, the mapping task is redistributed and a greater number of parameters ('interpolation' weights) will share the map approximation task. Thus, the relevance of the less significant inputs (with smaller variances in the dataset) will increase continuously. The mapping starts from discovering the rough features of the input-output relationship by a hypersurface in few dimensions, and pursues by increasing the attention paid to details by gradually extending the dimensionality of this hypersurface. This process continues until the quality of the representation is optimal, *i.e.*, a necessary-and-sufficient number of parameters describe without redundancy the interpolation, by giving appropriate relevance to each input. It should be pointed out that not the feature extraction property of the PCA transformation is the source of map smoothing here, because data compression was not employed.

The training algorithm of the employed NN implies least-squares solutions of an overdetermined equation system at each iteration (training epoch) :

$$[\mathbf{A}\ ,\ f_1(\mathbf{A}\cdot\mathbf{W}_{ih})]\cdot\begin{bmatrix}\mathbf{W}_{io}\\ \mathbf{W}_{ho}\end{bmatrix} = f_2^{-1}(\mathbf{B})\ , \tag{1}$$

where $\mathbf{A}$, $\mathbf{B}$ represent the input, and output training sets, respectively, f_1 and f_2, the nonlinear activation functions for the hidden and output nodes, $\mathbf{W}_{ih}$ is a randomly generated, fixed coefficient matrix, and $\mathbf{W}_{io}$, $\mathbf{W}_{ho}$ are the matrices containing the unknowns of the problem, *i.e.*, the input-output and the hidden-output interconnection weights, respectively. The least-squares solution – via singular value decomposition – of this regression equation represents an additional regularization of the mapping, equivalent to a maximum likelihood or 'perfect *a priori* ignorance' Bayesian estimation. Also, a new regularizing feature was introduced, *i.e.*, a 'cooling' procedure for the hidden nodes' nonlinearity. The steepness of f_1 is increased progressively – by a logarithm-based function – while training the network, allowing the regression to 'discover' first the rough features of the error surface, and to locate smoothly its details.

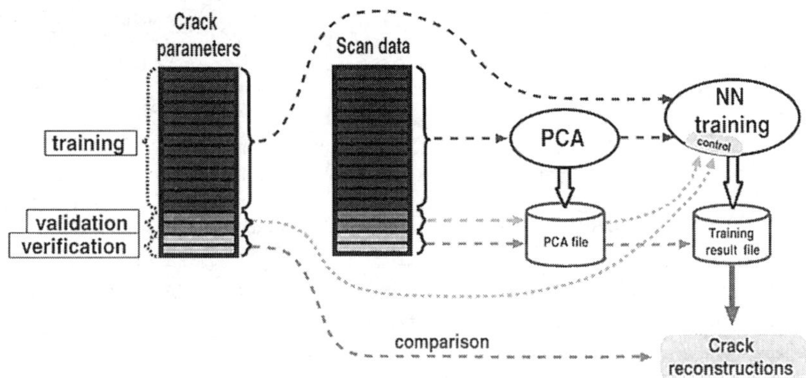

Fig. 1. Data flow through the basic inversion algorithm.

Fig. 1 depicts the data flow through this *basic* mapping algorithm. In the next Section, an additional processing step based on aperture shifting will be introduced to complete the proposed procedure. The I-O pairs of the initial database are partitioned into training, validation and verification sets. The 'PCA' and 'NN' modules are presented with the training data and issue the two files for later use. The PCA file is used for the transformation of both the validation, and verification sets. The validation set is used only to control the training optimality, by monitoring the currently achieved estimation error. The reconstructions are conducted on the verification set, and, if available, the corresponding 'correct' shapes are compared with the estimated ones. The latter set is the equivalent to the *in situ* scan data (the 'fresh' data). At this stage, the previously recorded files are employed, and only elementary operations are necessary for data transformation and propagation through the trained network. These two steps are very fast and can be performed in real-time with small computation requirements during the actual testing. Note also that such a mapping approach allows a continuous upgrading of the estimations' quality by enlarging the training set and by off-line - or even on-line - retraining the network. Being based on learning an input-output mapping from a set of examples (the training set) or fitting in a least-squares sense a hypersurface to this set in view of acquiring good generalization properties in unpopulated points, the whole procedure is equivalent to a *statistical regression*.

3. The shifting aperture approach

The rationale for the introduction of the new mapping method based on aperture shifting is detailed elsewhere[17,18]. Here, the basic points will be outlined for completeness.

Let us consider a complete B-scan along a probing line, as shown in Fig. 2.

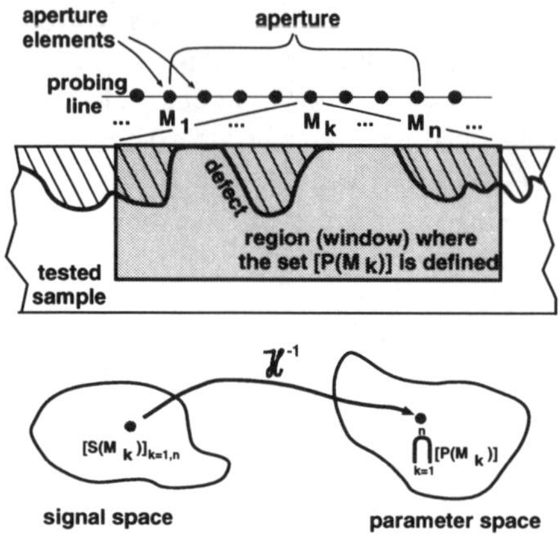

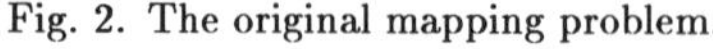

Fig. 2. The original mapping problem.

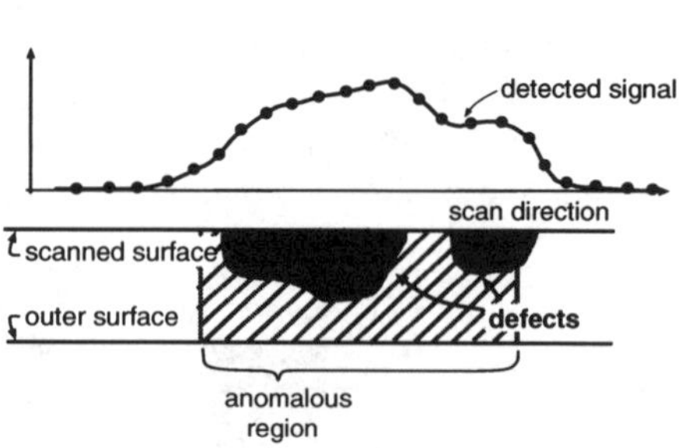

Fig. 3. An anomalous-region-based mapping: 'full' scans are mapped onto defects enclosed by the anomalous region boundary.

The scan consists in the discrete probe readings, or views. One can define an *aperture* as a fragment of a full-scan, *i.e.*, a fixed number n of successive views, or *aperture elements.* The signal $S(M_k)$ sampled at each aperture element M_k represents the result of a 'convolution' between the probe's PSF (centered in M_k), and the defect's shape function. If the flaw is defined by a set of shape parameters, let $[P(M_k)]$ denote this set only for the maximal flaw region which affects $S(M_k)$. A mapping procedure whereby one tries to regress the relationship $\mathcal{H}^{-1}$ between the n views and the whole set of crack parameters which affect this aperture:

$$\mathcal{H}\left[\bigcap_{k=1}^{n}[P(M_k)]\right] = [S(M_k)]_{k=1,n} \tag{2}$$

is highly frustrating, due to three reasons: (i) the contribution of each flaw parameters to each probe reading depends on their relative farness, as a consequence of the attenuation. This makes the 'far' parameters less significant to the view. (ii) the far parameters are not only less illuminated-read by any aperture element, but they are seen by fewer probe views, fact that reduces again their significance to the aperture readings. (iii) the number of parameters more-or-less affecting the aperture can be prohibitively large. An approach based on defining a maximal, anomalous region where the crack is supposed to be confined is only a particular case of the above described attempt. While partially solving the two latter listed problems, an anomalous-region-based mapping (see Fig. 3) introduces further difficulties related to flaw location, reconstruction of unconnected flaw profiles and suitability for process automation.

All these facts suggest the main point of the approach proposed hereafter. The sketches from Fig. 4 illustrate the underlying idea, which borrows from the synthetic aperture technique employed in ultrasonic and radar imaging in that the reconstruction is focused only on the flaw parameters highly affecting a given aperture. In Fig. 4-a), the 1-D probe's PSF is designated symbolically by a bell-shaped curve, whose lateral opening corresponds to the active area. The sensitivity characteristics of the probe used here agree fairly with this model. The figure shows the set of sensitivity profiles for a given aperture of six elements. It is straightforward to imagine equivalently the

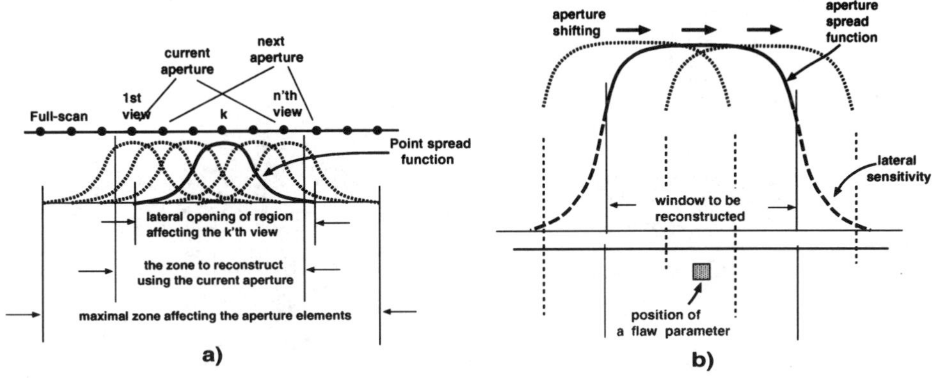

Fig. 4. Qualitative analysis of individual scan readings, a). *Aperture PSF* and data fusion through aperture shifting, b).

resultant sensitivity of the whole aperture to the flaw parameters as the superposition (in a first approximation) of the individual spread functions.

Clearly, this *aperture spread function* – see Fig. 4-b) – will present a smooth central peak, reflecting the larger relevance given to the parameters observed by most of the views. The sensitivity profile obtained in this way presents a larger contrast between the central, highly weighted object zone, and the lateral queues. The flaw parameters situated close to the periphery of the maximal zone illuminated by the aperture – being observed with low weights and by much fewer elements – will contribute with considerably lower relevance than their central analogues to the total amount of information encoded in the aperture data. We can now lift all three barriers discussed above, by trying to find the mapping toward a window which has a *smaller* opening than that of the maximal affecting region. Although this inverse mapping will not be injective – due to the lateral cuts – it is less ill-conditioned than the initial one. Fig. 4-b) depicts the second aspect of the approach. Three successive apertures are represented by their sensitivity profiles, which are cut to the extent of the window which has to be approximated by the mapping. One can notice that a given flaw parameter is seen by several apertures, thus it will be approximated several times as an output of the learned mapping. Obviously, the best accuracy will be expected from the aperture centered above it. This suggests the idea of giving different weights of confidence to these successive approximations of the same object, according to how far from the center of the current aperture it is located. This technique, which is basically a *data fusion* procedure, will enhance the estimation reliability for each parameter, by accounting for the unconsidered lateral sensitivities as well. The weights were chosen after several trials, in order to minimize the global estimation error for the tested sets.

Fig. 5 depicts the described shifting aperture mapping approach: each aperture from an initial full-scan is mapped onto a corresponding object window, and both are simultaneously shifted. Obviously, the optimal openings and shifting step of these scan and object fragments will depend on the frequency, scanning step, probe construction and material parameters. It was found[17,18] that equal openings – of about the same spatial extent as the probe's active area – and a shifting step equal to the scanning step ensure a good conditioning for the problem. After cutting and shifting the raw

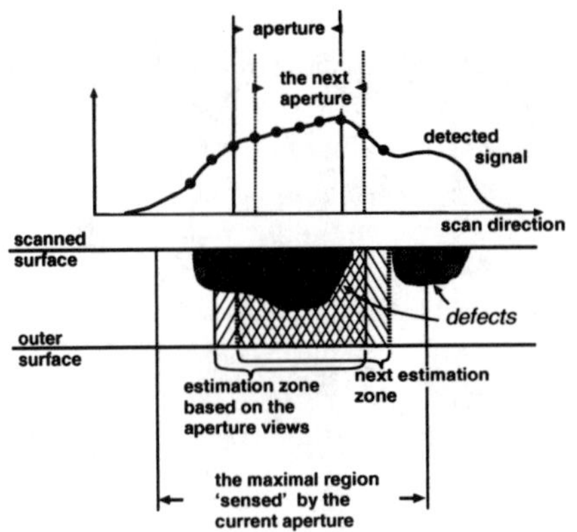

Fig. 5. Schematic of the shifting aperture strategy for flaw reconstruction.

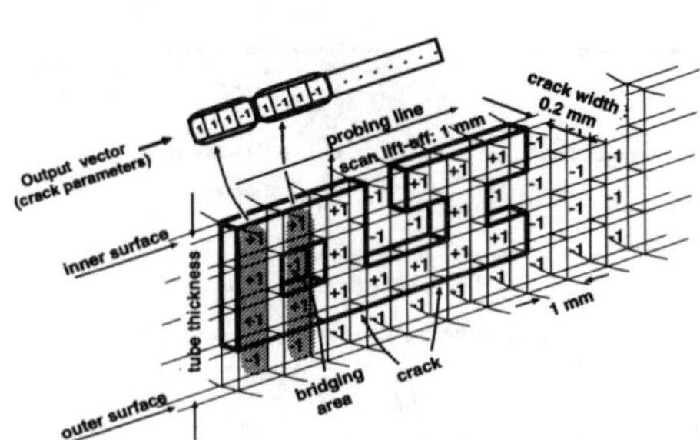

Fig. 6. The crack parametrization and the construction of output vectors.

full-scans, the resulted database will be sent to the basic inversion algorithm – as was depicted in Fig. 1 – to pass through the PCA, and NN modules.

The object profile was binary-parametrized at cell level, as shown in Fig. 6. The cells belonging to the flaw are denoted by the $+1$ level, while the unflawed cells receive -1 level. The figure shows also the manner of constructing the output (crack shape) vectors of the database. Between each pair of scan points, the in-depth cell values are concatenated.

4. Validation

Various validation steps were performed, including reconstructions by using noise-free or noise-polluted synthetic scan data, or laboratory data. In all cases only synthetic scans were employed for regressing the map, as obtained from the 3D FEM-BEM coupled analysis code, based on the $A - \phi$ method[19] that we are currently using in our analyses. The simulated cracks were treated as groups of fixed width (0.2 mm) void elements.

The quality of reconstructions from noisy inputs was significantly enhanced when introducing a characteristic technique for NN approaches, *i.e.*, *data jittering*, or training with noise. The aim of this procedure is multifold: (*i*) to increase the number of available training cases; (*ii*) to 'smooth' (regularize) the mapping; (*iii*) to increase the robustness of the final map against noisy inputs.

4.1. Validation for synthetic data collected by an 'intelligent' probe in a tube geometry

In a first testing stage, 106 full B-scans were simulated over a same number of quasi-randomly shaped inner-surface-breaking cracks in a tube specimen. The sample had the same geometrical and material properties as the actual INCONEL 600 tubing used in the steam generator (SG) tubing of pressurized water nuclear reactors (PWRs). The same optimized probe[20] as for the former work[14] was employed here to collect the scan data. Comparative inversion and data statistics studies positively acknowledged the potential of this probe arrangement to provide well behaved data for inversion purpose.

Both single-frequency (100 kHz), or dual-frequency (100 and 400 kHz) scanning sets were used, and it was noticed that the latter case brought significant improvements only when reconstructions from noise-polluted data were aimed. For the present, enlarged family of defects it turned out that a satisfactory reconstruction needs both the in-phase and quadrature components of the signal. Each complete scan consisted of 26 probe readings along the probing line, in 1 mm steps. The best mapping was obtained by taking apertures of 11 elements and estimation windows of 10 mm opening. From the total of 106 complete, 26-points scans over a set of highly dissimilar flaws were formed in this way $N = 106 \times (26 - 11 + 1) = 1696$ such input-output vector pairs. From this maximal database, validation and verification sub-sets were extracted prior to any subsequent module of the algorithm, and used only in the testing phase.

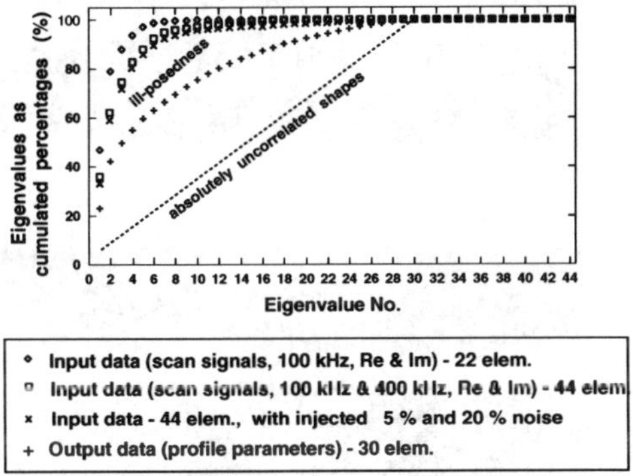

Fig. 7. Cumulated percentage distributions of correlation matrix eigenvalues.

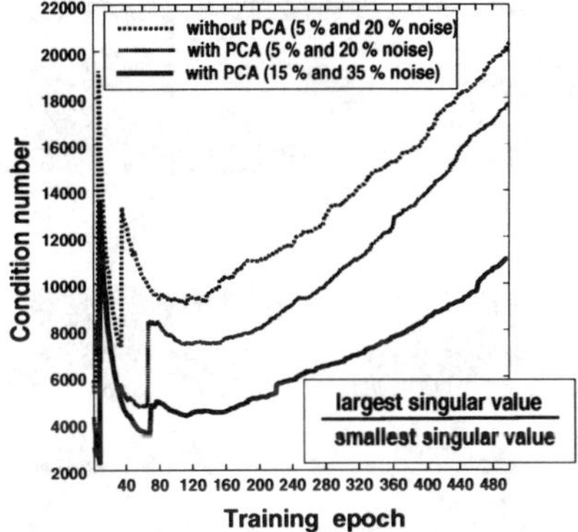

Fig. 8. Evolution of the regression condition number.

Fig. 7 depicts comparative cumulated percentage distributions of the correlation matrix eigenvalues for various data sets. If one compares the degree of correlation for the output data with that for the input data at 100 kHz, an important difference can be noticed. One can measure the degree of problem's ill-posedness by the *area* confined between the two distributions, or the difference between their starting values. The bigger is the difference of steepness between the two patterns, the stronger the ill-conditioning of the regression. Obvious improvements, even though not significant in this representation, can be noticed when dual-frequency data are used, or when two levels of artificial noise are appended to the original input set (jittering): the area between the curves decreases, this suggesting a better conditioning. Note that the correlation noticed in the shape parameters (the departure from a straight line) is due to the fact that the database did not contain, for example, single-cell isolated voids, thus reflecting the essence of the problem: we are searching for *cracks* as compact exceptions in a *sane material.*

A clear image of the improvements brought by PCA transformation and by data jittering is given by the curves from Fig. 8. They represent the condition number of the regression matrix $[\mathbf{A}\ ,\ f_1(\mathbf{A} \cdot \mathbf{W}_{ih})]$ from eq. (1). During training the network, the gradually increasing resolution of the map has the effect of decreasing the condition number for the SVD minimization module. The figure shows that without performing PCA transformation, this conditioning remains the worst all over the training period. The introduction of larger noise levels in the data is further beneficial for this parameter.

The performance measure of the regression was given by the *average deviation* of the reconstructed binary parameters for the validation, or verification sets. For this parameter we obtained values corresponding to an average per-cell deviation of less than 7 % from the range (= 2) of the cell binary parameter. However, a 'good' average deviation value did not correspond always to a good estimation of the parameters which define the flaw peculiarities and was sometimes just an indication of a higher contrast between the 'black' (unflawed, -1) and 'white' (flawed, +1) cells. As it was chosen to *not threshold* the outputs, the visual examination of the reconstructed shapes became equally important.

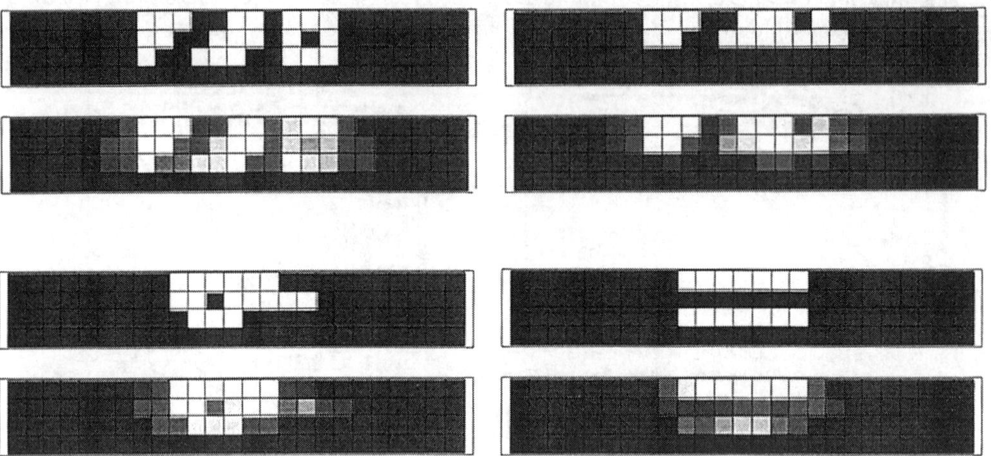

Fig. 9. Reconstructions *from* noise-free data. The mapping was obtained by *training* with single-frequency (100 kHz), unjittered signals. For each estimation, the correct image is given above.

In Fig. 9, four reconstruction examples are presented in gray-level images. The training was stopped after about 250 epochs, when a minimal error was obtained for the selected validation sets. One can notice that fair reconstructions are obtained for the electrical bridging areas, for the multiple successive cracks, and for the laterally buried profiles, that so far eluded most other inversion approaches. Each image represents a 25 mm long profile which is obtained by a weighted superposition of 16 successive windows.

In the same setup, the algorithm was tested on signals containing artificial uncorrelated (white) noise. The results were not satisfactory when the map was obtained from noise-free training sets. When the data jittering technique was employed and the original signal collection was expanded three times to contain additional noise levels, the map became very robust. Fig. 10 gives some examples, obtained after about 400 training epochs.

4.2. Validation for experimental data collected by a pancake coil in a plate geometry

The inversion algorithm was tested also for data obtained by laboratory measurements. Due to the fact that only a small number of experiments were available, the database used to construct the inverse map contained *only synthetic data*, computed by the same

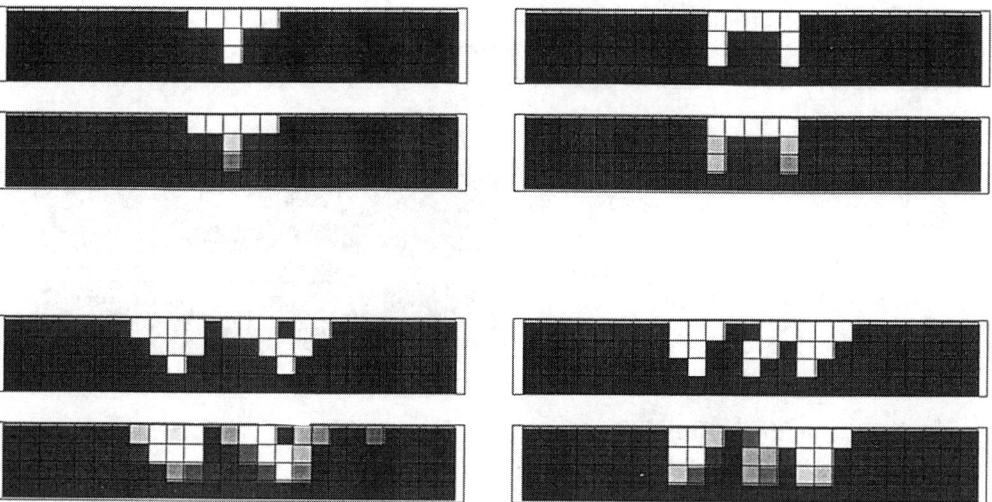

Fig. 10. Reconstructions from noise-polluted scan data. Artificial white noise, 25 % from the signal maximum. The training was done by using dual-frequency (100 and 400 kHz) scans, and by adding two levels of noise (15 % and 35 %) in the initial database (jittering). For each estimation, the correct image is given above.

code as in the previous investigations. The step No. 6 of the benchmark problems proposed by JSAEM[21] consists in 2-D scanning the surface of a conducting plate with a pancake coil. The plate is made from the same material and has the same thickness as the SG tubing in PWRs. The scans were performed on each of the plate faces, thus making available both inner, and outer data for each crack. The energizing frequency is 300 kHz.

The present attempt differs from the previous one by trying to find a *single* map which is able to invert signals of *both* inner, and outer cracks. The database, thus, contained both IC and OC simulations, and the inversion of the measurement data did not require any assumption concerning to which face of the sample the crack is breaking (*i.e.*, no IC-OC pre-classification was necessary). The depth direction was uniformly divided in five 20 % thick layers. For the construction of the database, a total of 110 cracks (half were ICs, and half OCs) were B-scanned in a succession of 21 points, in 1 *mm* steps. The training was controlled by a validation set, and the final regressed map was tested directly on the experimental set. The inherent coarseness of the mesh obviously resulted in a discrete approximation of the reconstructed crack shapes.

For deciding the aperture opening, the previously deduced criterion was used, *i.e.*, this was taken in correlation with the probe's active range. Accordingly, apertures of six, 1mm-step readings were cut and shifted from each complete scan, and the estimation windows covered corresponding zones of same opening (5 mm). Further studies[18] clearly showed that the present problem is significantly worse conditioned than the former, and that the principal reason is the quality for inversion of the employed probes. Data jittering was employed also in this case: the initial database was enriched with three new, noise-polluted sets (5 %, 20 % and 25 % white noise). An acceptable map was obtained after 700 training epochs.

The final map was, then, used to invert the experimental data. The relative percentage deviation per crack cell parameter was 7.1 % from the total range. Figs. 11, 12 depict the departures of the numerical model from the real situations and the reconstruction results for ICs, and OCs, respectively, for two of the analyzed cases. Each reconstructed window is formed by the weighted fusion of 16 successive approximations.

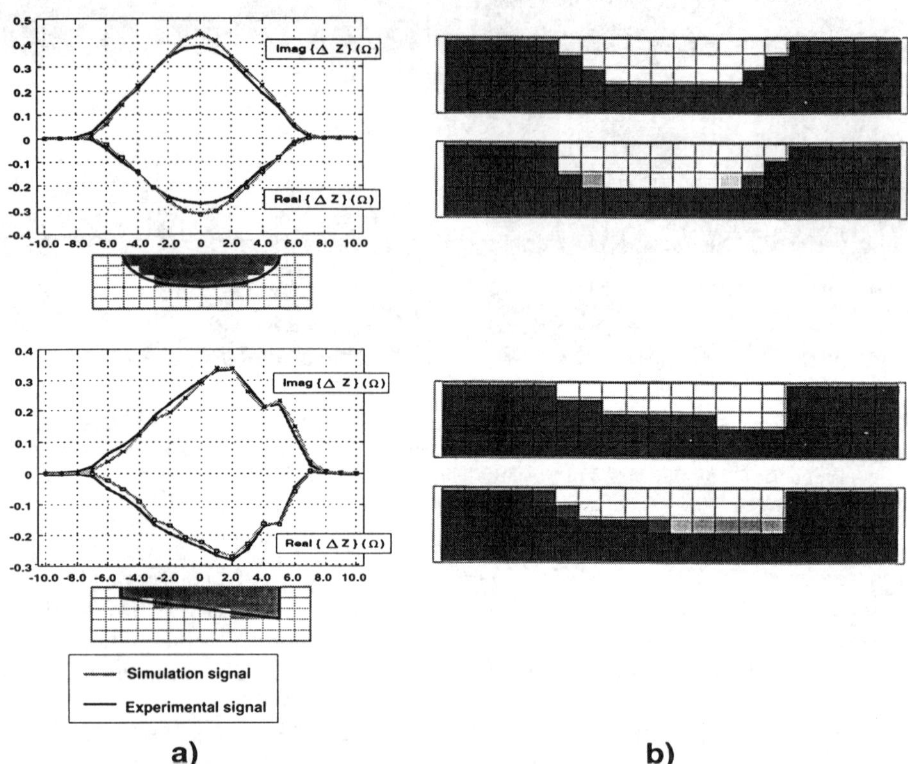

Fig. 11. Inner cracks. a) Comparison showing the departure of the numerical model from the real situation; the mesh allows just a 'best fit' approximation (shaded area) of the actual crack shape (thick-line curves); the corresponding numerical, and experimental scans are compared. b) The 'best fit' approximations (above), along with the corresponding reconstruction results.

As mentioned before, the resolution of the mesh in the crack region allows just a 'best fit' approximation for the shape of the manufactured cracks. These best fits, along with their original counterparts are represented below each corresponding plot of experimental, and simulated scans. For outer cracks, the relative deviations are significant and even for ICs, they translate into a spatially-correlated noise component. In some cases, inaccurate signal offset is noticed for the experimental scan. A common quality can be noticed for all reconstructions: the maximum depth of the crack is well estimated. Note that in the construction of the training set, no crack depth was favored. As expected from the quality of the experimental data, the IC reconstructions are superior to OCs. The main features of the notches were, however, well estimated in all cases. Besides the correlated character of the noise, two other factors contribute to the reconstruction inaccuracies: the insufficiency of the training cases, and the quality of the probe. Nevertheless, taking into account the huge dimensionality of the problem's state space, the presented results are more than one could expect from a database of only 110 cracks.

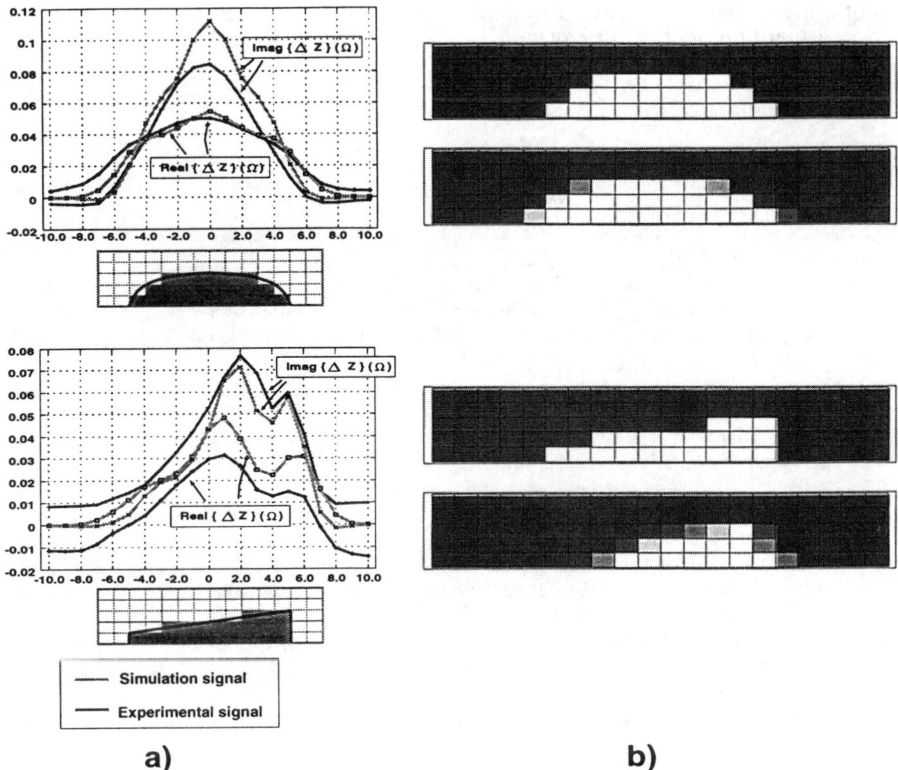

Fig. 12. Outer cracks. a) Comparison showing the departure of the numerical model from the real situation; the mesh allows just a 'best fit' approximation (shaded area) of the actual crack shape (thick-line curves); the corresponding numerical, and experimental scans are compared. b) The 'best fit' approximations (above), along with the corresponding reconstruction results.

5. Conclusions

A fast and reliable inversion method was proposed, based on statistically regressing the signal-to-shape inverse map. Various regularization techniques are proposed, in order to cope with the ill-posedness. The introduction of a new approach in ECT consisting in a *shifting aperture* mapping significantly enhanced the method, allowing good reconstructions from noise-polluted synthetic data, or experimental data. The main roles of the aperture shifting technique are: to reduce the ill-posedness of the map; to elude the problems caused by an approach based on anomalous region assumption; to reduce the dimensionality of the mapping problem and multiply the number of available cases. The profile of the tested object can be reconstructed dynamically, in nearly real-time by a weighted fusion procedure. The algorithm's capability of reconstructing multiply connected and unconnected crack domains is important, as this encourages the attempts to tackle the problem of natural cracks.

It should be noted that the resulted map has to be accurate only in a critical region of the problem space. Theories of fracture mechanics and results of experimental tests should be combined for deciding the critical dimensions and shape classes of propagating cracks. Based on these results, a minimal database can be constructed. The proposed mapping algorithm can be easily extended to 3-D reconstructions from 2-D scans.

References

[1] S. Barkeshli, D.J. Radecki and H.A. Sabbagh, "On a linearized inverse scattering model for a three-dimensional flaw embedded in anisotropic advanced composite materials", *IEEE Geosc. Rem. Sens.*, Vol. 30, **1**, 1992, pp.71-80.

[2] H.A. Sabbagh and R.G. Lautzenheiser, "Inverse problems in electromagnetic nondestructive evaluation", *Int. J. Appl. Electromagn. Mech.*, **3**, 1993, pp.253-261.

[3] V. Monebhurrun, D. Lesselier and B. Duchêne, "Eddy current nondestructive evaluation of a 3-D bounded defect in a metal tube using volume integral methods and nonlinerized inversion schemes", in *Studies in Applied Electromagnetics and Mechanics, Vol.14*, R. Albanese *et al.*, Eds., IOS Press, 1998, pp.261-270.

[4] S.J. Norton and J.R. Bowler, "Theory of eddy current inversion", *J. Appl. Phys.*, **73**(2), 1993, pp.501-512.

[5] J.R. Bowler, "Inversion of eddy current data to reconstruct flaws by an optimization process", *Int. J. Appl. Electromagn. Mech.*, **4**, 1994, pp.277-284.

[6] J.R. Bowler, "Pulsed eddy current inversion for the determination of crack shape", in *Studies in Applied Electromagnetics and Mechanics, Vol.12*, T. Takagi *et al.*, Eds., IOS Press, 1997, pp.263-269.

[7] Z. Badics, *et. al*, "Rapid flaw reconstruction scheme for three-dimensional inverse problems in eddy current NDE", in *Studies in Applied Electromagnetics and Mechanics, Vol.12*, T. Takagi *et al.*, Eds., IOS Press, 1997, pp.303-309.

[8] Z. Chen, K. Miya and M. Kurokawa, "Reconstruction of Crack Shapes Using a Newly Developed ECT Probe", in *Studies in Applied Electromagnetics and Mechanics, Vol.14*, R. Albanese *et al.*, Eds., IOS Press, 1998, pp.225-232.

[9] NDT Abstracts - Neural networks in nondestructive testing, *NDT&E Intern.* Vol. 30, **5**, 1997, pp.321-334.

[10] L. Udpa and S.S. Udpa, "Eddy current defect characterization using neural networks", *Mater. Eval.*, **48**(3), 1990, pp.342-347.

[11] R. Albanese *et al.*, "Analysis of metallic tubes with ECT and neuro-fuzzy processing", *Int. J. Appl. Electromagn. Mech.*, **9**, 1998, pp.325-338.

[12] S. Mandayam, L. Udpa, S.S. Udpa and W. Lord, "Wavelet-based permeability compensation technique for characterizing magnetic flux leakage images", *NDT & E International*, **30**(5), 1997, pp.297-303.

[13] F.C. Morabito, "An intelligent network for defect profile reconstruction in eddy current applications", in *Studies in Applied Electromagnetics and Mechanics, Vol.13*, Kose and Sievert, Eds., IOS Press, 1998.

[14] R.C. Popa and K. Miya, "A data processing and neural network approach for the inverse problem in ECT", in *Studies in Applied Electromagnetics and Mechanics, Vol.14*, Albanese *et al.*, Eds., IOS Press, 1998, pp.297-304.

[15] I.T. Jolliffe, *Principal Component Analysis*, Springer-Verlag, New York, 1986.

[16] C.L.P. Chen, "A Rapid Supervised Learning Neural Network for Function Interpolation and Approximation", *IEEE Trans Neural Networks*, **7**, 1996, pp.1220-1230.

[17] R.C. Popa and K. Miya, "Approximate inverse mapping in ECT, based on aperture shifting and neural network regression", under review at *J. Nondestr. Eval.* since February, 1998.

[18] R.C. Popa, *Optimized crack detection and inversion in eddy current testing*, Ph.D. Thesis, Nuclear Eng. Res. Lab., University of Tokyo, 1998.

[19] F. Matsuoka, "Calculation of a three dimensional eddy current by the FEM-BEM coupling method", *Proc. IUTAM on electromagneto-mechanical interactions in deformable solids and structures*, Elsevier Science, 1987, pp.169-174.

[20] R.C. Popa, K. Miya and M. Kurokawa, "Optimized eddy current detection of small cracks in steam generator tubing", *J. Nondestr. Eval.*, **16**(3), 1997, pp. 161-173.

[21] T. Takagi *et al.*, "Electromagnetic NDE research activities in JSAEM", in *Studies in Applied Electromagnetics and Mechanics, Vol.12*, Takagi *et al.*, Eds., IOS Press, 1997, pp.9-16.

Electromagnetic Nondestructive Evaluation (III)
D. Lesselier and A. Razek (Eds.)
IOS Press, 1999

Sizing and Localization of Cracks Using FEM and a Constrained Stochastic Algorithm

Fabrice ZAOUI, Claude MARCHAND, Adel RAZEK
Laboratoire de Génie Electrique de Paris (LGEP)
URA CNRS 0127 – SUPELEC – Universités Paris XI et Paris VI
Plateau de Moulon, F 91192 Gif-sur-Yvette Cedex

Abstract. This paper deals with the use of a constrained genetic algorithm (GA) associated with finite element modeling for solving inverse problems in eddy current testing. 2D sizing and localization of unknown inner cracks (shape and position) are performed using elementary geometrical models (EGM). These EGM can change and move in a pre-defined search domain. The inverse problem will attempt to find the EGM size and position whose electromagnetic field best matches the image of an unknown defect.

1. Introduction

Non-destructive evaluation (NDE) concerns the techniques involved in the tests of materials, pieces, or devices without modifying their characteristics. NDE deals with the precise characterization of eventual defects (physical properties, orientation, location, and size...). This is done for insuring a high level of quality in production phases (industrial steel for example), and reliability during periods of activities (plane transportation, gas pipelines...). Possible human and financial costs are highly dependent upon these considerations of quality and security. NDE is based on the observation of disturbed physical phenomena for detecting the presence of defects in specimens.

Techniques used in NDE are various (radiography, thermography, ultrasonic, eddy current testing, and microwave imaging...). Eddy current testing (ECT) is an efficient tool for dealing with shallow defects in conducting materials. ECT uses induced currents due to a variable exciting magnetic field. The distribution of induced currents is disturbed by the presence of any inhomogeneity. Thus it is the observation of the corresponding resultant magnetic field that allows the flaw detection. This is done with the solution of an inverse problem based on the optimization of a suitable error functional between field measurements and predicted values issued from theory [1,2].

Among the existing optimization algorithms, stochastic approaches are well known for their robustness. The genetic algorithm (GA), founded upon the survival of the fittest in the principle of evolution [3], is one of the most famous and promising tools in the optimization procedures [4], and specifically in NDE [5]. GA explores the search space to locate the global solution without being dropped into local optima. With GA non-differentiable or multi-modal functions can be easily studied because only values and not derivatives of the objective function are required.

In classic GA, the variables coding uses a binary scheme. This representation can be ill adapted for multidimensional high-precision numerical problems. In fact, for variables with large domains, the number of digits in the binary representation scheme has to be high enough to keep a good precision. This involves a significant search space where the classic GA, with acceptable computation times, performs poorly. Moreover, considering non-trivial constraints (i.e. not only the domain constraints) with the binary scheme is difficult. A priori constraints can be numerous when dealing with parametrical shapes moving in a space for the defect reconstruction (size and localization).

In this presentation, a GA based on a floating-point representation is used as an optimization tool in the inverse problem solution. Floating point representation is faster and provides higher precision especially with large domains for variables. This tool is based on the GENOCOP program developed by Michalewicz [6]. With this tool, linear constraints are easily implemented.

2. Problem Statement

The solution of the inverse ECT problem is searched by iteratively solving the forward problem, see Fig. 1. The defect modeling uses parameters P_i for updating the current solution toward the real defect.

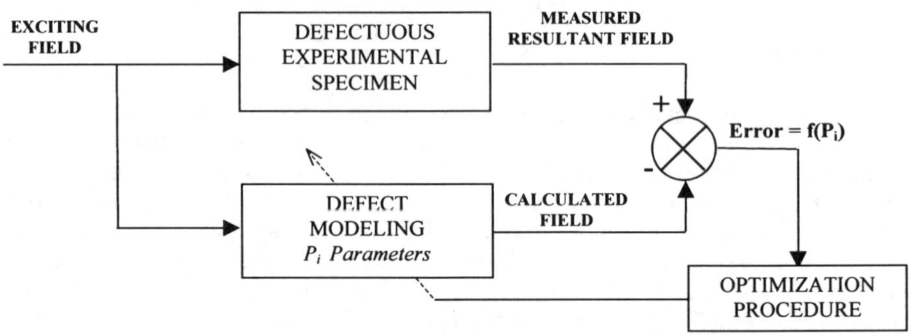

Fig. 1: Inverse problem approach in ECT

A 2D finite element method (FEM) is used for the numerical modeling of the defect. FEM is suitable for treating complex geometries encountered in inspection [7]. FEM consists in the division of the geometry into smaller domains called finite elements. All the finite elements constitute the mesh. The magnetic field is only calculated in mesh pre-defined points. In the rest of the working space, values are interpolated with polynomial approximations. Then, the numerical precision is highly dependent upon the mesh definition.

For the considered problem, the defect is an unknown inner object. No particular hypothesis is made according to the position or shape. Thus for reconstructing a complicated real defect with effectiveness [8,9], geometrical parameters Pi can not be easily established. Only an approximation on the true profile can be searched when dealing with a simultaneous localization.

Instead of working with an invariant mesh and changing materials properties considered in the finite elements modeling, the inverse problem is solved by updating the finite element mesh using elementary geometrical models (EGM). This choice avoids defining the size of the finite element in the binary approach (i.e. constant mesh). Furthermore it also avoids a very difficult optimization procedure implementation when treating with the size and inner localization of non-trivial profiles.

To present the method, two EGM are considered as shown in Fig. 2. The associated parameters and constraints for the optimization algorithm are also represented. These constraints define the complexity of EGM deformations. The aim is to find the right model location and size which best matches the actual defect position and shape.

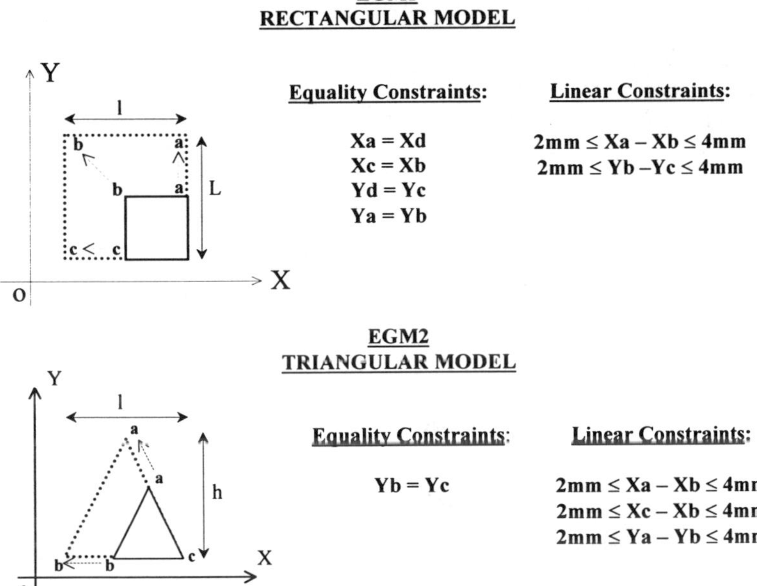

Fig. 2: Two EGM and associated variables and constraints for the reconstruction problem

Other constraints (the domain constraints) depend on the geometrical search space. The search space is rectangular and is a part of the magnetic plate specimen as represented in Fig. 3 and Fig. 4. Then all domain constraint dimensions for P_i parameters describing EGM are 18mm for x-values and 10mm for y-values.

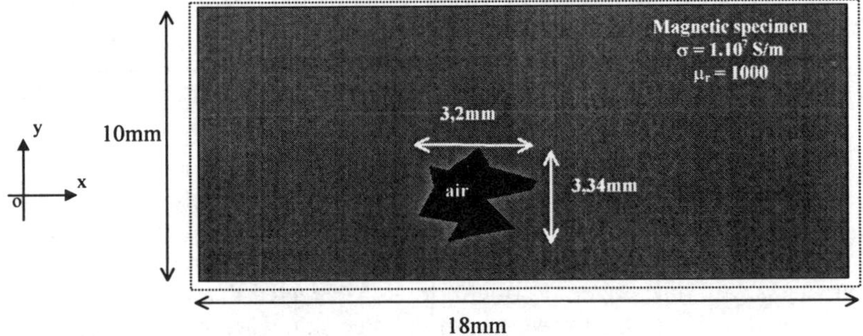

Fig. 3: An unknown defect located in the search space (dotted lines)

An inner defect with air permeability and conductivity is supposed to be located in this search space.

3. Numerical Modeling

The numerical model governing ECT process is issued from Maxwell's equations and is given by:

$$-\frac{\partial}{\partial x}\left(\frac{1}{\mu}\frac{\partial A(t)}{\partial x}\right)-\frac{\partial}{\partial y}\left(\frac{1}{\mu}\frac{\partial A(t)}{\partial y}\right)=J(t)-\sigma\frac{\partial A(t)}{\partial t} \qquad (1)$$

A: vector magnetic potential
J : applied AC current density in the exciting coil
μ: magnetic permeability
σ: conductivity

The mesh of the studied device uses first order triangular finite elements as shown in Fig. 4.

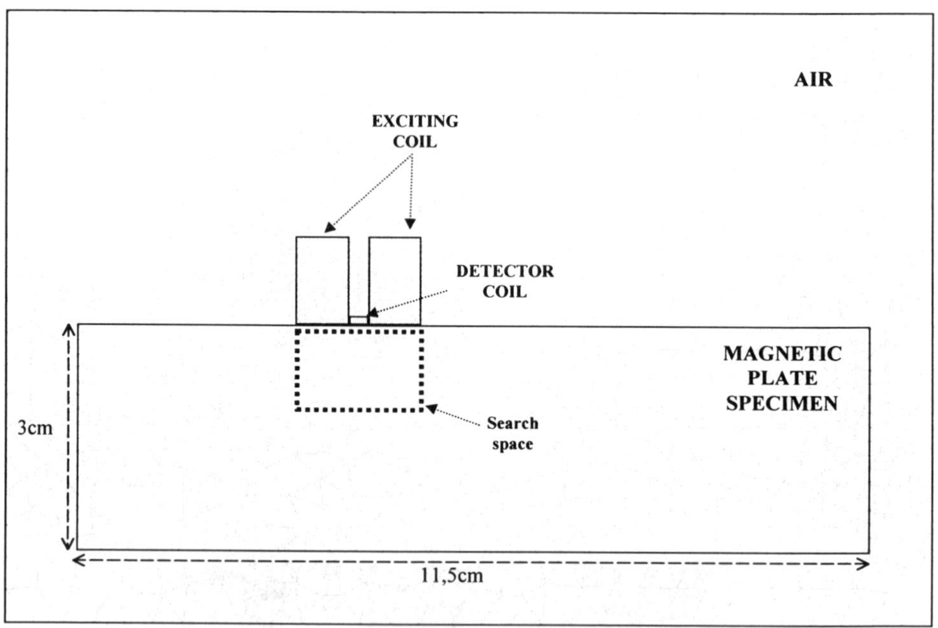

Fig. 4: Triangular mesh of the studied geometry (2970 elements, 1536 nodes)

The mesh is based on the Voronoï algorithm. Whatever the EGM located in the search space for matching the actual defect, the mesh variation during the iterative procedure is always progressive keeping a sufficient mesh density, see Fig. 5. Therefore, the defect geometry is well discretized. Furthermore it is expected that calculated magnetic values may not be mainly due to a different numerical precision when re-meshing.

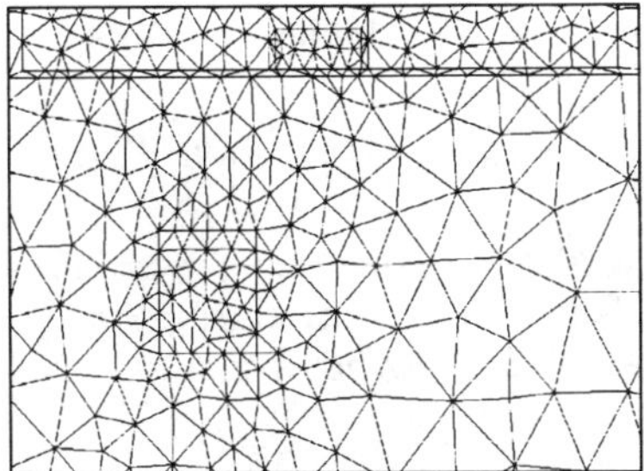

Fig. 5: a mesh example for EGM1

4. Optimization

4.1 Genetic Algorithm: GENOCOP

A numerical inverse problem solution based on FEM has serious accuracy problems in the evaluation of the estimates for the derivatives of the objective function *F*. Hence optimization algorithms based on the derivative calculations (i.e. deterministic algorithms) are not easy to implement. The mesh variations can affect the convergence properties of the first order difference usually used for the derivative approximation.

This derivative computation difficulty suggests the use of direct evaluations based methods. Among such methods, stochastic algorithms are widely known for their robustness. GA is probably the most popular stochastic algorithm due to the possibility to deal with a population of points. GA is based on the computer simulation of the natural evolution. This simulation renews the current population of individuals (i.e. EGM vertices) to tend toward the global optimum of the objective function.

The chosen genetic algorithm uses a floating-point representation. For the considered problem, this representation involves a higher precision in the design variables of the defect. The convergence is faster than with a binary representation. Finally the program is suitable for constrained EGM evolutions. GA used in this work is essentially based on the program GENOCOP III [6].

With the classic operators used in a floating-point representation (uniform mutation, boundary mutation, non-uniform mutation, simple and arithmetical crossover,...) GA works with two different populations P_r (i.e. reference population) and P_s (i.e. search population) where mutual influences exist. These influences depend on the individual development.

4.2 Procedure

The number of design variables is shown in Tab. I.

Tab. I: optimization variables

VARIABLES	EGM1	EGM2
X_1	X_a	X_a
X_2	X_b	X_b
X_3	Y_b	X_c
X_4	Y_c	Y_a
X_5		Y_b

The chosen goal function $F(X_1 ... X_5)$ used for the minimization procedure is:

$$F = \sqrt{(Bx - Bx_0)^2 + (By - By_0)^2} \quad (2)$$

where the magnetic flux densities are calculated above the specimen under test within the detector coil, see Fig. 4. B_0 is the flux density corresponding to the unknown defect and B the one corresponding to the current EGM configuration (size and location).

5. Applications

5.1 Examples

Tests have been done with the FEM code for different locations and shapes of defects. Fig. 6 presents the test cases and associated magnetic flux densities.

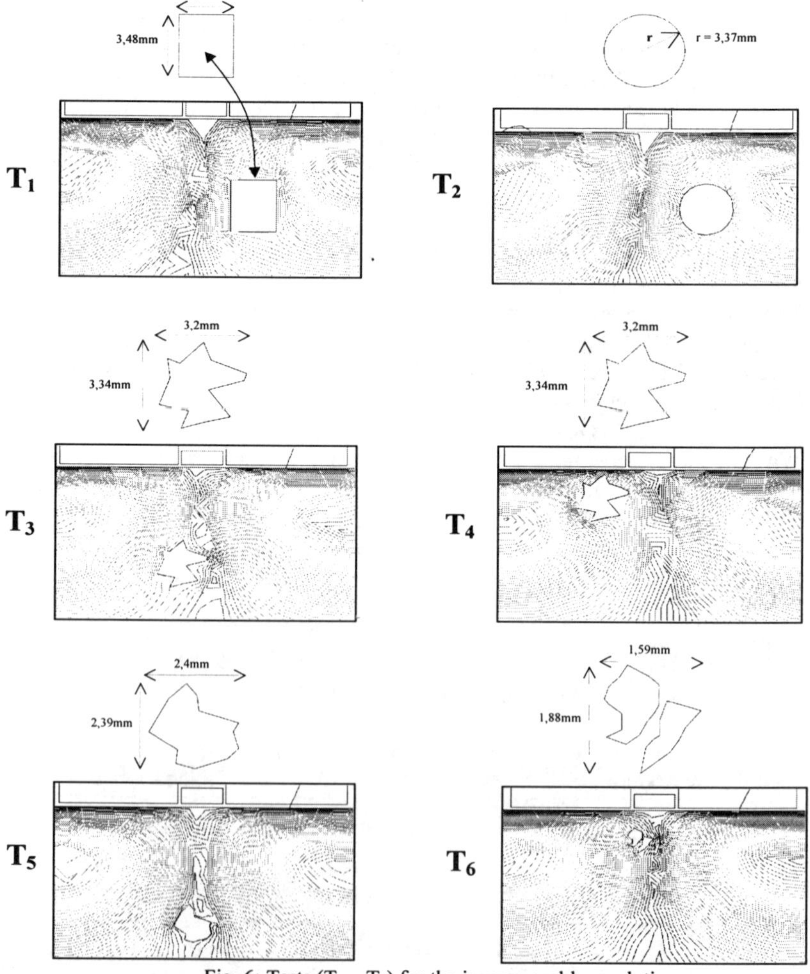

Fig. 6: Tests ($T_1 ... T_6$) for the inverse problem solution

Except for the particular test case T_6, it is supposed that considered positions and sizes belong to the feasible solution space of the objective function F, as mentioned in the definition of constraints. T_6 groups two small profiles presenting a phantom electromagnetic image. Moreover, T_6 dimensions are out of bounds of the problem constraints. T_1 and T_2 are simple shapes (T_1 is a shape like EGM1) and permit a preliminary test of GA behavior. T_3, T_4, and T_5 have no special characteristics.

Moreover, for simplicity, only defects presenting air properties are taken into account. The exciting current frequency is constant for all simulations involving a skin depth of 1cm.

5.2 GA Results

Calculations have been done with EGM1 and EGM2 for all the T cases. Fig. 7 shows some results for optimizations with EGM1 (P_r = 10 and P_s = 30). Given that GA quickly finds the global solution zone, a re-definition on domain constraints has been done on case T_4 during the process. This has involved a computed result in fewer times.

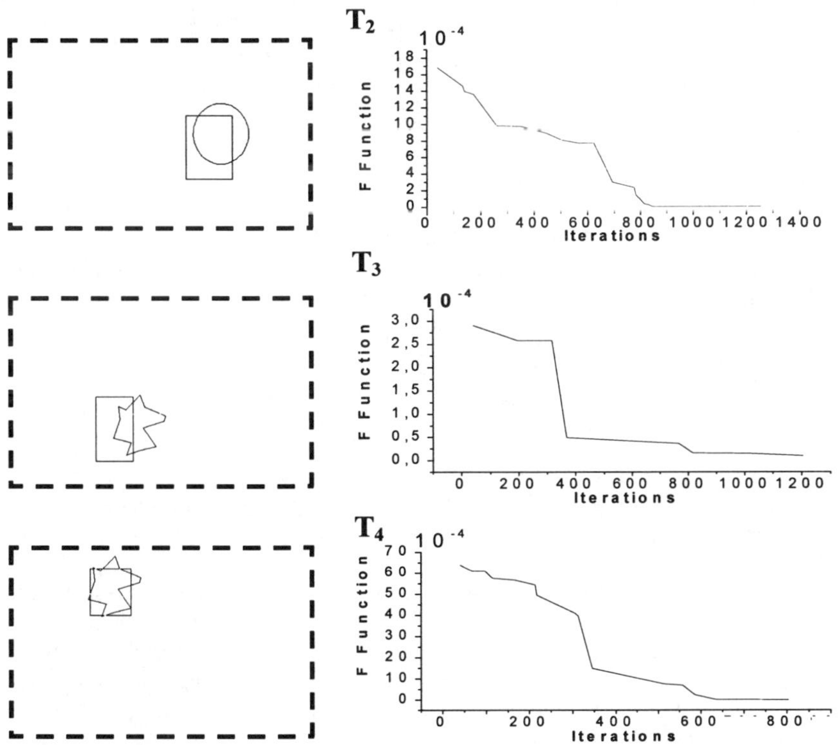

Fig. 7: Some EGM1 results for three tests
(Search space in dotted lines)

Tab. II presents some significant results for all the T test cases with EGM1 and EGM2. Population sizes, F results, EGM dimensions and errors on positions are indicated for demonstrating the sizing and localization capacities.

Tab. II: EGM responses for *T* cases

Test cases	EGM	Pr	Ps	F	Errors on localization	Sizing results
T1*	1	10	30	$4,5.10^{-8}$	$\Delta x = 0,23$mm $\Delta y = 0,06$mm	l = 2,1mm L = 3,3mm
T2	1	25	60	$4,8.10^{-5}$	$\Delta x = 0,78$mm $\Delta y = 0,66$mm	l = 2,8mm L = 3,2mm
T2	2	10	30	$2,6.10^{-6}$	$\Delta x = 4,2$mm $\Delta y = 0,24$mm	l = 3,8mm h = 3,1mm
T3	1	10	30	$8,4.10^{-6}$	$\Delta x = 1,5$mm $\Delta y = 0,18$mm	l = 2,3mm L = 3,56mm
T3	2	10	30	$7,4.10^{-6}$	$\Delta x = 2$mm $\Delta y = 1,2$mm	l = 3,3mm h = 3mm
T4*	1	10	30	$3,7.10^{-6}$	$\Delta x = 0,2$mm $\Delta y = 0,32$mm	l = 2,5mm L = 2,5mm
T4	2	25	60	$6,6.10^{-6}$	$\Delta x = 0,25$mm $\Delta y = 0,15$mm	l = 2,87mm h = 2,5mm
T5	1	10	30	$1,01.10^{-4}$	$\Delta x = 2,3$mm $\Delta y = 0,63$mm	l = 2,6mm L = 3,11mm
T6	2	10	30	$3,9.10^{-7}$	$\Delta x = 0,03$mm $\Delta y = 2,14$mm	l = 3,15mm h = 3mm

*: *With a domain re-definition during the process*

Results show the effectiveness of the proposed stochastic approach for simultaneously determining sizes and positions of unknown inner profiles. As GA works with a population of points, the possibility to drop in a local optimum is minimized.

Finally, as for all stochastic methods, the convergence phase is highly dependent upon the optimization parameters (population sizes, mutation and crossover probabilities and so on…).

6. Conclusion

This paper presents a stochastic optimization approach for solving inverse problems in magnetodynamic NDE using simple shapes. The accuracy on the reconstructed shapes and positions are mainly due to the complexity of chosen EGM and the numerical precision of FEM computations.

It is possible to deal with more complicated shapes (hexagon, octagon…) and deformations for EGM. Limitations are the required calculations times when working with numerous variables. If GA is an efficient global optimization procedure, each genetic individual (a position and a shape of EGM) of the current working population involves a FEM computation. For more rapidity, one can think to mix the genetic algorithm with other optimization methods [10].

A new mesh leads to a different numerical precision for consecutive calculations. This can affect or destroy the convergence phase. The solution is to re-mesh keeping a high density of elements or to work with a constant mesh. In this last case, affecting appropriate physical values to the finite elements will do all EGM evolutions. With a well-defined mesh, errors on F calculations, due to poor approximations on working physical values, can be considerably reduced. Moreover, an invariant mesh reduces the calculation time spent in the inverse problem loop. This calls for enhancing the proposed stochastic approach by defining elaborated EGM in a binary representation .

Even if stochastic methods with FEM remain costly in terms of calculations, computer speed evolutions promise an efficient reconstruction in the future NDE activities.

Acknowledgements

This work is supported by the European Commission in the frame of the INCO-COPERNICUS – MANODET project (Contract no. ERBIC15CT960703).

References

[1] "Inverse Problem Methodology and Finite Elements in the Identification of Cracks, Sources, Materials, and their Geometry in Inaccessible Locations", *S. Ratnajeevan H. Hoole et al.*, IEEE *Trans. Mag.*, Vol. 27, No. 3, May 1991, pp. 3433-3443

[2] "Solution of Inverse Problems by Using FEM and Structural Functions", S. P. Papazov, V. D. Borshukova, IEEE *Trans. Mag.*, Vol. 31, No. 6, Nov. 1995, pp. 4297-4305

[3] "Genetic Algorithms in Search, Optimization & Machine Learning", D. E. Goldberg, Addison Wesley, 1989

[4] "Ancillary Techniques for the Practical Implementation of GAs to the Optimal Design of Electromagnetic Devices", G. F. Üler and O. A. Mohammed, IEEE *Trans. Mag.*, Vol. 32, No.3, May 1996, pp. 1194-1197

[5] "Microwave Imaging Using a Genetic Algorithm", S. Caorsi, M. Pastorino, Electromagnetic Nondestructive Evaluation (II), R. Albanese et al. Eds., *Studies in Applied Electromagnetics and Mechanics 14*, IOS Press, Amsterdam, 1998, pp. 233-242

[6] "Genetic Algorithms + Data Structures = Evolution Programs", Z. Michalewicz, Springer, 3rd Edition, 1996

[7] "A Data Processing and Neural Network Approach for the Inverse Problem in ECT", R. C. Popa, K. Miya, Electromagnetic Nondestructive Evaluation (II), R. Albanese et al. Eds., *Studies in Applied Electromagnetics and Mechanics 14*, IOS Press, Amsterdam, 1998, pp. 297-304

[8] "Iterative Algorithms for Electromagnetic NDE Signal Inversion", *M. Yan et al.*, Electromagnetic Nondestructive Evaluation (II), R. Albanese et al. Eds., *Studies in Applied Electromagnetics and Mechanics 14*, IOS Press, Amsterdam, 1998, pp. 287-296

[9] "Crack Shape Characterization in Eddy Current Testing", *H. Fukutomi et al.*, Electromagnetic Nondestructive Evaluation (II), R. Albanese et al. Eds., *Studies in Applied Electromagnetics and Mechanics 14*, IOS Press, Amsterdam, 1998, pp. 305-312

[10] "A Hybrid Technique for the Optimal Design of Electromagnetic Devices Using Direct search and Genetic Algorithms", *O. A. Mohammed and G. F. Üler,* IEEE *Trans. Mag.*, Vol. 33, No. 2, March 1997, pp. 1931-1934

Electromagnetic Nondestructive Evaluation (III)
D. Lesselier and A. Razek (Eds.)
IOS Press, 1999

Cracks Detection and Recognition by Using of Multi-frequency Signal Processing and Neural Networks

Tomasz CHADY, Masato ENOKIZONO, Takashi TODAKA, Yuji TSUCHIDA
Faculty of Engineering, Oita University, 700 Dannoharu, Oita 870-1192 Japan
Ryszard SIKORA
Department of Theoretical Electrotechnics , Technical University of Szczecin, al. Piastow 19, 70-310 Szczecin, Poland

Abstract. In this paper authors propose a new system for multi-frequency testing of conducting plates (magnetic and nonmagnetic materials). Precise crack imaging is achieved by using of spectrograms obtained from an eddy-current probe multi-frequency response and application of a neural network.

1. Introduction

A considerable effort is underway to develop new eddy current systems for detecting and recognizing cracks. Both frequency-domain and time-domain techniques are developed. In this paper authors propose a new concept of multi-frequency system. A complex signal containing selected sinusoidal components is used for testing. In case of such signal an information content is greater than in a case of single-frequency signal. On the other hand the proposed technique is far faster than swept-frequency measurements. Time required for acquire the complex signal is the same as for single- frequency signals.

2. System description and operation

In order to achieve precise crack recognition, a new computer-controlled system for eddy current multi-frequency nondestructive testing was constructed. The experiences from the previous system [1] have been taken into account by the authors and significant improvements have been done. The new system can operate at any set of frequencies between 5Hz and 8MHz and is capable of making measurements with various sensors and techniques.

It consists of the following main components:

- eddy current sensor,
- X-Y-Z scanning device,
- high performance data acquisition system (amplifier, anti-alias filter and A/D converter),
- excitation subsystem (wave synthesizer, and high speed power amplifier),
- pentium based computer with GPIB and VXI interface boards,
- computer program for instruments control and data processing.

A block scheme of the system is shown in Fig. 1. The most important element of the system is a computer equipped with specially prepared software.

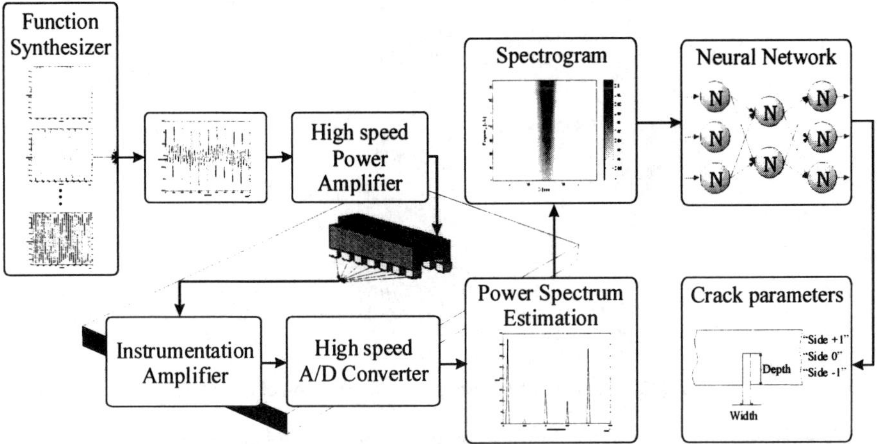

Fig. 1 Schematic view of the system

The computer controls all elements of the system through the VXI and GPIB interface and it performs such tasks as digital signal processing of incoming signal and crack recognition.

An eddy current sensor shown in Fig. 2 provides the signal, which is used for crack detection and identification. The sensor consists of an exciting coil and a set of search coils. All coils are wounded on ferrite core shown in Fig. 2. The exciting coil has 200 turns (ϕ0.2mm), and each of the fourteen search coils has 100 turns (ϕ0.04mm). Multiple search coils enable to perform at once measurements in many points.

A function synthesizer supplies excitation coil of the sensor through the high-speed power amplifier. A complex signal containing selected harmonic components is used in order to achieve information from number of frequencies.

The analog signals obtained from the search coils are converted into digital form by the data acquisition system. This system is a VXI device and consists of a multiplexer, a computer-controlled amplifier, and a high performance A/D converter. The high speed (20 megasample per second) and excellent dynamic range (32 bit resolution) of the converter are crucial for the performance of the whole system. The speed is necessary to achieve proper signal representation. The resolution is also necessary because generally the changes of the signal caused by a defect are very small in comparison with the level of background signal.

Noise and other artifacts are introduced by several internal and external sources, including transducer and instrumentation. In order to remove influence of noises and to extract the amplitude of main components from the measurements, signal-processing algorithms have been applied. Estimation of power spectra is useful in variety of applications, including the detection of signals buried in wide-band noise [2]. Authors have also used this technique for

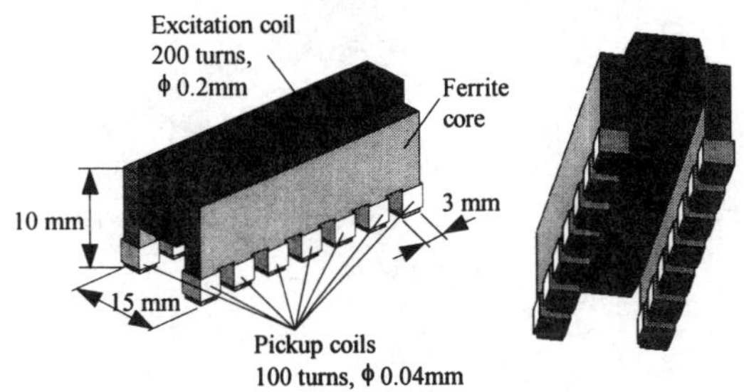

Fig. 2 View of the eddy current sensor

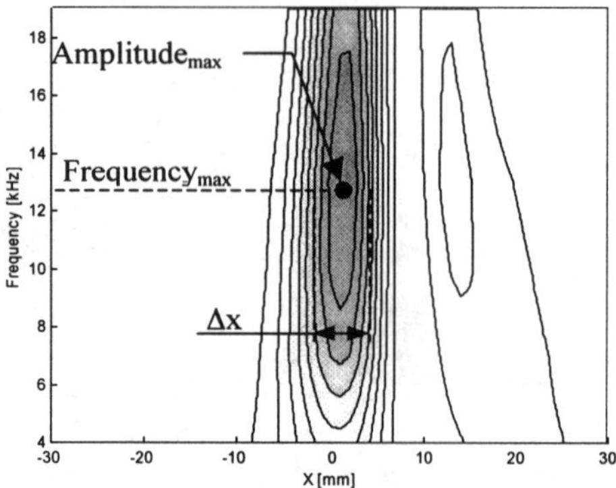

Fig. 3 Example of a spectrogram and definition of parameters

amplitude estimation of the each frequency component. Then a digital low-pass infinite impulse response filter processes the resulting signals in order to remove high frequency fluctuations. The applied anti-causal filter performs zero-phase digital filtering by processing the input data in both the forward and reverse directions. After all mentioned operations, the resulting signals were used for creation of spectrograms. Example of the spectrogram is shown in Fig. 3. The spectrogram is a two-dimensional display of the relative amplitude of the frequency components of a signal from the single search coil over the sensor position. The relative amplitude is a difference between the current amplitude and the amplitude measured for the fault free plate divided by the amplitude measured for the fault free plate. After analysis of spectrogram a set of selected parameters (values at the maximum of the spectrogram: "Amplitude$_{max}$", "Frequency$_{max}$", and "Δx", see Fig. 3) is calculated and is passed to the input of a neural network used for crack recognition.

3. Neural network for a crack recognition

A conventional (fully connected) feed-forward network was used to solve the inverse problem of crack identification. The schematic view of the network is shown in Fig. 4.

In the proposed net there are two layers with connections to the outside world: an input plane consists of 3 units, which receives parameters extracted from the spectrograms and 3 output neurons which gives the information about crack width, depth and position (same side, opposite side, inner). The main design parameter of the network is the number of hidden neurons, which guarantee the desired degree of accuracy. There is no established rule to strictly determine how many neurons are needed to obtain a given degree of accuracy [3]. After some experiments it was found that the minimal number of hidden units is 2.

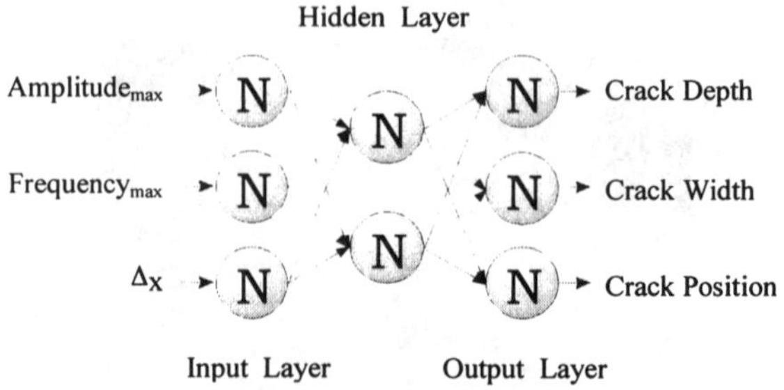

Hidden Layer

Amplitude$_{max}$ → N → N → Crack Depth

Frequency$_{max}$ → N → N → Crack Width

Δx → N → N → Crack Position

Input Layer   Output Layer

Fig. 4 Architecture of the neural network used for a crack recognition

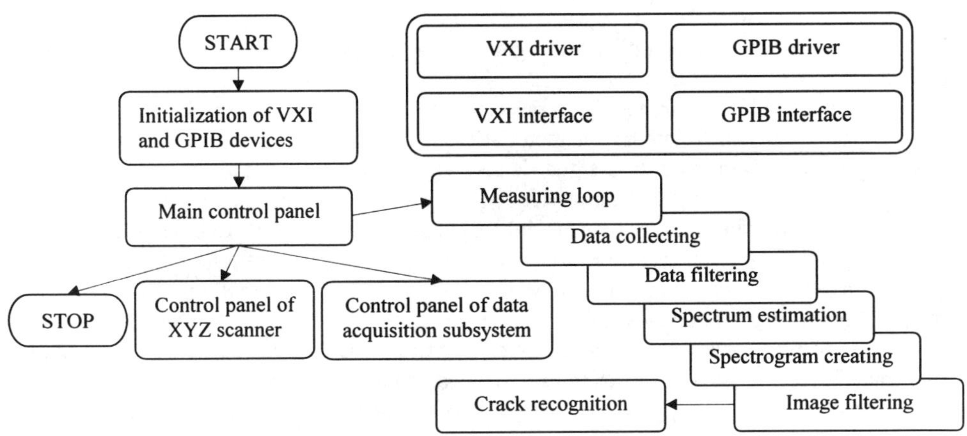

Fig. 5 Simplified flowchart of the program

4. Software structure

The Windows™ based software controls operation of all instruments as well as the data acquisition process.

The program can be divided for the following five main modules:

- Main control panel.
- Interface to GPIB and VXI devices (two DLL libraries).
- Control panel of data acquisition and excitation subsystems.
- Control panel of the XYZ scanner.
- Module for digital signal processing and power spectrum estimation.
- Module for spectrograms creation and data visualization.

Module for a crack recognition (interface to a neural network toolbox).

A simplified flowchart of the program is shown in Fig. 5. One of the most important parts of the software is the control panel (Fig. 6) of the data acquisition and excitation subsystems.

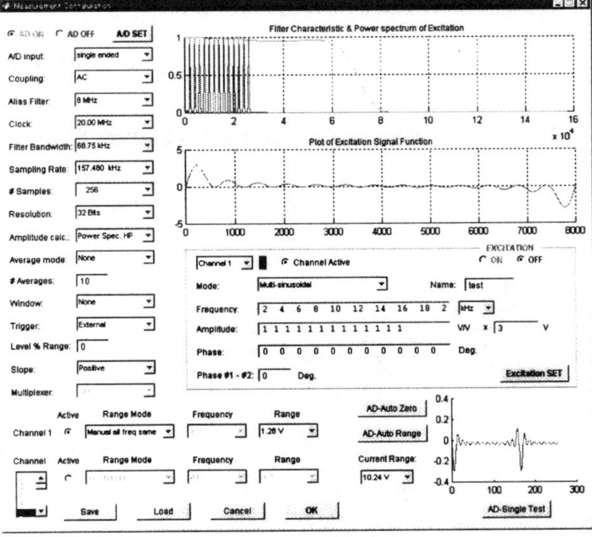

Fig. 6 Control panel of the data acquisition and excitation subsystems

In this panel one can define nearly all parameters of the measuring process (e.g. input filter bandwidth, sampling rate, amplification gain, triggering parameters, type of excitation signal and many others).

In order to simplify the process of parameters selection two auxiliary plots are provided. In the top graph a plot of frequency characteristic of the input filter is shown. In the same plot a power spectrum of an excitation signal is also shown. This plot enables proper selection of A/D parameters even by an inexperienced user. The next plot contains a view of the excitation signal. This plot is very useful, especially for the multi-frequency mode. In the multi-frequency mode a user has to define frequency and amplitude of each frequency component. Beside the multi-frequency mode one can select also a single frequency, a step function and a swept-frequency mode.

The measuring process can be monitored be using various kind of plots. An example of four main available plots is shown in Fig. 7. These windows show respectively:

- Left-top - view of a signal from the search coil after digitalization and filtering
- Right-top - plot of a signal spectrum
- Left-bottom - spectrogram
- Right-bottom - plot of amplitude of each frequency component

Some additional plots are also available (e.g. three-dimensional spectrogram). All graphs are updated on-line during the sensor movement.

Because of user-friendly graphic interface and auxiliary plots less than one hour is enough to teach a new person how to measure and use the system.

The software can be also easily extended. Many techniques (signal and image processing, neural computing, FEM, fuzzy logic, genetic algorithms, wavelet analysis) can be

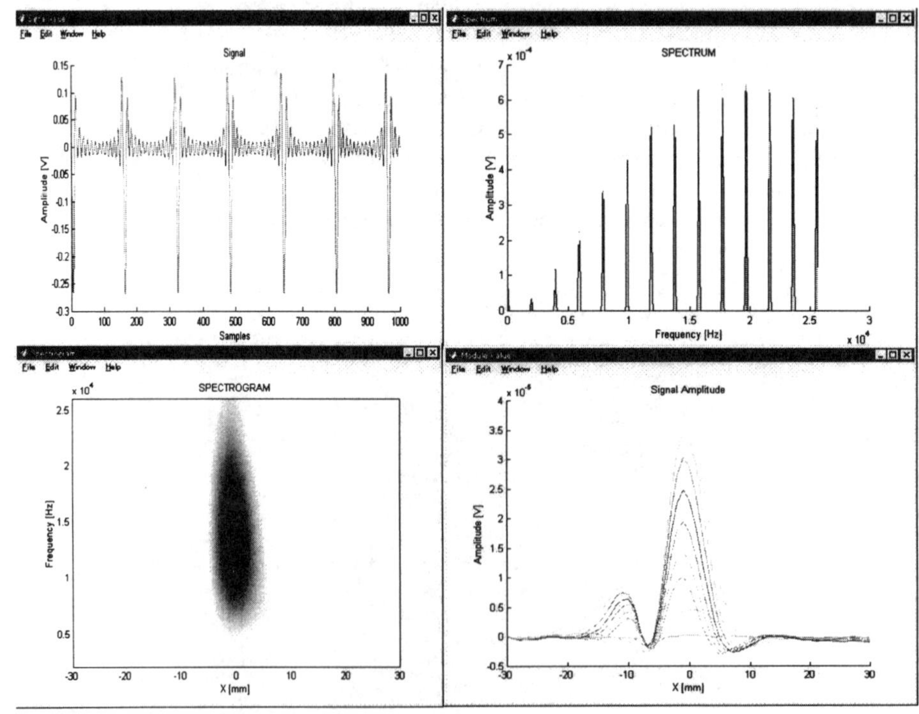

Fig. 7 Four main plots offered for a user during measuring process

immediately applied. Thanks to the VXI and GPIB interfaces prepared by the author a majority of measuring instruments can be controlled from the system directly.

5. Experimental results

In order to demonstrate advantages of the system a set of experiments was carried out. The test specimens consisted of a set of 5 mm thick plates of SUS304 (Fig. 8a) and SS41 (Fig. 8b) each containing a single rectangular crack. The dimensions of the test specimens were sufficiently large that measurements can be made without any edge effects. Each crack was located at the center of the plate and aligned so that the slot axis was perpendicular to the major axis of the plate.

For the SUS304 plates the width of the crack was 5 mm, the length was 10 mm, while the depth was from 1.0 mm up to 4.0 mm.

For the SS41 plates the width of the crack was 0.6 mm and 1 mm, the length was 120 mm, while the depth was from 1.0 mm up to 4.0 mm.

The measurements were done by scanning the probe along the line perpendicular to the slot axis in steps of 1 mm. The measurements were made using the multi-frequency excitation signal consists of sinusoidal signals having frequencies from 2kHz up to 26kHz for the SUS304 plates and from 5Hz up to 500Hz for the SS41 plates.

The spectrograms for the SUS304 plates are presented in Fig. 9 and for the SS41 plates in Fig. 10 respectively. The plot shown in Fig. 11 summarized all results of the measurements made for SUS304 specimens. From this figure it is possible to observe that these data enable very precise determination of a crack position and depth and can be used successfully for cracks identification.

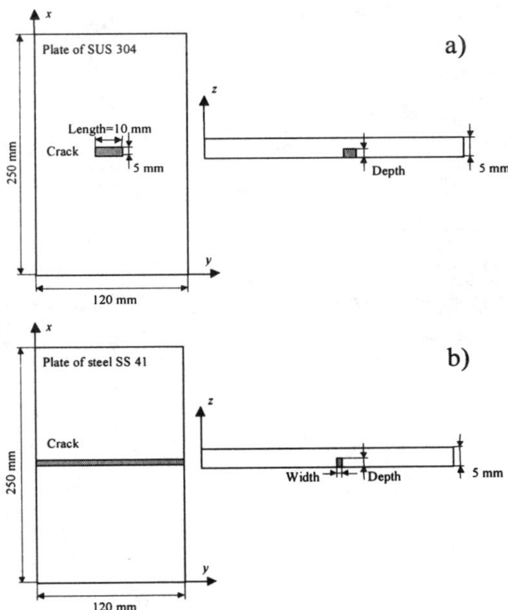

Fig. 8 Test specimens

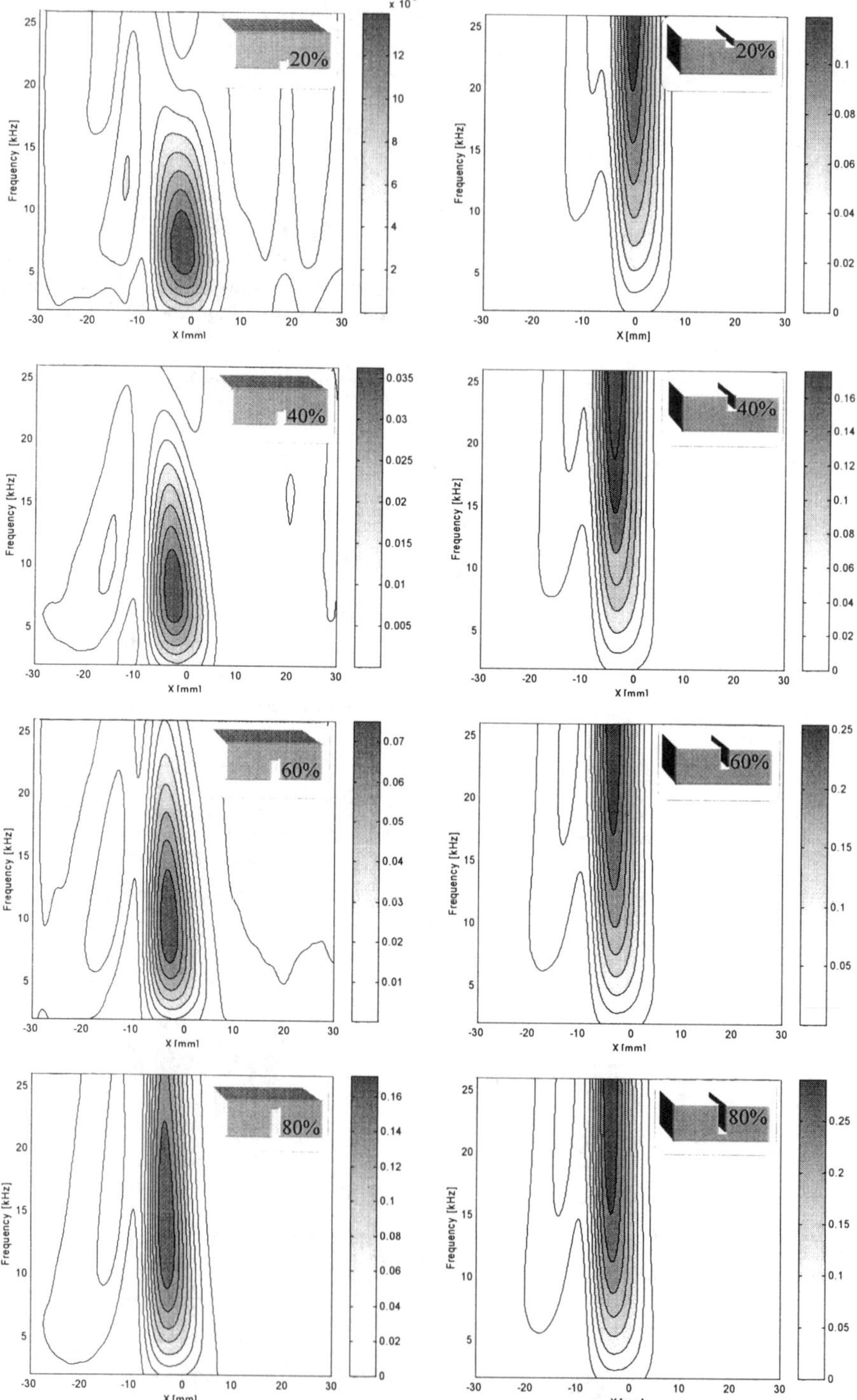

Fig. 9 Spectrograms for cracks having different depth placed on the same side and opposite side of the plate (SUS301)

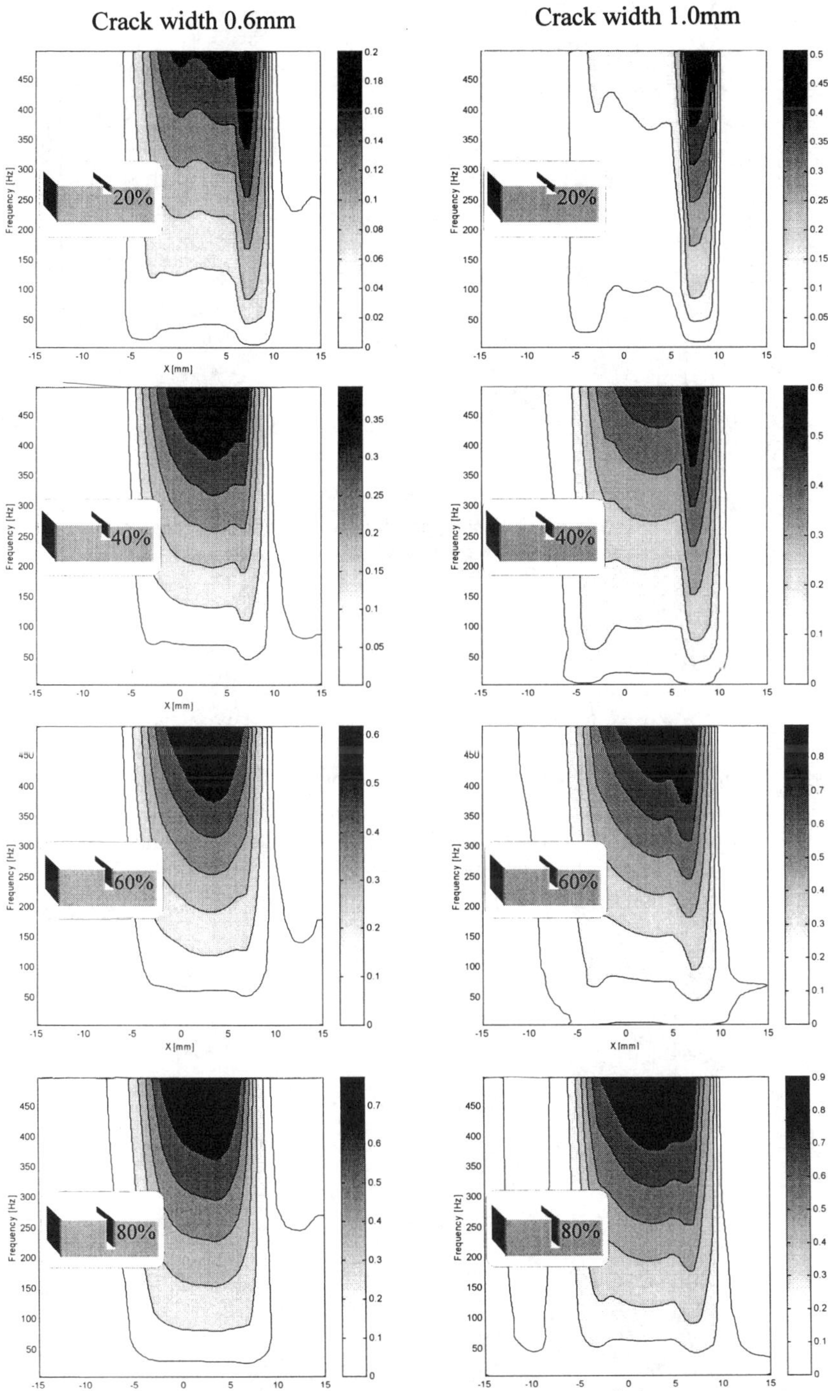

Fig. 10 Spectrograms for cracks having different depth placed on the same side of the plate (SS41)

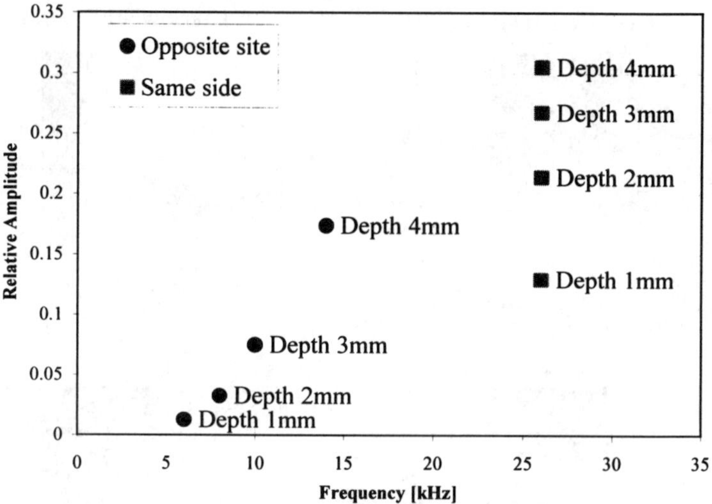

Fig. 11 Parameters of spectrograms for cracks having different depth placed on the same side and opposite side of the plate (SUS301)

6. Training of the neural network

The neural network used for crack recognition was trained with the achieved data from measurements performed for SUS304 plates. All the data were divided into three parts: training set, validation set and the test set. The learning should be stopped at the minimum of the validation set error. When learning is not stopped, overtraining occurs and the performance of the net on the whole data decreases. After finishing the learning phase, the net was finally checked with third data set, the test set.

Because simple backpropagation is very slow, improvement techniques [3] such a

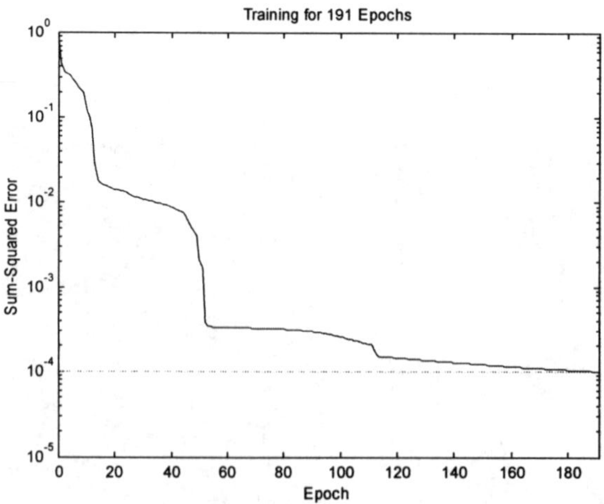

Fig. 12 Estimation error for training patterns vs. iterations (epochs)

momentum and adaptive learning rate and an alternative method to gradient descent, Levenberg-Marquardt optimization, have been tested. After experiments the Levenberg-Marquardt method was selected as the most sophisticated rule. The starting values of the network parameters were found by using a random optimization method.

Fig. 12 shows how the sum of squared errors over all training vectors converged as the training epochs occurred. It took only 180 epochs for the network to achieve error less than the goal of 10^{-4}. The resulting network has a sum-squared error of 10^{-4}. All calculations were performed using a neural network simulator on a personal computer with 300 MHz Pentium processor. The learning process took only four seconds.

7. Conclusions

The principal conclusion drawn from the experiments is that the proposed multi-frequency method and the system seem to be very promising for the evaluation of crack properties.
The offered advantages by the system can be summarized as follows:

- The key feature of the proposed multi-frequency ("multi-sinusoidal") method is the wide bandwidth and high speed of operation it provides. As a result of using multi-frequency excitation the response to several different frequencies can be measured at once. Since the penetration depth depends on the frequency of excitation, information from a range of depths can be obtained within single measurement.
- Testing frequencies can be freely selected according to user requirements.
- User can also decide amplitude of each frequency component.
- Only useful frequency components are used as a sensor excitation.
- Information obtained from the range of depths and collected in one spectrogram enables the precise flaw depth and shape identification.
- The same instrumentation as for single frequency method is sufficient for our technique.
- Various kinds of probes can be used and the system can operate at any frequency from 2Hz to 8MHz,

Application of the neural network for identification of cracks has also some additional advantages:

- High speed of operation (only a few milliseconds is necessary to calculate a crack shape).
- Flexibility (other types of eddy current transducers may be modeled by simply additional training).
- Short time of learning (less than 10 seconds).
- Noise tolerance.
- Small requirements (only results of measurements made for selected test specimens are needed).
- Simplicity (the proposed network contains only a few neurons).

References

[1] T.Chady, M.Enokizono, "Eddy Current Testing of Stainless Plates by Using Matrix Sensor", Proceedings of the 7th Magnetodynamics Conference, April 24, 1998, pp.107
[2] S. Kay, Modern spectral Estimation, Englewood Cliffs, NJ: Prentice Hall, 1975.
[3] J. Hertz, Introduction to the theory of neural computation, Addison Wesley: Reading, 1991

Electromagnetic Nondestructive Evaluation (III)
D. Lesselier and A. Razek (Eds.)
IOS Press, 1999

A Wavelet Neural Network Processor of Eddy Current NDE Data

Francesco Carlo MORABITO[1], Antal GASPARICS[2]

[1] *Ass. EURATOM/ENEA/CREATE, DIMET, Univ. degli Studi di Reggio Calabria Via Graziella, Loc. Feo di Vito, I-89128 Reggio Calabria, Italy*
[2] *Research Institute for Materials Science, H-1525 Budapest, P.O. Box 49, Hungary*

Abstract. In this paper, some useful properties of both Neural Network and Wavelet Decomposition processing which have a potentially relevant impact on eddy current NDE data analysis are investigated. The analysis regards both synthetic and experimental data, which are the output of researches based on fluxset probe carried out in the co-operation of the Research Institute for Material Science and the Technical University of Budapest, Hungary. The outcome of this work is that the use of wavelet processing can be of interest for pre-processing of the data as well as for selecting features of the experimental readout at the different levels of resolutions. Neural Networks are of high interest for generalising the conclusions to actual NDE applications.

1. Introduction

One of the most promising industrial developments of electromagnetic non-destructive evaluation regards the detection and characterisation of cracks and flaws in metallic materials. Along this way, the typical inspected specimens are thin plates and tubes. The identification procedure calls for modern signal processing techniques to be applied in order to overcome various practical difficulties. In this paper, we will show how important tools like Wavelet Transform Analysis (WTA), Neural Networks (NN) can be of help in analysing ECT-NDE data. The proposed results will regard both simulated and experimental data. Firstly, the algorithms are tested on synthetic data derived from an analytical solution of the direct problem. Finally, experimental data related to the MANODET FluxSet type sensor are analysed to motivate our investigations [1].

The ECT-NDE system outputs signals depending on the geometry of the inspected specimens, on the probing system, and that are affected by various sources of inaccuracies, e.g., the imprecise contour and shape of the defect, the mis-positioning of the inspection system, the imprecise determination of the lift-off, and the measurement noise. NN models have been proposed for facing the challenging requirements of ECT-NDE [2-3]. In this paper, we propose use a Modular Neural Network (MNN) structure in order to improve the accuracy in the crack parameters estimation.

Wavelet analysis can be useful in processing ECT-NDE data from two opposite perspectives. The first one concerns the de-noising of signals, which means that small details can be omitted in the signal without affecting its main features. Secondly, the filters used during signal decomposition can produce details that highlight the presence of defects and indicate its presumable location. Both aspects are analysed in the following.

2. The Benchmark Problems and the Wavelet Transform

We shall use two benchmark problems: the first one is the characterisation of a small circular hole in a planar, ideally conducting, thin metallic plate. The aim of the procedure is to associate a non-linear mapping between a set of properly selected magnetic measurements taken in the vicinity of the specimen and the set of geometric quantities defining the hole (i.e., two co-ordinates representing its location on the plate and its radius). This mapping is implemented by means of a NN. Both the database of training and testing have been generated through an analytical procedure [2]. The possibility of using a numerical generator is evidently important for allowing a variety of different situation to be accounted for. The second test problem aims to analyse the experimental data proposed in [1], by comparing our results to the simple straightforward processing of the measurement data there presented.

In both cases, the signal picked-up by the sensors shows a typical peaked behaviour which lends itself to be analysed by means of a wavelet pre-processing system. The wavelet transform has recently emerged as an exciting new tool for statistical signal processing. The wavelet domain provides a natural setting for various applications involving real-world signals, including ECT-NDE [4]. The remarkable properties of the wavelet transform lies basically in the representation coefficients of the analysed signal mapped in the WT domain. The wavelet transform is an atomic decomposition that is able to represent 1-D time (or space) signals in terms of shifted and dilated versions of a prototype bandpass wavelet function (mother wavelet), and shifted versions of a lowpass scaling function. The wavelet functions are both time and frequency localized. For a wavelet centered, let's say, at time zero and frequency fo, the corresponding wavelet coefficient, w(j,k), measures the energy signal content around 2^j k and frequency 2^{-j} fo. In wavelet-based signal processing, we process the 1-D signal by operating on its wavelet and scaling coefficients. The wavelet transform has several attractive properties that make it natural for NDE data processing. The main properties are:

1) *Locality*, each atom is localized in both time and frequency domain;
2) *Multiresolution*, wavelet atoms are compressed and dilated to analyse the signal at a nested set of scales;
3) *Compression*, the WT of real-world signals tend to be sparse.

Two additional more rarely claimed properties are of interest in our problems:

1) *Clustering*, if a particular wavelet coefficient is large/small, then adjacent coefficients ae very likely to also be large/small;
2) *Persistence*, large/small values of wavelet coefficients tend to propagate across scales.

By exploiting these properties, we are able to locate important clues in our data, which are related to the presence of defects and to their characterisation [3, 4]. In our experiment, for example, the forward wavelet transform of the raw data is computed and the map of the transform coefficients scanned to find the peak value at the highest resolution level. This coefficient indicates the presence of the crack around the corresponding location. The comparative analysis of the coefficients at the various levels of resolution allows us to estimate its presumable centre and size. This information can be used for further processing. It is always possible to estimate the size of the cracks starting from the measurement of the distance between the peaks, however, this is really reliable just on ideal cases. Indeed, in practice, the presence of white noise influences the readout of the measurement system, and we may have a group of peaks among which it is difficult to select the correct one. Other kinds of correlated noise may in practice affect the raw data, like, for example, the mis-positioning of the sensor, the lift-off noise and the surface roughness of the specimen. In particular, in the case of "other side" cracks, surface roughness may give rise to false detection comparable to potential real cracks.

As an example, Fig. 1 reports the analysis based on the wavelet transform of a typical pattern of fluxset-type raw data. The signal related to the presence of two cracks parallel with the axis of the OD20% and OD40% cracks are reported. The multi-resolution analysis allows to look through the signal at different scales, thus allowing to separate the different effects on the measurements of different sources (e.g., noise, lift-off variations, presence of flaws, The reading of the highest resolution levels is often sufficient to decide about the presence of defects in the plate specimen. It is here evident that the WT is able to locate and roughly characterise the cracks, by means of the measuring of the distance between the peaks and the deepness of the valleys. It is to be noted explicitly that the power of the tool is related to the possibility of separating in both frequency and space the different contributions to the final signal. This means that we are able to select a frequency (*scale*) level on which the only measurable effect is related to the presence of the crack (second row from the top in Fig. 1) while the other components of the measured signal are taken into account at the various other frequency levels. In the example, the measurement noise could be read at the fourth row from the top. ...). Fig.2 depicts the typical distribution of the WT coefficients for the problem at hands. The analysis in the WT domain consists in forcing to zero some subset of the transform coefficients. The inverse transform computed starting from the remaining nonzero coefficients allows to obtain a signal "cleaned" from the various noises. Without inverse transforming the data, we are yet able to determine the presumable location and length of the crack by scanning the shift axis at a selected dilation level. An additional advantage of the WT approach is related to the possibility of making different choices for the mother wavelet. In this paper, we have used the Daubechies 4, however, the different properties of the mother wavelets allows to best match the characteristics of our problem. For example, in [4] it is shown that a suitable choice of the mother wavelet allows to filter the lift-off noise from the data.

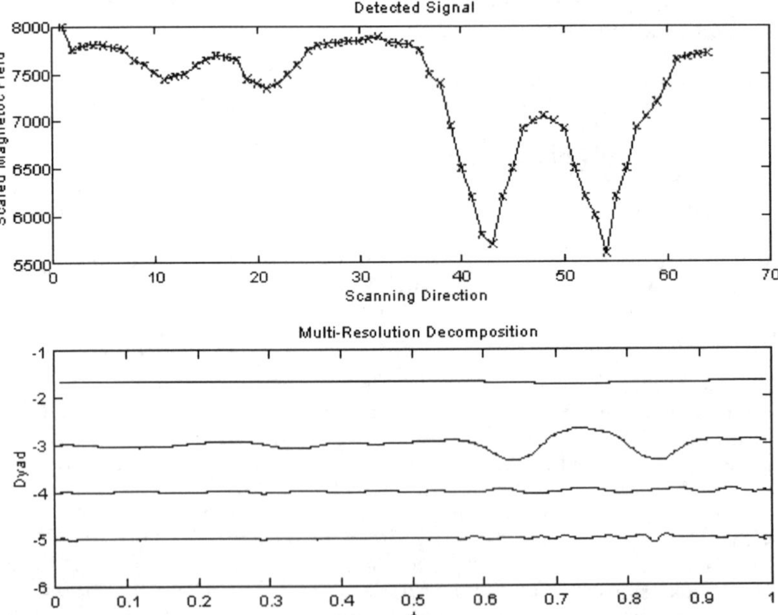

Fig.1: The readout of the measurement system when the probe is scanned along a line parallel to the axis of two cracks of different width, and the multi-resolution decomposition of the signal (Daubechies 4).

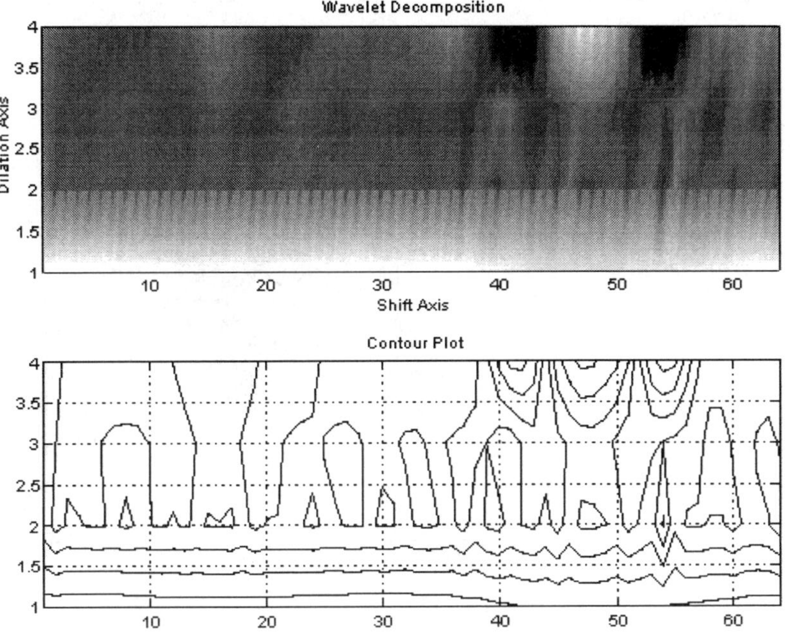

Fig.2. Image and Contour plot of the wavelet coefficient at the different levels of resolution. It is highlighted the detection capability of the WT.

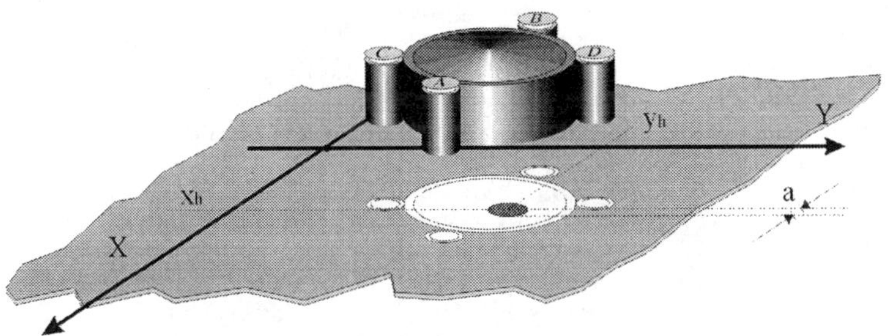

Fig.3: A drawing of the inspection system for the analytical benchmark problems.

3. The ECT-NDE Analytical Problem

The analytical test problem is the characterization of a small circular hole in a planar, ideally conducting, thin metallic plate. The corresponding direct problem can be formulated in terms of Maxwell equations and an analytical solution is already available [2]. The solution of the inverse problem consists in characterising the crack in terms of location and size, starting from the reading of a pattern of measurements (or some linear combinations of them). The non-linear mapping between the two sets is implemented by a NN. We generate a database of patterns for the training and the testing of the NN through an analytical procedure. When the plate is excited by a time varying external magnetic field, a reaction current is induced in the plate, which perturbs the original field. The presence of a hole modifies the reaction current map, and thus the magnetic field. We suppose that the hole size is small enough with respect to exciting system dimensions and with respect to the excitation field wavelenght. Further details of the proposed test case can be found in [2].

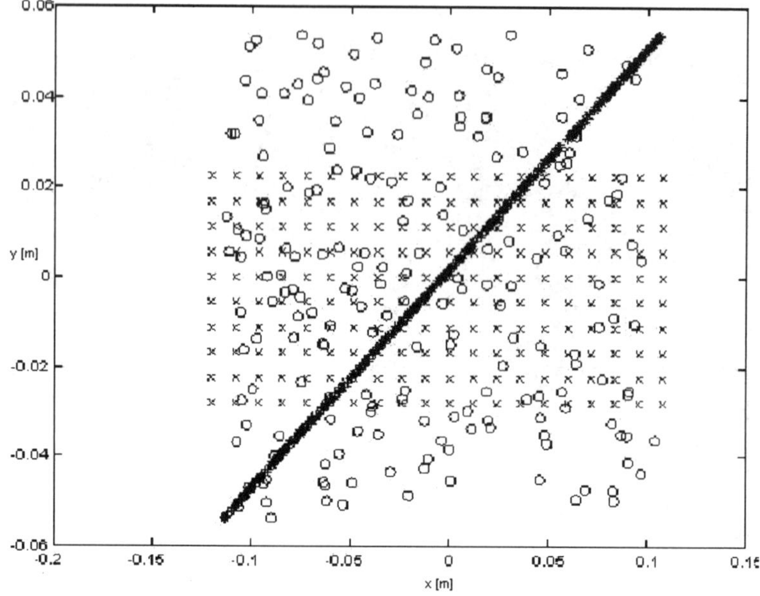

Fig. 4: The distribution of the training database (o), and of the two test database (x, *). The first block of test patterns has a fixed radius value (0.5 mm).

The inspection procedure is simulated through an exciting-receiving apparatus that moves all over the plate, and the diagnostic system consists of two differential probes located around the inducing solenoid. The region of the plate under inspection is sized 40x20 cm. Fig. 3 depicts the exciting solenoid and the pick up sensors configuration. The magnetic flux measured by the probes is computed by supposing an uniform field within each receiving loop. The measurements assumed as inputs of the identification procedure are the differences of flux within the receiving probes (Ψ_A -Ψ_B), and (Ψ_C -Ψ_D).

The proposed configuration is used to generate a suitable database for training a NN model in order to have an identification model of the characterisation procedure. The selected NN topology is a multilayer perceptron with 16 inputs (i.e. the set of simulated differential measurements at 8 different positions), 10 hidden nodes and 3 outputs (centre co-ordinates, x_h and y_h, and radius, a). Two different database have been generated for testing (the distributions of the three database are shown in Fig.4).

In Fig. 4, the axes correspond to the x- and y- direction of the plate specimen and are quoted in meters. In particular, we have trained the NN on a set of patterns (circles in the figure) corresponding to hole locations which have been randomly generated within the inspection area. The test databases have been selected in order to test the generalisation and extrapolation abilities of the resulting model. The information gathered by means of the WT procedure applied to each incoming pattern during a fresh inspection procedure is used to drive a gate which switches on a specialised NN trained on patterns competing to that area [5]. The resulting processor is a modular system whose performance are compared in Table I to those of the basic NN processor.

Table I reports the resulting average errors of the NN on the two different test database. Since the first test database has been generated by fixing the radius of the hole, the corresponding estimation error has not been computed (symbol "/" in the Table).

Tab.1 – Root Mean Squared Physical Error [mm].

	Crack Parameter	Wavelet + NN	16-12-3 NN
Test 1st block	x_h	0.3	1.7
	y_h	0.4	2.0
	a	/	/
Test 2nd block	x_h	0.9	2.1
	y_h	1.1	2.3
	a	0.26	0.46

4. The Experimental ECT-NDE Problem

The proposed approach has been preliminary tested on an experimental database related to the experimental activity carried out in Hungary [6]. In the proposed application the operational frequency of excitation of the eddy currents is 20 kHz. The exciting-receiving apparatus is described in a different paper proposed by one of the authors in this E'NDE conference. The distance between the exciting coil and the surface of the specimen (lift-off) was more than 1 mm. Some parameters of interest concerning the plate and the artificial crack are reported in Fig. 5 and in the attached Table. The middle of the probe has been moved along four lines perpendicular to the axis of the crack. This explains the different appearance of the readout with respect to Fig.1.

The scan-lines are at 1, 3, 5 and 7 mm along x-axis. Each line includes 61 data that we pad to 64 measurements to match the dyadic structure of WT. The cracks lie in the middle line of the plate on the other side (OD15%) or on the same side (ID15%) with respect to the measuring apparatus. Each crack is 9 mm long and occupies 15% of the plate thickness.

Fig. 6 shows the Mother Wavelet used in our experiment (Daubechies with 4 vanishing moments). Fig. 7 reports the readout of the measuring systems. For our analysis we use a differential set of measurements, obtained by considering two successive scan-line. The denoising procedure allows to filter out unwanted contributions to our signal, which reduces to have a non redundant representation of the original signal filtering out the high-frequency noise.

The remaining rapid variations in the signal can easily be related to the presence of cracks in the specimen.

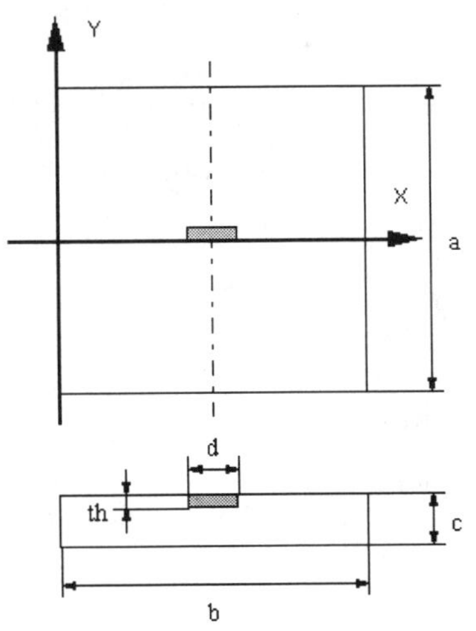

Parameter	[mm]
a	80
b	80
c	1.25
d	9
th 15%	0.1875
th 10%	0.1250

Fig.5: Schematic drawing of the ECT test specimen and its reference parameters.

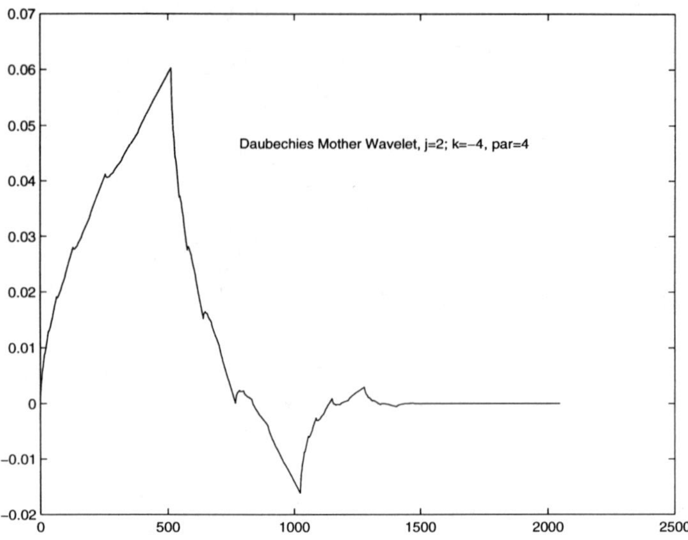

Fig. 6: The Daubechies_4 Mother Wavelet for two selected scaling and translation parameters.

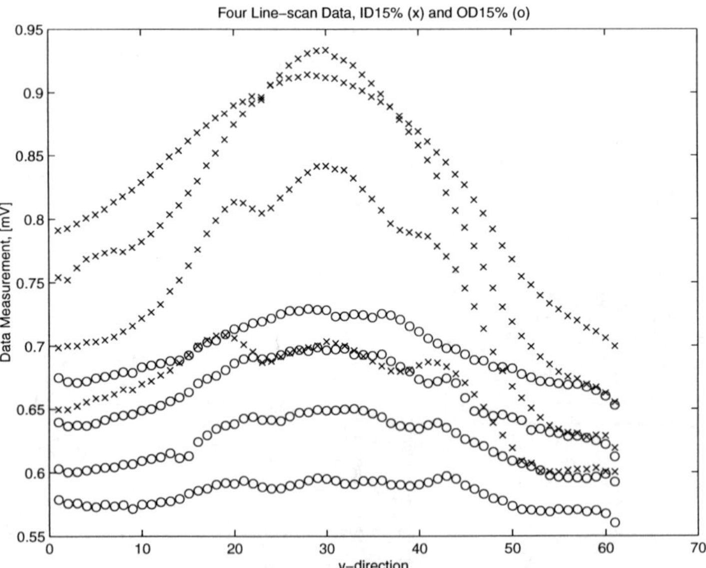

Fig. 7: The readout of the measuring system when the probe is scanning along four lines perpendicular to the axis of the crack. The ID15% and OD15% cases are considered.

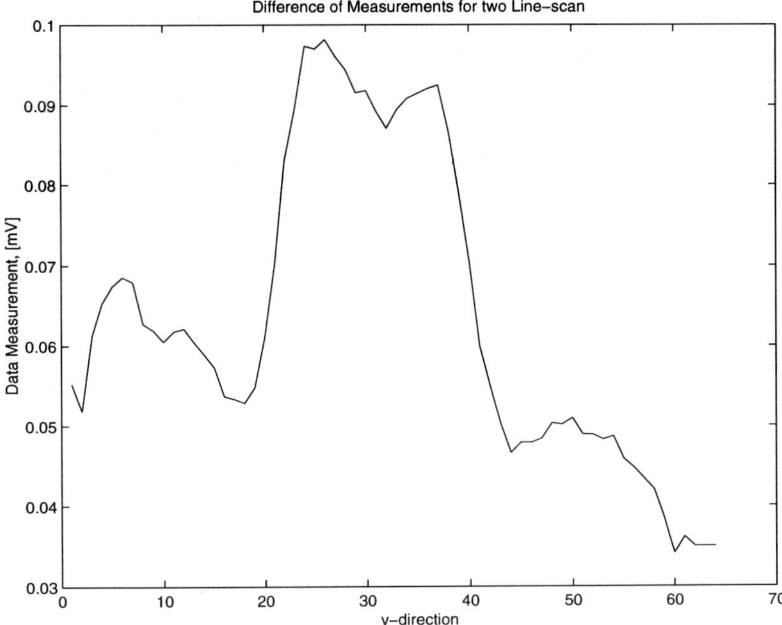

Fig. 8: The Wavelet Transformed pattern, i.e., the difference between two line scan of data

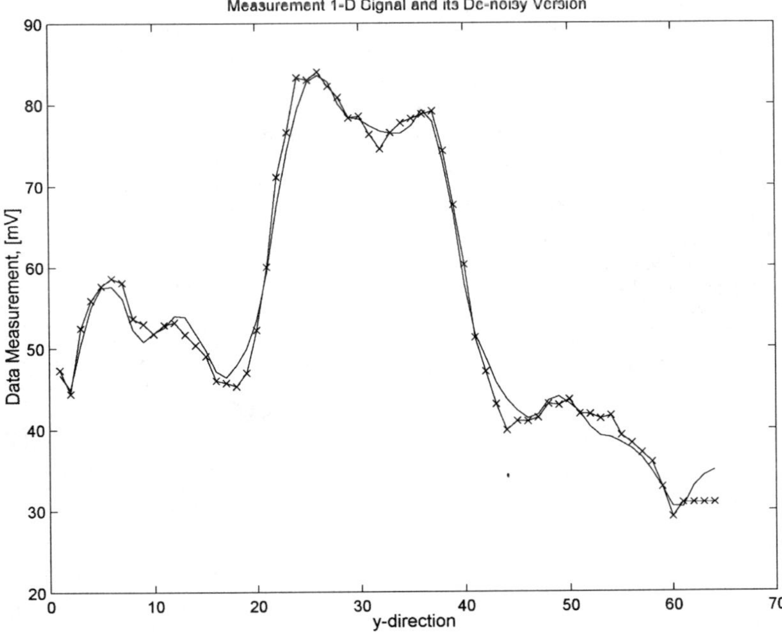

Fig. 9: The effect of the denoising procedure on the differential data of Fig. 8. Crosses indicates original data. Denoisy data are basically a smoothed version of the original ones.

5. Conclusions

The use of the wavelet approach in ECT-NDE data processing is motivated by its suitability to cope with data with sharp valleys or peaks. Furthermore, the wavelet transform yields an opportunity to filter out several sources of noise, namely, mis-positioning of sensors, which are typically encountered in this kind of analysis. Indeed, the WT approach allows a frequency-space representation of the energy content of the measured signal. Since such disturbances like mis-positioning and lift-off variation of the sensor are represented by band-pass signals, the filtering of the related contribution is possible and it is practically carried out by zeroing the proper set of wavelet coefficients. As a difference with Fourier harmonic analysis, the wavelet approach allows to operate in the space-frequency domain. One difficulty of the WT approach is related to the selection of the optimal mother wavelet. In the presence of a database approach, like the one used in the NN framework, one can use the available data to construct a super-mother wavelet as a linear combination of admissible mother wavelet, where the coefficient values should be estimated from the data.

This work was supported by the INCO-COPERNICUS Project PL 964037 of the European Commission and by the Italian MURST.

References

[1] J.Pavo and A.Gasparics, "Proposal for Benchmark Problem qualifying Some Aspects of the Performance of ECT Probes", Proceedings of the E'NDE'97 Symposium, R. Albanese et al., Eds., Electromagnetic NonDestructive evaluation (II), IOS Press, **14** (1998) 337-342.

[2] F.C.Morabito, R.Albanese, E.Coccorese, R.Martone, G.Rubinacci, "Identification of circular holes in thin plates using a neural network approach", Proceedings of the International Symposium on Nonlinear Electromagnetic Systems, ISEM, Cardiff, UK, September 1995, in: Nonlinear Electromagnetic Systems, Edited by: A.J.Moses and A.Basak, IOS Press Series Studies in Applied Electromagnetics and Mechanics, Vol.10, March 1996;

[3] A.Fomisano, R.Martone, F.C.Morabito, "Wavelet Tools for Improving the Accuracy of Neural Network solution of Electromagnetic Inverse Problems", Paper presented at the 11th Intl. Confer. on the Computation of Electromagnetic Fields, COMPUMAG, November 2-6, 1997, Rio de Janeiro, Brazil, will appear on IEEE Trans. on Magnetics, September 1998;

[4] G.Chen, A.Yamaguchi, K.Miya, "A Novel Signal Processing Technique for Eddy-Current Testing of Steam Generator Tubes", IEEE Transactions on Magnetics, **34**, 3 (May 1998) 642-648;

[5] F.C.Morabito, M.Campolo, "A novel neural network for NDT applications", Proceedings of the Third International Conference on Computation in Electromagnetics, Bath, UK, (10-12 Apr '96) 364-9, London, UK, IEE Conf. Publ. **420** (Apr 1996);

[6] A.Gasparics, C.S.Daroczi, G.Vertesy, J.Pavo, "Improvement of ECT Probes based on Fluxset type Magnetic Field Sensor", in Electromagnetic Nondestructive Evaluation,(II), R.Albanese et al Eds., IOS Press, Studies in Applied Electromagnetics and Mechanics, **14** (1998) 146.

Electromagnetic Nondestructive Evaluation (III)
D. Lesselier and A. Razek (Eds.)
IOS Press, 1999

Neural Network Identification of Defects in RFEC Technique

P. BURRASCANO, E. CARDELLI, D. FAZZIOLI, S. RESTEGHINI
Istituto di Energetica, Università di Perugia, Via G. Duranti 1/A-4, 06125 Perugia

Abstract. In this paper we discuss the use of the Probabilistic Neural Network (PNN) for the classification of the defects detected in tubes via the remote field Eddy Current (RFEC) inspection technique. The defects are divided into two groups: the class of "deep" defects, whose depth is larger than the width, and the class of "surface" defects, whose width is greater than the depth. The neural network is trained to classify each defect to one of the two classes. Each defect is represented by means of the phase response of the probe system, obtained by the numerical simulation of the geometry using a Finite Difference approach in the frequency domain (FDFD). Results obtained show that the proposed artificial network offers reliable classification results.

1. Introduction

The RFEC technique is a non-destructive test (NDT) technique, especially suitable for the inspection of metallic tubes and pipes using a probe system placed into the tube [1].

This probe system consists of two coils located at a distance of more than two diameters from each other, in order to operate in the so called remote field zone, and physically connected together, to allow them to move as one unit. One coil (source coil) is driven with low frequency ac current. The second coil (pick up coil) is connected to a nanovoltmeter (see Fig. 1). The voltage induced in the pick up coil and its phase are used for the detection of defects. An important aspect is the data analysis for the discrimination and classification of the defect shape. Results obtained using a neural network technique to address this problem are described below.

2. Numerical Simulation

The data used for training and testing the neural network have been obtained using numerical simulation techniques employing finite-difference methods in the frequency domain [2]. The main features of the algorithm used in this code are briefly summarised.

In this case the problem has axisymmetric geometry. Consequently the magnetic vector potential (A) has thus the only A_θ component.

The governing equation used in the numerical analysis is:

$$\oint_{\gamma}\left(\nabla \times \dot{A}\right)\cdot d\gamma = \iint_{\Gamma}\mu\dot{J}\cdot d\Gamma \tag{1}$$

where the dot indicates phasor notation, Γ is a suitable surface in the plane r, z and γ is its boundary. The value of $\dot{J}$ in equation (1) is:

$\dot{J} = 0$ in air,

$\dot{J}$ equal to a given value J_0 in the probe coil, and

$\dot{J} = -j\omega\sigma \dot{A}_\theta$ in the tube.

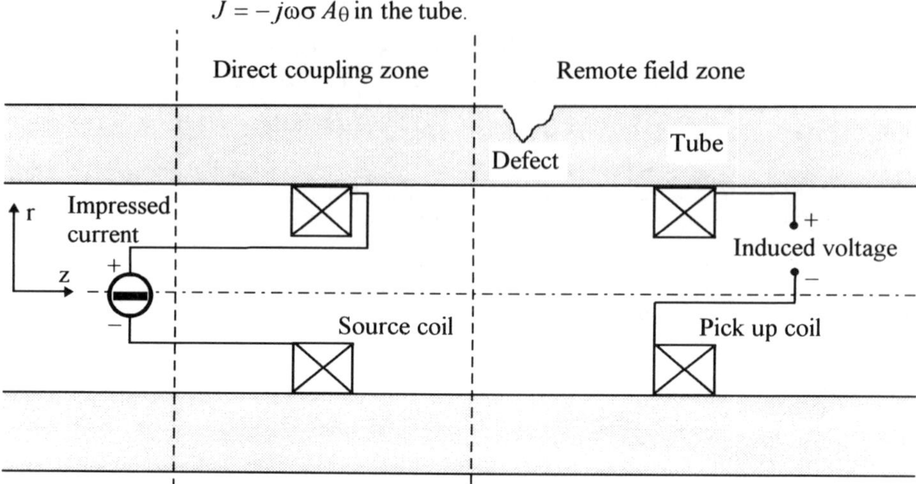

Fig. 1 - Scheme of the RFEC technique.

We use the boundary condition $\dot{A}_\theta = 0$ along the axis of symmetry. We also imposed the Dirichlet condition $\dot{A}_\theta = 0$ at a suitable distance from the tube.

To numerically solve the equation we have used a finite difference scheme. We have decomposed the domain under analysis in an axisymmetric regular grid. Δr and Δz denote the spatial steps along the directions r and z respectively. The nodes of the grid are indexed by means of the indexes i and j. The grid consists of rectangular elements whose vertices coincide with the nodes of the grid. We have built the grid in order to have only one type of material (conductor, air, etc.) in each element. Consequently the probe and the tube are bordered by nodes (see Fig. 2). The values of the magnetic vector potential are considered in the nodes of the grid, and the integration path γ connects centres of nearby elements in a clockwise direction.

Following this scheme the following finite difference equation can be written:

$$rA_\theta^{i,j} = \frac{\frac{\Delta r}{r\Delta z}\left(rA_\theta^{i,j+1} + rA_\theta^{i,j-1}\right) + \frac{\Delta z}{\Delta r}\frac{2}{2r+\Delta r}\left(rA_\theta^{i+1,j}\right) + \frac{\Delta z}{\Delta r}\frac{2}{2r-\Delta r}\left(rA_\theta^{i-1,j}\right) + \mu J_0 \frac{\Delta r \Delta z}{2}}{\frac{2\Delta r}{r\Delta z} + \frac{\Delta z}{\Delta r}\left(\frac{2}{2r+\Delta r} + \frac{2}{2r-\Delta r}\right)} \tag{2}$$

for the interface between probe coil and air (fig. 1), and:

$$rA_\theta^{i,j} = \frac{\frac{\Delta r}{r\Delta z}\left(rA_\theta^{i,j+1} + rA_\theta^{i,j-1}\right) + \frac{\Delta z}{\Delta r}\frac{2}{2r+\Delta r}\left(rA_\theta^{i+1,j}\right) + \frac{\Delta z}{\Delta r}\frac{2}{2r-\Delta r}\left(rA_\theta^{i-1,j}\right)}{\frac{2\Delta r}{r\Delta z} + \frac{\Delta z}{\Delta r}\left(\frac{2}{2r+\Delta r} + \frac{2}{2r-\Delta r}\right) + j\omega\mu\sigma\frac{\Delta r\Delta z}{r}} \qquad (3)$$

for the interface between air and the tube (fig. 1).

Equation (2) is also valid inside the probe coil and in air, taking into account the proper values of the assigned current J_0 while the equation (3) is valid inside the tube.

The form of the equations (2) and (3) allows the use of the successive relaxation method.

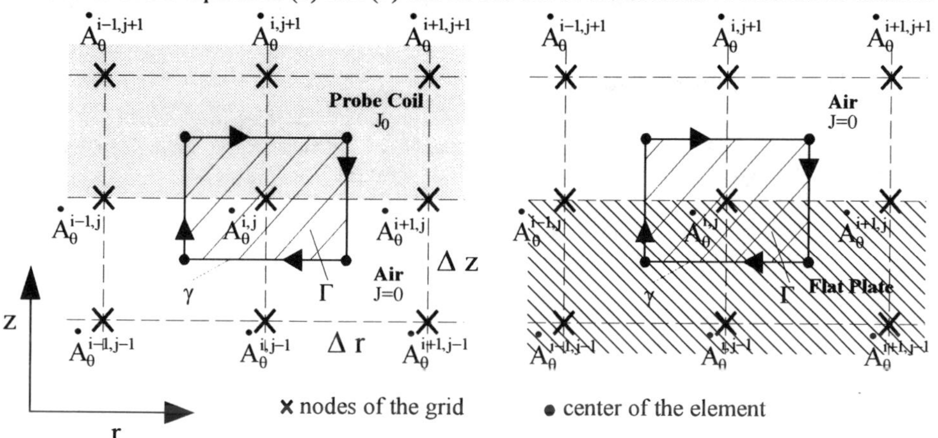

Fig. 2 - Discretization used for the FD algorithm.

Once the values of the magnetic vector potential are determined, the induced voltage in the pick-up coil is given by the expression:

$$\dot{V} = 2\pi \sum_{i=1}^{n} \dot{A}\theta_i r_i \qquad (4)$$

where $\dot{A}\theta_i$ and r_i are the magnetic vector potential and the radius of the i-th turn of the coil, respectively. We have suitably approximated (4) using the values of $\dot{A}\theta_i$ computed at the nodes of the grid.

Taking into account the real shape of defects, the axisymmetric geometry seems to be not realistic. However numerical simulations of three-dimensional geometries with defects having the same cross section, but located circumferentially only along few degrees, have given similar output responses, consuming a large computation time. So we have assumed the axisymmetric geometry to be valid to test the proposed neural network approach.

For the numerical calculation we have used a PC with Pentium II INTEL© 300 MHz processor and 128 MB RAM. The numerical code is written in C++ language and it is compiled in a 32 bit Windows environment. The code uses double precision variables for the real numbers and "complex" precision variables for the complex numbers. The double precision variable supports 64 bits. We have used a rectangular mesh with $\Delta r = 1\,\text{mm}$ and

$\Delta z = 0.5\,\text{mm}$ (see also fig. 2) with 44225 nodes and 42658 unknowns. The convergence criterion used was:

$$\frac{\max\limits_{\forall i,j}\left\| A_n^{i,j} \right| - \left| A_{n-1}^{i,j} \right\|}{\max\left| A_n^{i,j} \right|} \leq 1\text{E}-7$$

where the subscript n indicates the iteration number.

The memory used was approximately 2.5 MB and the total CPU time needed was about 1200 sec.

3. The Probabilistic Neural Network

The neural network we have used for classification of defects is the Probabilistic Neural Network, which is based on a Bayes classification strategy. Each classifier based on Bayes strategies minimises the "expected risk" of misclassification. If a two-category classification is considered, the problem is to decide whether the unknown pattern $\boldsymbol{X}_t = [X_{t1}, X_{t2}, \dots, X_{tp}]$ belongs to class θ_A or to class θ_B. The Bayes decision rule can be written as:

$$\begin{aligned} d(\boldsymbol{X}) &= \theta_A \quad \text{if} \quad h_A l_A f_A(\boldsymbol{X}) > h_B l_B f_B(\boldsymbol{X}) \\ d(\boldsymbol{X}) &= \theta_B \quad \text{if} \quad h_B l_B f_B(\boldsymbol{X}) > h_A l_A f_A(\boldsymbol{X}) \end{aligned} \tag{5}$$

where

$f_A(\boldsymbol{X})$ and $f_B(\boldsymbol{X})$ are the probability density functions for category A and B, respectively;
l_A is the loss function associated with the decision $d(\boldsymbol{X}) = \theta_B$ when the truth is A;
l_B is the loss function associated with the decision $d(\boldsymbol{X}) = \theta_A$ when the truth is B;
h_A is the a priori probability of occurrence of patterns from category A;
$h_B = 1 - h_A$ is the a priori probability of occurrence of patterns from category B.

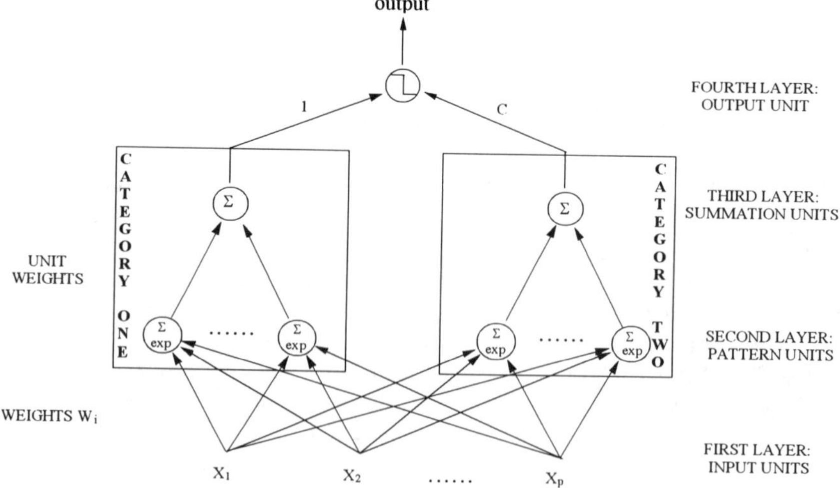

Fig. 3 - The Probabilistic Neural Network

The Bayes decision surface relative to the two-category problem we are considering is the boundary between the region in which $d(\boldsymbol{X}) = \theta_A$ and the region in which $d(\boldsymbol{X}) = \theta_B$.

The main problem in using Bayes strategies for classifications is the estimation of the probability density functions relative to each class. This task is usually accomplished by using a set of training patterns with known classification. A consistent estimate of a multivariate probability density function can be obtained by using the product of univariate kernels. In the particular case of the Gaussian kernel, the multivariate estimate of the probability density function of category A can be expressed as:

$$f_A(\boldsymbol{X}) = \frac{1}{\sigma_{PNN}^p \sqrt{(2\pi)^p}} \frac{1}{m} \sum_{i=1}^{m} exp\left[\frac{-(\boldsymbol{X} - \boldsymbol{X}_{Ai})^T (\boldsymbol{X} - \boldsymbol{X}_{Ai})}{2\sigma_{PNN}^2}\right] \tag{6}$$

where

i is the current pattern index;

m is the number of training patterns from category θ_A;

$\boldsymbol{X}_{Ai}$ is the i-th training pattern from category θ_A;

σ_{PNN} is a smoothing parameter;

p is the dimensionality of the input space.

In using expression (6) it is implicitly assumed that X_{Ai1}, X_{Ai2}, ... , X_{Aip} are independent identically distributed random variables [3, 4].

It is worth noting that expression (6) is the sum of multivariate Gaussian distributions centred at each training sample; moreover, although Gaussian kernels have been used in (6), $f_A(\boldsymbol{X})$ is not restricted to be Gaussian. The smoothing parameter σ_{PNN} defines the amount of interpolation between the different modes of $f_A(\boldsymbol{X})$ corresponding to the location of the training patterns. Varying σ_{PNN} also affects classification results: the classifier can be shown to be similar in effect to the nearest neighbour classifier as $\sigma_{PNN} \rightarrow 0$, and to the K-nearest neighbour classifier for increasing values of σ_{PNN}.

The Probabilistic Neural Network is a direct implementation of the Bayes classifier described above [3, 4] and its structure consists of four layers (see Fig. 2). The input layer consists of simple fan-out units. In the second layer there is one unit per training pattern; each pattern unit performs a dot product of the input pattern vector $\boldsymbol{X}$ with weight vector $\boldsymbol{W}_i$, $Z_i = \boldsymbol{X} \cdot \boldsymbol{W}_i$, and then performs the non-linear operation:

$$exp\left[\frac{(Z_i - N^2)}{(\sigma_{PNN}^2)}\right]$$

which, assuming that both $\boldsymbol{W}_i$ and $\boldsymbol{X}$ are normalised to N, is equivalent to each exponentiation of formula (6). In the third layer, the summation units (one for each category) simply sum the outputs from all pattern units of the relative class. In the two-category case, there is one output unit in the fourth layer, which implements the decision of equations (5). It is a two input unit which produces a binary output with only a single variable weight C, defined as:

$$C = -\frac{h_B l_B}{h_A l_A} \frac{n_A}{n_B}$$

where n_A and n_B are the numbers of training patterns from category θ_A and θ_B, respectively. If the numbers of training samples from the different categories are in proportion to their a priori probabilities and class losses l_i do not reflect any bias in the decision, C may simplify to -1.

The network is defined (trained) by setting the $\boldsymbol{W}_i$ weight vector in one of the patter units equal to each $\boldsymbol{X}$ pattern in the training set and then connecting the pattern unit output to the appropriate summation unit.

4. Results

Several tests were performed in order to verify the performance of the proposed neural classifier. The results reported in the paper refer to defects produced in a 28 mm inner diameter and 6 mm wall thickness tube. The tube is made of non-magnetic material. In the numerical simulation we have used the values $\mu_r = 1$ for the relative magnetic permeability and $\sigma = 5E6$ for the electrical conductivity. The distance between the centres of the two coils is 112 mm. Each coil consists of 400 turns and is 4 mm thick and 10 mm long. The source coil is driven with a 50 Hz sinusoidal current having a current density of 3E6 A/m^2.

We have considered axis-symmetric defects having rectangular cross sections (see Fig. 4); for each defect the behaviour vs. the relative defect-probe position was considered. Fig. 5 shows the behaviour of amplitude and phase responses for a family of defects sharing the same width. We have chosen this configuration because it appears to be critical from the classification point of view. The amplitude and phase responses of different defects are in this case very close.

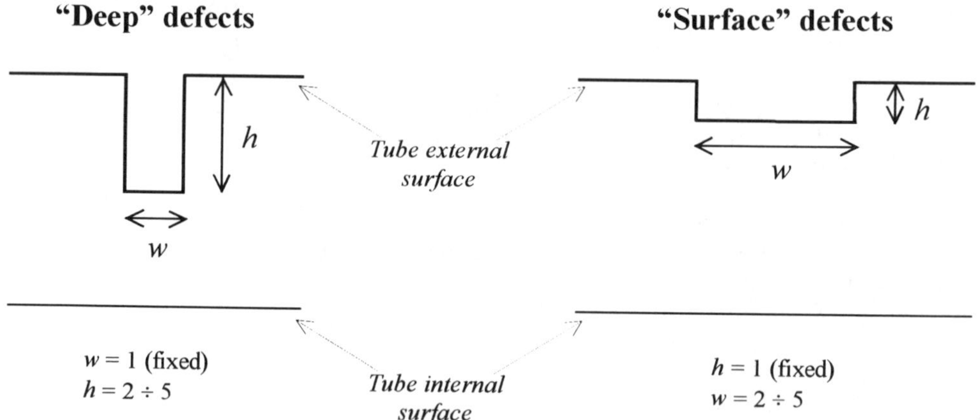

Fig. 4 - Graphical explanation of the defects used as training samples for the PNN.

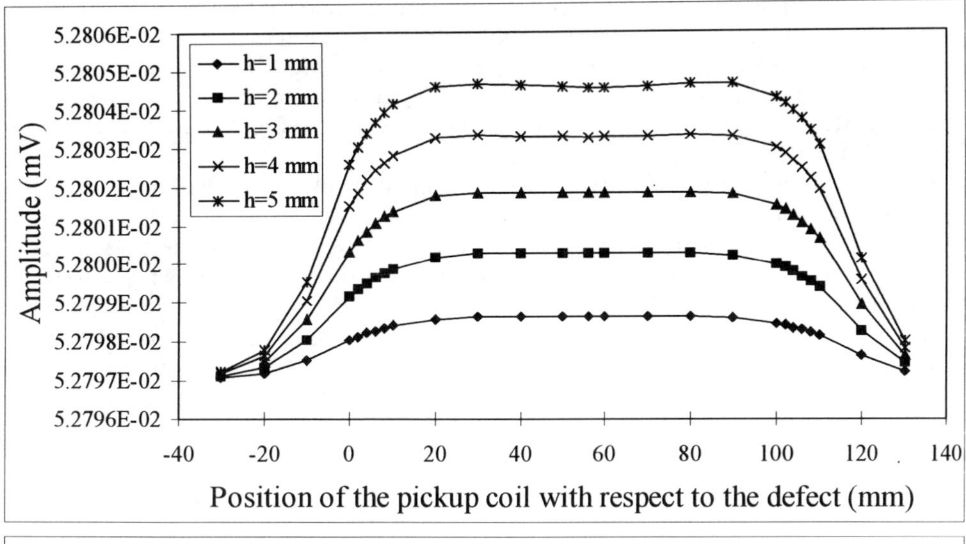

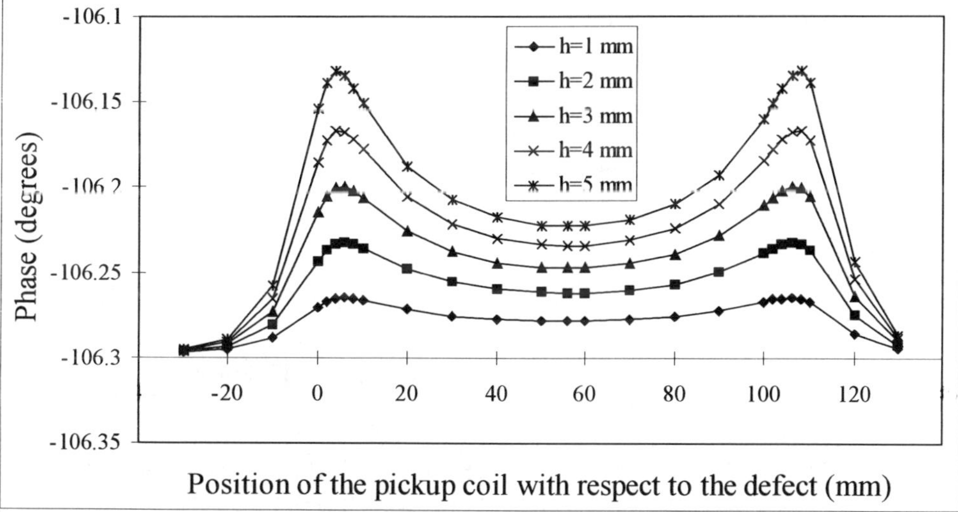

Fig. 5 - Amplitude and phase response computed by FDFD for a class of deep defects having the same width of 1 mm.

A preliminary analysis of the data has shown that the phase response is more informative than the amplitude response for the classification problem considered. In order to collect the training samples, we have taken into account only the phase response of the defects. The input vector consist of 25 samples of the signal (those from -20 mm to 130 mm) plus an additional dummy element which has been used to normalise the vector. In order to work with small-norm vectors, we centred the phase samples by subtracting their mean value. We then choose a value of 0.4 for the vector normalisation.

We define "deep" defects as those having a value of the ratio between height (*h*) and width (*w*) greater than two. We assign such defects to class 1. In the same way, we define "surface" defects as those having a ratio w/h greater than two, and assign them to class 2. We used these regular defects for training the neural network: in particular, we considered four defects from class 1 with w fixed (1 mm) and h with integer values ranging from 2 to 5 mm, and four defects from class 2 with h fixed (1 mm) and w ranging from 2 to 5 mm. Thus, on the basis of the PNN description given in section 3, the resulting network consists of eight pattern units in the second layer, four for each class.

Fig. 6 shows the classification performance vs. the network interpolation parameter σ_{PNN} for the training and testing set.

The curves show that on the training samples the network provided excellent classification results for a wide range of σ_{PNN}. A 100% classification performance was obtained with the training set when σ_{PNN} is less than 0.05.

The PNN was defined on the basis of a training set consisting of regular defects; in defining the testing set we simulated a number of irregular defects (non-rectangular cross sections), to take into account that the real defects have, in general, not regular borders. The defects we used as testing samples are represented in Fig. 7. Most of them belong to the "deep" class because defects belonging to this class are the most dangerous ones (pitting, cracking, etc.).

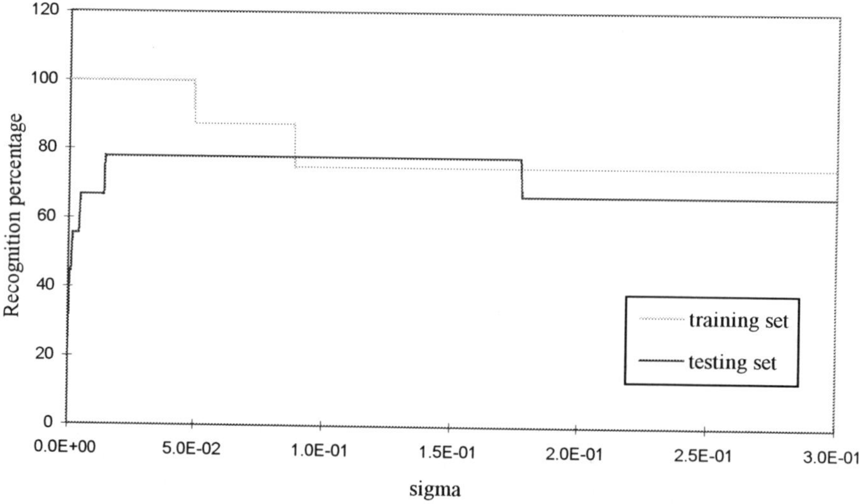

Fig. 6 - PNN performance: the curves represent the correct classification percentage of the network, trained over the regular defects, for the defects of training and testing set, respectively.

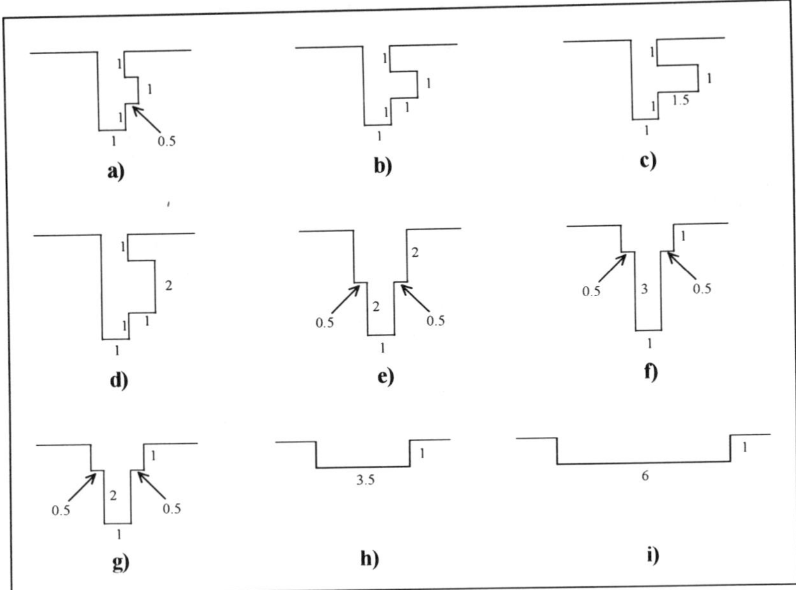

Fig. 7 - Graphical description of defects used as testing samples. Defects a) – g) are of deep type, defects h) and i) are of surface type. Dimensions are expressed in mm.

The performance of the PNN over the irregular shaped testing set described above is reported again in Fig. 6. The curves report the percentage of correctly classified test defects vs. σ_{PNN}. The PNN has been able to correctly classify seven of the nine test defects, achieving a recognition percentage of 78%. The two defects that the network misclassified are e) and g), i. e., those from deep class which mostly resemble surface defects. However, defect f), which is similar to the previous defects, was correctly recognised by the network. If we evaluate the performance over the whole set of defects (training defects plus testing defects), the PNN has an overall recognition capability greater than 88% (over the appropriate interval of the parameter σ_{PNN}). We observe that the σ_{PNN} interval for which the PNN achieves the best performance is [1.5E-2, 5E-2]. Such values are indeed small, due to the small differences in the phase curves shown in Fig. 5. Similar classification results were obtained in the presence of random gaussian noise with diagonal covariance matrix added to the test samples (standard deviation of the random noise up to 2E-2).

5. Conclusions

In this paper we have presented the application of a PNN for the classification of the defects in metallic tubes detected by means of the RFEC technique. We have used for the training and the testing of the PNN the responses in amplitude and in phase calculated by a FDFD numerical method.

Among several possible geometries and cases we have chosen a special case of "critical data" where the amplitude and the phase of the induced voltages for different defects have very close values. In this case we have used defects with rectangular cross section called "regular" defects, to define the PNN parameters. We then used the resulting network to classify of "irregular" defects that are either "surface" defects or "deep" defects.

The PNN technique was able to classify more than 88% of defects. The proposed neural network appears to be a promising method to apply to other eddy-current NDT techniques.

References

[1] T. R. Schmidt, The Remote-Field Eddy Current Inspection Technique, *Mater. Eval.* vol. 42, Feb 1984.

[2] M. Angeli, P. Burrascano, E. Cardelli, S. Resteghini, R. Sikora, Numerical Analysis of Eddy Current Non Destructive Testing (JSAEM Benchmark Problem #1 – Circular Plate), *Proc. of the E'NDE Conference*, Reggio Calabria, September 1997, in *Electromagnetic Non destructive Evaluation (II)*, IOS Press, 1998.

[3] D. F. Specht, Probabilistic Neural Networks, *Neural Networks*, vol. 3, n. 1, Jan 1990, pp. 109-118.

[4] D. F. Specht, Probabilistic Neural Networks and Polynomial Adaline as Complementary Techniques for Classification, *IEEE Trans. on Neural Networks*, vol. 1, n. 1, March 1990, pp. 111-121.

[5] P. Burrascano, Learning Vector Quantisation for the Probabilistic Neural Network, *IEEE Trans. on Neural Networks*, vol. 2, no. 4, July 1991, pp. 458-461.

Innovative Modalities and Applications

Electromagnetic Nondestructive Evaluation (III)
D. Lesselier and A. Razek (Eds.)
IOS Press, 1999

Nondestructive evaluation of A533B Steel by Hall Sensor

Sinnichi Shoji[†], Koji Yamada, Y.Tanaka, Y. Uno, Y. Takeda, Sprapedi and Satoru Toyooka and Yoshihiro Isobe*

Graduated School of Saitama University, Urawa, Saitama 338-8570, Japan
**Nuclear Fuel Industries, Ltd., Sennan-gun, Osaka, 590-0481, Japan*

Abstract: Nondestructive evaluation of low alloy steel (A533B) sample has been investigated using a GaAs Hall sensor for magnetic leakage flux detection and using a small pick-up coil for Barkhausen jump detection. A sensitive measurement of leakage magnetic field over the sample was performed in zero external field, after a sample polarization in 1 kOe external field. The residual stresses formed by 0.1 % or less deformation and the magnetic anisotropies in the Lüders band were found tilted at angles between 25-45 degree to the sample direction. It fits well with numerical simulations by FEM based on a model of different residual magnetizations, depending on local residual stresses. As an indicator of the material fatigue, average fields of discontinuous magnetizations are considered. They are functions of residual stresses and angles against the applied tensile stress direction.

1. *Introduction*

Nondestructive evaluations (NDE) of iron-based material are very important for the environmental problems of the earth. Various methods of NDE for iron based material have been investigated on low alloy steel by observing ultrasonic propagation, X-ray diffraction, Vickers hardening or magnetic noises [1], etc. Recently, by using superconducting quantum interference device (SQUID), magnetic leakage flux of strained samples has been intensively investigated for NDE [2,3]. In the present paper, we performed experiments on the leakage flux observation using a semiconductor Hall sensor instead of SQUID. We reached a high spatial resolution of degradation inhomogeneity as small as 100 μm due to a small size of the Hall element and a short lift-off distance from the element to the sample surface. Further, we developed a residual stress evaluation by processing noises observed from the sample surface. This evaluation method is independent of the leakage noise magnitude by normalizing the total data with an integration up to the observation

[†] On leave from Mitec Co. Ltd., Sendai, Miyagi 981-1105, Japan

time. The validity of the evaluation processes and the physical meaning are examined in the discussion. Finally, we simulated the leakage flux distribution by adopting a model the of the leakage flux generation by different residual magnetizations at each point due to residual stresses after a polarization in a strong magnetic field. The simulations are compared with the experimental results.

2. Experimentals

The steel of A533B is composed of Fe, Mn, Ni, Mo and C with total impurities less than 3 Wt %. The ingot of the specimen was cut and shaped in a standard loading test sample as shown in Fig.1. These were quenched by water-cooling after heating 2 hr at 860-890°C and annealed by air cooling at 650-665°C for 2hrs. After this heat treatment, the residual magnetization and coercivity of this material were 1.3 T and 800 A/m, respectively.

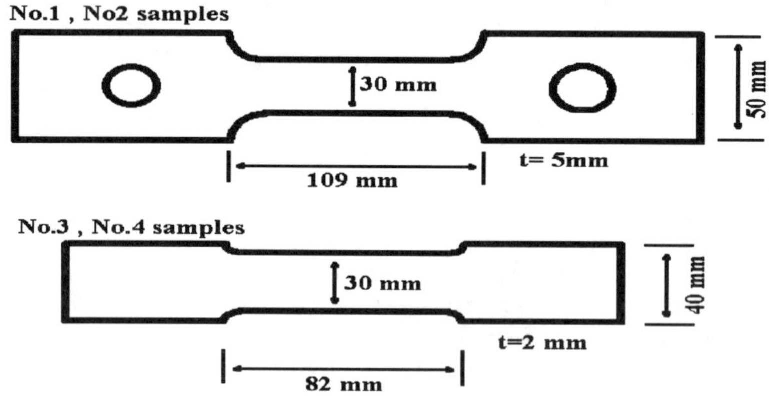

Fig. 1 Sample shapes and their sizes

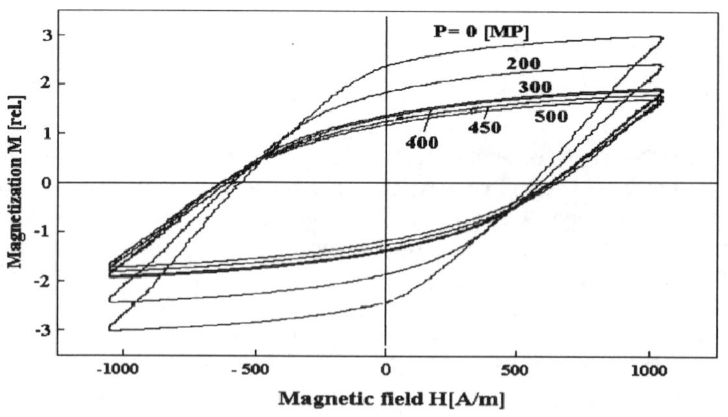

Fig.2 The magnetization curves for differently strained samples

We applied tensile stresses of 540 MPa to No.1 and No.3 samples, and of 660 MPa to No.2 and No.4 samples, respectively. After unloading of these stresses by 2hrs, we performed all the measurements. Fig.2 shows the magnetization curves for differently strained samples in a weak field range less than ±1 kA/m for a thread shaped sample of 1mm radius at room temperature. Residual magnetizations and coercivities decreased with applied tensile stresses up to 500 MPa. These decreases are enhanced for a thread shaped sample due to uniform residual stresses at each point compared with those in a thick plate sample of 5mm (thickness) x 30mm (width). Thickness d of a samples was measured directly by a digital caliper with 1μm accuracy over the surface. The residual strain ε of stretched rectangular sample is calculated as

$$\varepsilon = -2\ln(\frac{d}{d_0}) \quad , \qquad (1)$$

where d_0 denotes the thickness before the stretch. Fig. 3 shows the distributions of the residual strain ε represented monochromatic patterns, where a point is brighter if larger strain occurs. The maximum residual strains of No.1, 2, 3 and 4 samples were 0.014, 0.14, 0.025 and 0.05, respectively. The Lüders bands inclined at about 45 degrees are observed in samples No.1 and No.3 samples with small strains less than 1%.

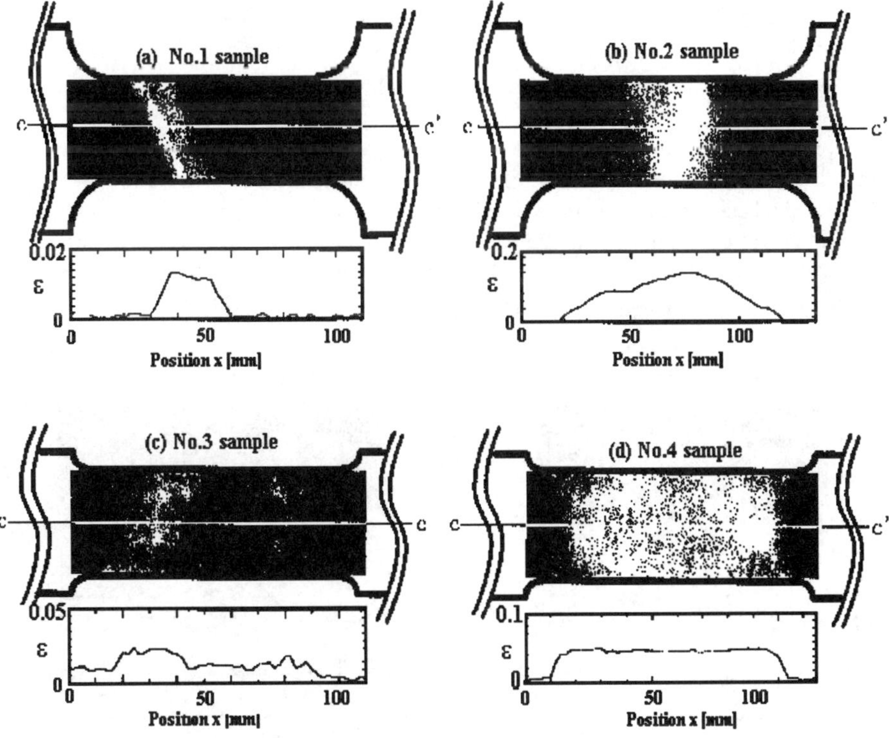

Fig. 3 Residual strain distributions of the samples

Further, in these two samples of No.1 and No.3, strains are almost limited and dominated within their Lüders bands. The curves in fig.3 show the plots of ε vs. position x along the axis c-c'. The residual stresses of σ_x, σ_y were measured by X-ray diffraction from α-Fe (211) plane at 157 [deg] of Cr-K_α line. Fig. 4 shows σ_x and σ_y respectively for samples No.1 and No.2 as a function of x along the same line c-c'. The residual stresses increased with increasing thickness deformations in the range less than 1 % and become slightly decreased with increasing deformations up to 8 %.

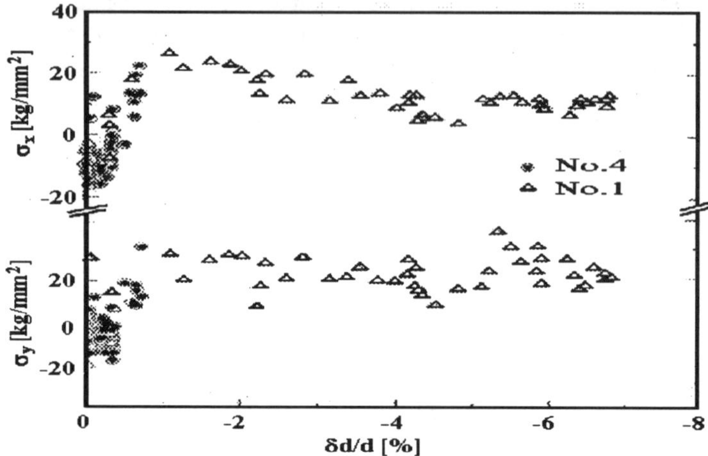

Fig. 4 Residual stresses of samples No.1 and No.2 observed by X-ray diffraction

both for σ_x and σ_y. In this experiment, the residual stresses in both x and y directions were negative even after a high temperature annealing. Observations of leakage flux distributions were performed by using a GaAs Hall sensor of a small size (120μm x 120μm), operated in a frequency at 10 kHz by a Lock-in amplifier where the leading wires to the Hall terminals were twisted with a small pitch of 1mm with copper mesh shield outside. We attained a high sensitivity down to 100 nT corresponding to 10 LSB (the Least Significant Bit) for a 12-bits A/D converter.

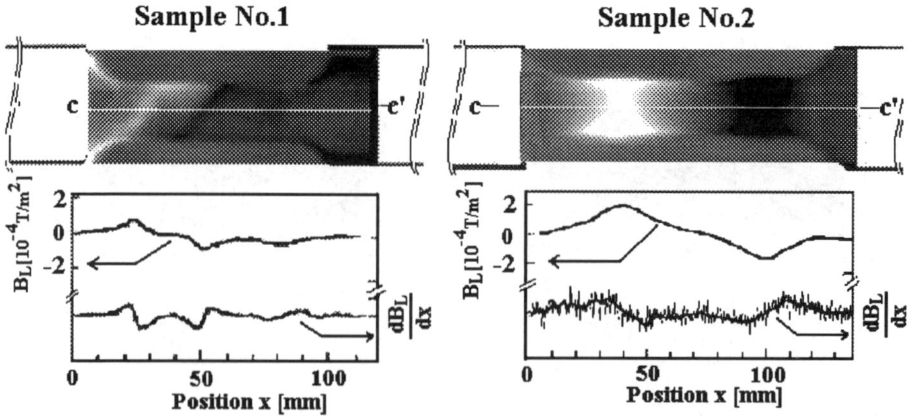

Fig. 5 Leakage flux density for samples No.1 and No2

A Hall sensor was mounted perpendicularly to the surface by a X-Y scanning table with different lift-off distances between 0.2 mm and 1.1 mm, and the element was scanned in a slow speed of 0.1 mm/s over 512x256 sample points for every 25 μm or 50 μm. The samples were polarized before the leakage flux measurements in a strong magnetic field of 80 kA/m. Fig.5 shows the distributions of leakage flux density B_L with 0.2 mm lift-off distance, where a point is colored in white for out-going flux from the sample, and colored in black for in-coming flux. The curves under these distribution patterns show the intensity of the leakage flux and the first derivative with x along the central line c-c'. As shown in this figure, the distributions of the leakage flux are close patterns with patterns of residual strains in Fig. 3. An obvious correlation exists between ε(x) and dB_L/dx.

3. Discussions

In this experiment, we observed the distribution of the magnetic leakage flux after an intentional full polarization in a strong magnetic field of 80 kA/m. With this condition, we show how the leakage flux occurs depending on the local material constant change. The leakage flux, in general, could be expressed or caused by the inhomogeneity of the local residual magnetization. The magnetic flux density $B(\boldsymbol{R})$ is given by the equation (2), where $\boldsymbol{R}$ denotes a measuring point as

$$B(\boldsymbol{R}) = grad\left\{ \iiint \frac{\rho_m(\boldsymbol{r})\cdot(\boldsymbol{R}-\boldsymbol{r})}{4\pi|\boldsymbol{R}-\boldsymbol{r}|} dV \right\}, \tag{2}$$

where, $\rho_m(\boldsymbol{r})$ denotes local magnetic moment at a position $\boldsymbol{r}$. Now, the configuration of local magnetic moments is derived as a function of the residual stress by adopting a model that the total magnetic energy ε_m at each point is composed of the anisotropy energy ε_A and the magnetostrictive energy ε_E as a function of stress α as follows.

$$\varepsilon_m = \varepsilon_A + \varepsilon_E \quad , \tag{3}$$

$$\varepsilon_A = K_1 \sum_{i \neq j} \alpha_i^2 \alpha_j^2 + K_2 \alpha_1^2 \alpha_2^2 \alpha_3^2 \quad , \tag{4}$$

$$\varepsilon_E = -\frac{3\sigma}{2}\left\{ \lambda_{100}\left(\sum_i \alpha_i^2 \gamma_i^2 - \frac{1}{3}\right) + 2\lambda_{111} \sum_{i \neq j} \alpha_i \alpha_j \gamma_i \gamma_j \right\} \quad . \tag{5}$$

Here, K_1 and K_2 denote the anisotropy constant of b.c.c (body centered cubic) Fe, α_i and γ_i the direction cosine of a magnetization and a residual stress, respectively. λ_{100} and λ_{111} denote magnetostrictive constants of <100> and <111> directions, respectively. Further, *i*, *j*=(x, y, z). The residual magnetization $\boldsymbol{M_R}$ of the whole sample is obtained by the minimum energy configuration of the local magnetic moments as

$$M_R = \iiint \rho_m(r) \cos\theta(r) dV \quad , \tag{6}$$

For an approximation for this material, we adopted a set of values of b.c.c. Fe for this material as

$$K_1 = 4.72 \times 10^4 [J/m^3],\ \ K_2 = -0.075 \times 10^4 [J/m^3],\ \ \lambda_{100} = 20.7 \times 10^{-6},\ \ \lambda_{111} = -21.2 \times 10^{-6}.$$

By using the finite element method, we obtained a numerical result for residual stress dependence of local residual magnetization as follows,

$$\frac{m_r}{m_s} = -0.0013\sigma + 0.86, \tag{7}$$

Here, $m_r = \rho_m \Delta V$ denotes the local residual magnetization and $m_S = \rho_S \Delta V$, the saturation magnetization. Both m_r and m_S are defined for a large volume of ΔV given by the observation area times thickness to determine σ by XRD. The top of Fig.6 (a) shows the thickness deformation observed by a digital caliper as a function of x for a sample No.1. The numerical simulations are shown in this figure for the residual magnetization (S-1), the leakage flux change with locally constant magnetization as a virtual case (S-2) and the leakage flux change with local fluctuations of the residual magnetization as a real case (S-3), respectively. Here, the residual magnetization at each position was calculated by (7) with each $\sigma(x)$. Fig. 6(b) shows the same quantities as those in Fig.6(a), for No.2 sample to which a strong stress was applied just before the destruction. Apparently, the leakage flux caused only by the thickness deformation with constant magnetization seems very small effects on the leakage flux generation in a deformation range less than 3 % and it is evident that the leakage flux change is given rise dominantly by the change of the residual magnetization with x due to different residual stresses.

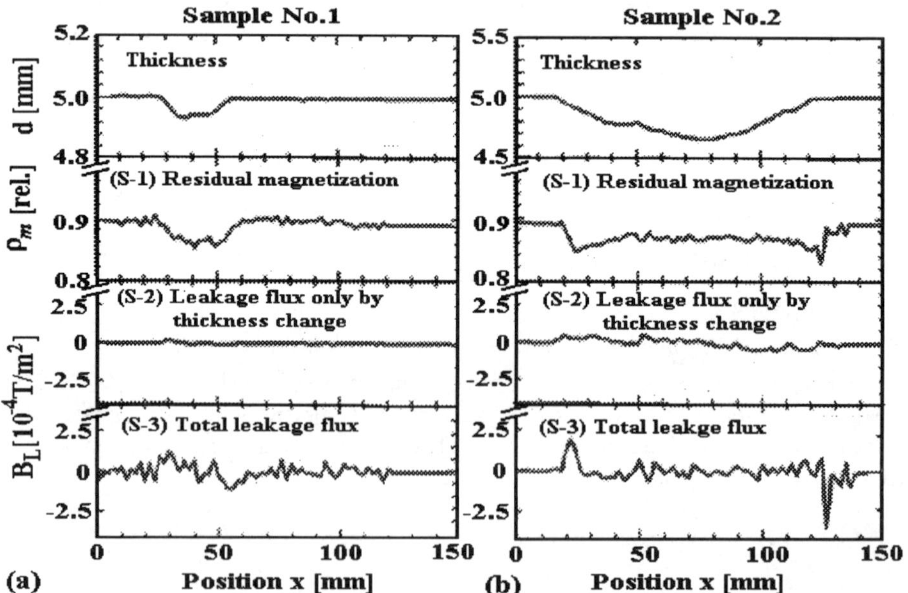

Fig. 6 The results of simulations for (a) slightly and (b) heavily deformed samples, respectively

4. Conclusions

The magnetic leakage flux in plastically deformed A533B steel has been observed for a deformation range less than several percent, after an intentional polarization of samples in 80 kA/m. We analyzed these phenomena by adopting a model that the magnetic leakage flux generations are caused by local residual magnetization decreases with increasing in residual stresses at each position. The results of the simulations explained rather accurately the profile of the leakage flux along the central positions of the samples. Therefore, the nondestructive evaluation of magnetic material can be performed by using a Hall element with a high spatial resolution as small as 100 μm. However, three dimensional analysis and the deformations along thickness direction must be researched for more precise evaluation of these material. A more detailed detection of the local deformation is possible by using the smaller size of Hall sensor and with the shorter lift-off distance than those adopted in this study.

Acknowledgments

The authors thank Prof. M. Uesaka and Prof. K. Miya, University of Tokyo, Dr. K. Ara, Japan Atomic Energy Institute, Prof. S. Takahashi and Prof. A. Chiba, Iwate University, and Dr. N. Kasai, Electro Technical Lab for valuable discussions.

References

[1] K. Yamada, S. Shoji, K. Yamaguchi and Y. Tanaka and R. Groessinger, Proc. 8th Int Symp. On Non-linear Electromagnetic Sytems (ISEM Braunschweig, Germany, June 1997), ISBN 90 5199 381 1. IOS Press, Amsterdam, 1997, pp153-156

[2] H. Weinstock, IEEE Trans. Magn. 27 (1991) 3231-3251

[3] A. Cochran, J. Macfarlane, L. Morgan, J. Kuznik, R. Weston, L. Hao, R. Bowman and G. Donaldson, IEEE Trans. Appl. Supercond. 4 (1994) 128-135

Electromagnetic Nondestructive Evaluation (III)
D. Lesselier and A. Razek (Eds.)
IOS Press, 1999

Application of the spherical harmonic model to identify the magnetic state of a system and to extrapolate its signature

L-L. Rouve*, J-P. Bongiraud*, P. Lethiec*, M. Legris**, J-L. Coulomb*
**Laboratoire du Magnétisme du Navire*
Domaine Universitaire
ENSIEG, BP 46
38402 St Martin d'Hères Cédex, France
***DGA-GESMA*
29240 Brest Naval, France

Abstract. In many applications and especially in the Navy, the magnetic field created by equipment must be known. Specific tools based on finite element method have been developed to compute this field when geometry and magnetic characteristics are well known. When it is not possible or when the temporal evolution of the studied magnetic state must be determined, measurements are made. But as all the space around the devices can not been filled with magnetic sensors, some solutions have been studied to identify the magnetic state of a system from near-field measurements. Then, the field can be extrapolated all around. This paper deals with the application of a simple but quite interesting solution to the problem: the spherical harmonic model.

This work has been performed for the Groupe d'Etudes Sous-Marines de l'Atlantique (G.E.S.M.A.) from the French Navy with the financial support of the Direction Générale de l'Armement (D.G.A.)

1. Introduction

Any magnetic source creates all around its own field distribution. This field distribution adds to the earth field and is known as a magnetic anomaly (with respect to the earth field). The determination of the anomaly generated by magnetic sources has always been a research topic, especially for the Navy. Most ships are made of magnetic sheets that are magnetized under the earth field or have on board magnetic devices that can generate an anomaly. The L.M.N. (Laboratoire du Magnétisme du Navire, Ship Magnetism Laboratory) has been working on the determination of magnetic anomalies created by Navy systems for fifty years. The L.M.N. has developed many specific tools to measure and compute such magnetic anomalies that are in the very low field range (where the common unit is nT).

The L.M.M.C.F. (Laboratoire de Métrologie Magnétique en Champ Faible, Laboratory of Magnetic Metrology in Low Field) has been built to control the earth field on a volume of 15m length and $4m^2$ section with a high precision and homogeneity (better than 5.10^{-4}).

Thanks to numerical acquisition and very precise magnetometers, really accurate measurements can be made.

On the theoretical point of view, two approaches have been studied. The first one is a direct approach that consists in computing the anomaly of a system by knowing its geometry, having a good idea of the magnetic properties of its materials and using the finite element method. When modeling such a complex structure as a ship, specific tools had to be developed [1]: thin surfaces with potential gap to take into account in the meshing the description of very thin but also very long sheets; the infinite box that makes it possible to apply the infinite magnetic conditions to a small finite element volume thanks to a mathematical transformation. The modeling of the magnetic effect of electric loops near thin sheets with an acceptable meshing has also been treated [2]. The second approach is based on an inverse problem approach and consists in determining the magnetic state of a complex device by only measuring the field around it. The source modeling by the spherical harmonic expansion has been studied. Combined with the surface normal equivalent dipole representation, it gave a new model, more complex but more efficient [3].

The present paper deals with the application of the simplest form of the model, the spherical harmonic model, to the determination of the magnetic state of a given structure and the extrapolation of its signature. The goal of this work is to provide the bases of the model [4], to specify its conditions of application and to present the experimental and theoretical implementation. Finally, the comparison of the measurements with the inductions extrapolated by the model will show the pertinence of such a model when correctly applied.

2. The spherical harmonic model

Let us define:

(Ω) the studied volume

(Ωi) the volume containing all the magnetic sources (magnetic materials and/or current distributions)

(Ωe) the complementary volume of (Ωi) in (Ω).

Let us consider the Maxwell equation at any point P of (Ω):

$$\mathbf{Rot}\ (\mathbf{H}) = \mathbf{j} \qquad (1)$$

where **H** is the magnetic field (A/m)

and **j** the current density (A/m^2).

The local law at any point P of (Ω) is:

$$\mathbf{B} = \mu_0\, \mathbf{H} + \mathbf{J} \qquad (2)$$

where **B** is the magnetic induction (T)

and **J** is the polarization (T).

If there is no current source in (Ωi) then at any point P of (Ω):

$$\mathbf{Rot}\ (\mathbf{H}) = \mathbf{0} \qquad (3)$$

That means that:

$$\mathbf{H} = -\,\mathbf{grad}\ (V) \qquad (4)$$

where V is the magnetic scalar potential.

At any point P of (Ωe), the induction satisfies:

$$\mathbf{B} = \mu_0\, \mathbf{H} \qquad (5)$$

Moreover:

$$\mathrm{Div}\ (\mathbf{B}) = 0 \qquad (6)$$

As a consequence, V verifies the Laplace equation at any point P of (Ωe):

$$\Delta V = 0 \qquad (7)$$

where Δ is the Laplacian operator.

If there is a current source in (Ωi) then the equivalent magnetization distribution of the current **Mc** is used as defined in the following equation:

$$\mathbf{j} = \mathbf{Rot}\,(\mathbf{Mc}) \quad (8)$$

That means that:

$$\mathbf{Rot}\,(\mathbf{H} - \mathbf{Mc}) = \mathbf{0} \quad (9)$$

So:

$$\mathbf{H} - \mathbf{Mc} = -\,\mathbf{grad}\,(V) \quad (10)$$

If (Ωi) is a reunion of simply connected elements, then at any point P of (Ωe):

$$\mathbf{Mc} = \mathbf{0} \quad (11)$$

(4) and (5) are verified and, as a consequence, (7) also.

As a first conclusion, if (Ωi) is simply connected then, at any point P of (Ωe), the magnetic scalar potential verifies the Laplace equation. That means that there is a harmonic expansion for V. As a remark, if in (Ωi) one element is not simply connected, then (Ωi) will be extended to the generally small volume where **Mc** is non zero. In any case, as it will be seen later, the spherical harmonic expansion will apply for an external volume really distant from (Ωi) where it is sure that the Laplace equation is verified.

Moreover, from (4) and (5), it comes:

$$\mathbf{B} = -\,\mu_0\,\mathbf{grad}\,(V) \quad (12)$$

In a general case, the induction **B** satisfies:

$$\mathbf{B}(P) = \frac{\mu_0}{4\pi}\iiint_{N\in\Omega i}\left(\mathbf{j}(N)\wedge\frac{\mathbf{NP}}{NP^3}\right)d\Omega + \frac{\mu_0}{4\pi}\iiint_{N\in\Omega i}\left(3(\mathbf{M}.\mathbf{NP})\frac{\mathbf{NP}}{NP^5} - \frac{\mathbf{M}(N)}{NP^3}\right)d\Omega \quad (13)$$

where **M** is the magnetization (A/m).

If current sources are replaced by their equivalent magnetization representation, then the induction **B** verifies at any point P of (Ωe):

$$\mathbf{B}(P) = \frac{\mu_0}{4\pi}\iiint_{N\in\Omega i}\left(3(\mathbf{M}.\mathbf{NP})\frac{\mathbf{NP}}{NP^5} - \frac{\mathbf{M}(N)}{NP^3}\right)d\Omega \quad (14)$$

where **M** is the magnetization of the magnetic material and/or of the current equivalence.

By using (12) and integrating (14), the magnetic scalar potential V becomes:

$$\mathbf{V}(P) = \frac{1}{4\pi}\iiint_{N\in\Omega i}\left(\mathbf{M}(N).\mathbf{grad}_N\left(\frac{1}{NP}\right)\right)d\Omega \quad (15)$$

The distance 1/NP can be developed with Legendre polynomials and V can be written as:

$$V(P) = \frac{1}{4\pi}\sum_{n=1}^{\infty}\sum_{m=-n}^{n}\frac{a_{nm}}{r^{n+1}}Y_n^m(\theta,\varphi) \quad (16)$$

where (r, θ, φ) are the spherical coordinates of P

a_{nm} are the spherical harmonic model coefficients

$Y_n^m(\theta,\varphi)$ are the spherical harmonics defined by:

$$Y_n^m(\theta,\varphi) = \sqrt{\frac{(n-|m|)!}{(n+|m|)!}}\,P_n^{|m|}(\cos\theta)\begin{cases}\sqrt{2}\cos(m\varphi), \text{ si } m>0\\ 1, \text{ si } m=0\\ \sqrt{2}\sin(|m|\varphi), \text{ si } m<0\end{cases} \quad (17)$$

$$P_n^m(x) = (-1)^m(1-x^2)^{\frac{m}{2}}\frac{d^m P_n(x)}{dx^m} \quad (18)$$

$$P_n(x) = \frac{1}{2^n.n!}\frac{d^n(x^2-1)^n}{dx^n} \quad (19)$$

(16) applies for $r \geq r_0$, where r_0 is the radius of the sphere including all the magnetic sources.

The spherical harmonic model coefficients a_{nm} have the following expression:

$$a_{nm} = \iiint_{N\in\Omega i}\left(\mathbf{M}(N).\mathbf{grad}_N\left(r_1{}^n Y_n^m(\theta_1,\varphi_1)\right)\right)d\Omega \quad (20)$$

where $(r_1, \theta_1, \varphi_1)$ are the spherical coordinates of N.
(20) shows that the coefficients a_{nm} only depend on the sources.
Moreover, one of the main interests of the model is that induction decreases as $1/r^{n+2}$ as written in the equation:

$$\mathbf{B}(P) = -\frac{\mu_0}{4\pi}\sum_{n=1}^{\infty}\sum_{m=-n}^{n} a_{nm}\,\mathbf{grad}\left(\frac{Y_n^m(\theta,\varphi)}{r^{n+1}}\right) \tag{21}$$

Since the coefficients a_{nm} are of the same order, a finite number of the first terms of (21) will be enough to compute the induction at a given distance because all the other terms will be negligible with respect to them.

The coefficients a_{nm} can be determined by measuring induction at several points around the sources. For measuring points whose coordinate r is larger than a given radius r_2, a good approximation for the induction can be computed with an order n_0.
From these measurements at a number NPT of points, a matrix system can be built.
Let us define :

$$\underline{bs_i} = \begin{bmatrix} B_{r_i} \\ B_{\theta_i} \\ B_{\varphi_i} \end{bmatrix} \tag{22}$$

which is the vector of the spherical components of the induction measured at the point n°i, with $i \in \{1,\ldots,NPT\}$.

$$[As_i] = -\frac{\mu_0}{4\pi}\begin{bmatrix} G_r{}_1^1(P_i) & G_r{}_1^0(P_i) & G_r{}_1^{-1}(P_i) & \ldots & G_r{}_{n0}^{-n0}(P_i) \\ G_\theta{}_1^1(P_i) & G_\theta{}_1^0((P_i) & G_\theta{}_1^{-1}(P_i) & \ldots & G_\theta{}_{n0}^{-n0}(P_i) \\ G_{\varphi 1}^{\,1}(P_i) & G_\varphi{}_1^0((P_i) & G_\varphi{}_1^{-1}(P_i) & \ldots & G_\varphi{}_{n0}^{-n0}(P_i) \end{bmatrix} \tag{23}$$

which is the matrix element associated to the point n°i, where $\left(G_r{}_n^m, G_\theta{}_n^m, G_\varphi{}_n^m\right)$ are the components of the vector:

$$\mathbf{G}_n^m(P_i) = \mathbf{G}_n^m(r_i,\theta_i,\varphi_i) = \mathbf{grad}\left(\frac{Y_n^m(\theta_i,\varphi_i)}{r_i^{\,n+1}}\right) \tag{24}$$

$$\underline{a} = \begin{bmatrix} a_{11} \\ a_{10} \\ a_{1-1} \\ \ldots \\ a_{n_0-n_0} \end{bmatrix} \tag{25}$$

which is the vector of $n_0 \times (n_0+2)$ coefficients to determine.
For each point n°i, we can write:

$$\underline{bs_i} = [As_i].\underline{a} \tag{26}$$

And for the NPT points:

$$[A].\underline{a} = \underline{B} \tag{27}$$

where:

$$[A] = \begin{bmatrix} [As_1] \\ \ldots \\ [As_{NPT}] \end{bmatrix} \tag{28}$$

whose row number is 3×NPT and column number is $n_0 \times (n_0+2)$

$$\underline{B} = \begin{bmatrix} bs_1 \\ \ldots \\ bs_{NPT} \end{bmatrix} \tag{29}$$

whose dimension is 3×NPT.

In (27), the unknown vector $\underline{a}$ is determined thanks to a least squares method based on a singular value decomposition of matrix $[A]$.

3. Measurements

The L.M.M.C.F. has been built to measure the magnetic signature of small and medium devices (whose weight is lower than 1000kg and whose maximum dimensions with measurement system are included in a parallelepiped of 15m x 2m x 2m) under any orientation and amplitude of the earth field. A system of coils (26m long and 8m diameter) makes it possible to control the earth field. The coils are supplied by 14 Kepco amplifiers driven by 3 high stability (lower than 10^{-5}) voltage references H.P. 3245A. A field closed loop system is planned to improve the speed of control of the current in the coils. For a general use, a 2.5m long and 1.6m wide trolley rolls over a 26m long aluminium rail track held by 12 concrete supports. The trolley is drawn by a notch belt; a second smaller notch belt driving an encoder gives a positioning accuracy better than 1mm. In the present work, a special displacement system has been built for the measured device and is going to be described. For the magnetic measurements, sensors are connected to a 128 channel-16 bit data acquisition system (H.P. E 1314B) driven by an H.P. computer 745i. Measurements can be made at full speed (1.5kHz-128 channels) or triggered by the encoder to precisely correlate the measured value with the measured point.

In this work, the measured device is included in a parallelepiped. This parallelepiped can be fixed on a rotating cage according one of its 3 principal directions (Figures 1 and 2). While rotating, the magnetic signature is measured by 13 tri-axial flux-gate magnetometers (ULTRA sensors) located in a vertical plane on a semicircle of 0.55m radius. All the

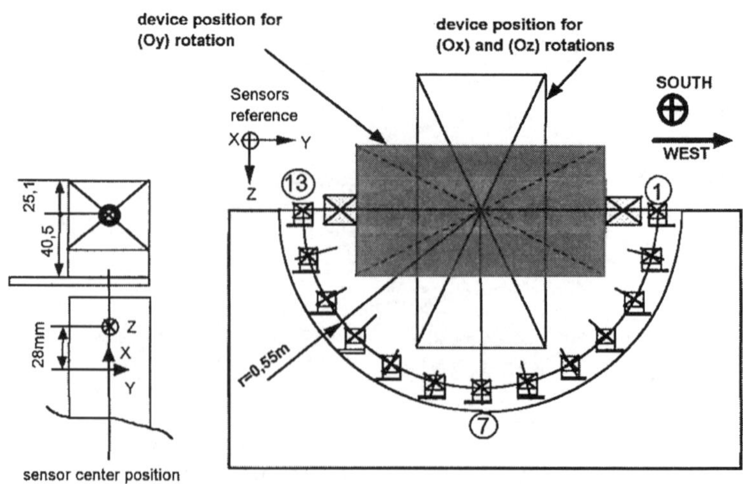

Figure 1: Measurement system principle

Figure 2: Measurement system

magnetometers are oriented in the same direction. As the sensors are not homocentric (there is a distance of 28mm between the centers of X/Y axes and Z axis), the sensors are placed with their median plane including the rotation axis of the parallelepiped. Thus, the 3 components of the induction can be measured on the same radius. A reference magnetometer is used to compensate the local field (earth field or controlled field) and makes it possible to measure a very weak anomaly. To be immune against possible misalignments between the measuring and the reference magnetometers and possible spatial variations of the local field, another difference is made between the differential measurement in the presence of the rotating device and the differential measurement without the device. The 50Hz magnetic pollution is numerically filtered. Finally, the global noise on the sensors is about 0.5nT. The measured device is drawn manually or by a pneumatic motor. The acquisition is triggered by an optic encoder at every 1 degree rotation. The system precision is around ±1mm.

4. Model identification

The first remark is specific to the measurement method of the magnetic signature. To be correctly represented by the model, the measured device magnetic state must not change during its rotation measurement. That means that the local field must not be perpendicular to the rotation axis; otherwise, a rotating field is applied to the system. As a consequence, measurements must be made under zero local field or under local field parallel to the rotation axis.

The used data are finally the induction components at points located on a sphere in a cartesian reference. The radius of the sphere must be chosen so as to have an order n_0 as small as possible to reduce the number of unknown values. At the same time, the signal-to-noise ratio must be as high as possible.

If the number of magnetometers is limited, the sensors must be put where the magnetic field will be expected to be the highest. In the present work, the system is over dimensioned. If it is under dimensioned, a more complex algorithm should be used that could compensate the loss of signal information by using probability about main magnetization orientation and amplitude [3].

Another problem is that the exact radius including all the magnetic sources of the studied device is not known. The device is as a black box whose magnetic state is to be determined. That means that the appropriate system order is not known a priori. The radius of the sphere including the studied parallelepiped is a superior limit of the exact radius but its value can require a high system order. As a consequence, the model coefficients have been computed for different orders. Then, the system degeneracy has been studied: if the model coefficients have been computed at the nth order, then the signatures used for the identification have been computed again with a smaller order than n. Then, the computation result is compared to the measurement to determine the sufficient system order.

One drawback of the spherical harmonic model is that it does not take into account the geometry of the studied device. For example, if the magnetic device is longer than wide, the maximum dimension will impose a distance for which the model will be applied at an acceptable order. This identification radius can be very important with respect to the smallest dimension of the device. As a consequence, in the direction of the smallest device dimension, the magnetic signal may be very weak at the distance of the identification radius. Moreover, it could be interesting to determine the magnetic induction near the smallest dimension or it could be necessary to put the sensors very close to the device. In this case, a model more adapted to the geometry has to be used: the first one is the elliptical harmonic model; the second one, more complex, is a combination between the harmonic model and the search of normal equivalent dipoles spread over the device surface [3].

5. Results

To estimate the spherical harmonic model, the following steps have been followed.

First, the device has been magnetically treated. For each of the 3 principal axes of the parallelepiped, the device has been stabilized in a polarization field of several hundreds A/m with a low frequency decreasing alternative field. Between each of the 3 stabilizations, the device has been demagnetized. The demagnetizing (and stabilizing) system consists of 8 squared coils supplied by Kepco amplifiers driven by a voltage reference controlled by computer. The maximum field is about 6000A/m. The field polarization is created by 4 squared coils concentric to the stabilizing ones, also fed in series by Kepco amplifiers. The homogeneity of the generated fields on the studied volume is better than 5%. The whole system is inside the simulator and demagnetizations can be made under zero local field.

Then, for each polarization state, the signature is measured on the 0.55m radius sphere, under zero local field. The model is applied and the system order determined.

To compare the model extrapolations to measurements and for internal needs, the induction component associated to the polarization axis is measured along the polarization axis thanks to an ULTRA magnetometer (Figure 3). This induction component is expected to be the highest. It is measured under zero local field.

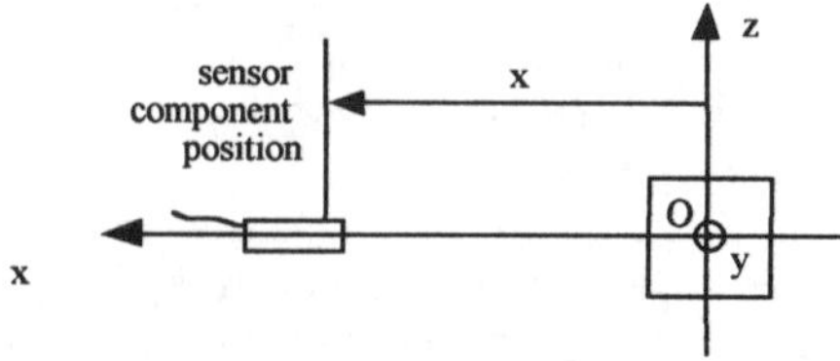

Figure 3: Measurement of the component induction along the polarization axis

The comparison between experiment and the model shows a very good agreement (Figure 4 to 6). The relative error is less than 10%. The abscissa corresponds to the distance between the center of the device reference to the position of the sensor component center along the polarization axis (see Figure 3 for (Ox) axis). Naturally, the extrapolation is given for a distance superior to the model identification radius. For security reasons, the real scale of induction is not given.

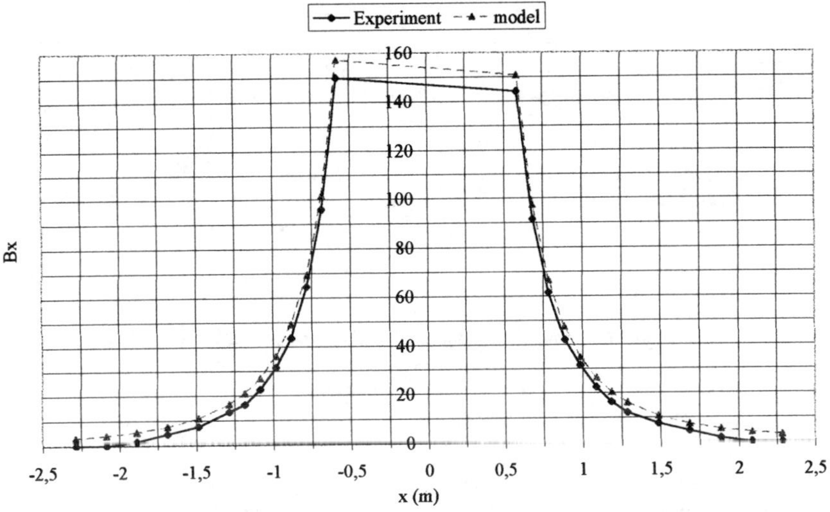

Figure 4: Component induction Bx along the polarization axis (Ox)

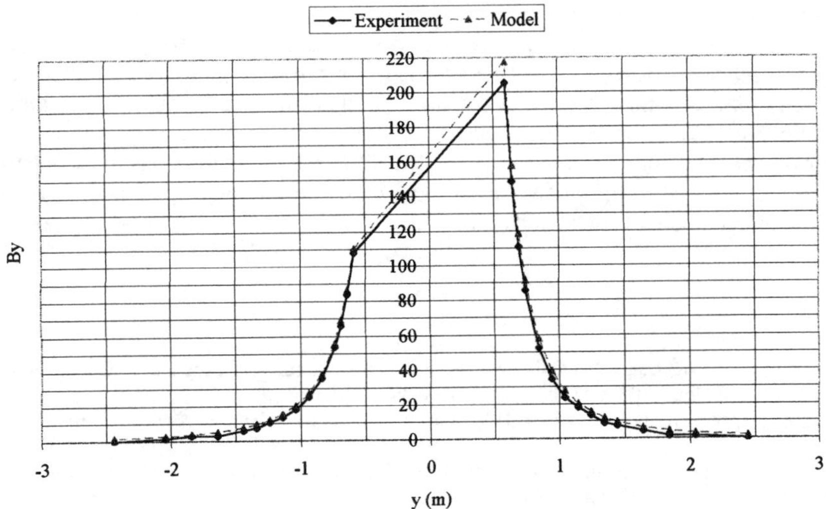

Figure 5: Component induction By along the polarization axis (Oy)

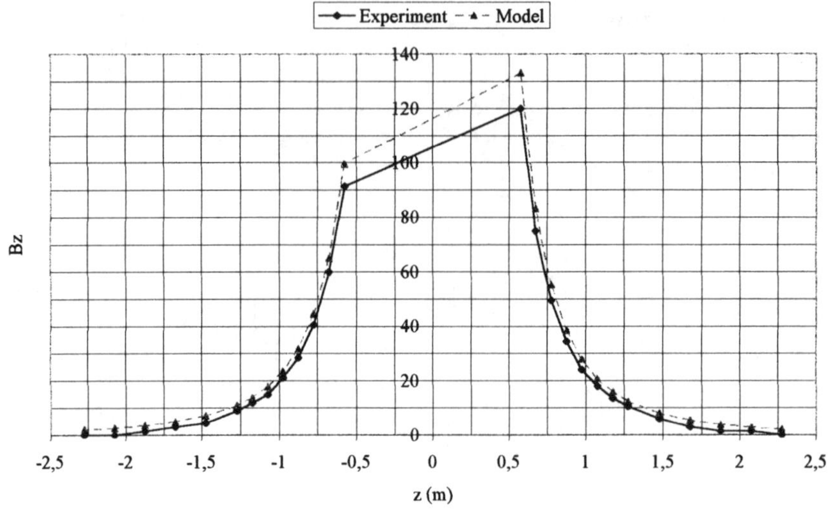

Figure 6: Component induction Bz along the polarization axis (Oz)

6. Conclusion

As a conclusion, the spherical harmonic model gives very interesting results for the identification of the magnetic state of a given system by near field measurements. The main advantages of the model are that the unknown coefficients only depend on the magnetic sources and that the farther the field is measured, the less expansion terms are required. That means that for identification, measurements must be made at a distance for which the best signal-to-noise ratio and the smallest system order are obtained at the same time. The theoretical minimum distance correspond to the radius of the sphere including all the system magnetic sources. Moreover, when identified, the model is able to compute the magnetic field only for a distance superior or equal to the identification distance. These minimum and identification radius can be drawbacks if the studied device has a non-symmetric geometry, if sensors must be very close to the measured system or if the field has to be determined at a very small distance from the device. In these cases, more complex models must be used to better take into account such needs [3].

References

[1] X. Brunotte, Modélisation de l'Infini et Prise en Compte de Régions Magnétiques Minces – Application à la modélisation des Aimantations de Navires, Thèse, INPG, Grenoble, 1991

[2] F. Ledorze, Modélisation des Effets de Boucles d'Immunisation dans les Navires, Thèse, INPG, Grenoble, 1997

[3] M. Legris, Identification de l'Etat Magnétique d'un Système Ferromagnétique à partir de Mesures de Champ Proche, Thèse, INPG, Grenoble, 1996

[4] E. Durand, Electrostatique, Tome 2, ISBN: 2-225-52107-7, Masson et Cie, Paris, 1966

Electromagnetic Nondestructive Evaluation (III)
D. Lesselier and A. Razek (Eds.)
IOS Press, 1999

Microwave nondestructive testing: A stochastic optimization approach

Salvatore Caorsi* and Matteo Pastorino**

**Department of Electronics, University of Pavia, Via Abbiategrasso 209, I-27100 Pavia, Italy*
***Department of Biophysical and Electronic Engineering, University of Genoa, Via Opera Pia 11A, I-16145 Genova, Italy*

This paper presents a stochastic optimization approach of microwave tomographic reconstruction that can be of interest in the light of microwave nondestructive evaluations. The approach is developed in the spatial domain and is based on the integral equations of the inverse scattering problem. Two-dimensional dielectric objects are reconstructed under transverse magnetic illumination conditions. After discretization of the continuous model, the problem is recast in an optimization problem. A *fitness* function is defined and minimized by applying a genetic algorithm. Reconstruction results are presented in the framework of the second order (nonlinear) Born approximation.

1. Introduction

In the field of nondestructive evaluations (NDE), some tomographic microwave methodologies begin to appear in the scientific literature among different methodologies [1], most of them now rather consolidated and of proven efficiency.

1.1 Microwave NDE

Microwave imaging approaches exhibit significant potential features to be considered for practical applications in the field of nondestructive evaluations. A recent overview of results obtained in the microwave NDE (with particular emphasis in the field of diagnostic of civil structures and materials) has been recently proposed in [2]. In this paper, the authors considered aspects related to multifrequency tomographic imaging, image enhancement by multifrequency field statistics and techniques for object recognition. In the approach taken by these authors the model assumes that the processes of propagation, reflection and detection of microwaves may be represented as a linear system.

In the field of civil engineering, microwave imaging techniques have been proposed for the evaluation of the structural integrity of road ways, buildings, and bridges [3]. Moreover, in the NDE field, microwave imaging approaches have been considered for on-line control of materials; in particular, detection of possible defects and measurements of physical quantities (e.g., moisture content) on conveyed products (textiles, paper, wood, etc.) [4]-[6]. In addition, the increased use of thick composites for both industrial and military applications represents a challenge to microwave NDE. The capabilities of microwaves to penetrate inside dielectric materials makes NDE techniques very attractive for interrogating such types of materials [7]. This investigation is performed on contact or noncontact basis using amplitude and/or phase information of the transmitted and/or reflected waves to create a two-dimensional (2D) or three-dimensional (3D) image of the specimen [7]. Furthermore, in the field of microwave NDT for civil engineering applications, a challenging task is also represented by the potential evaluation of "combined materials", such as honeycomb or mixtures of water and air [3] and, in aerospace application, MI has potential applications in the detection of cracks that may occur in aircraft fuselages, turbine blades, and steel-bridge members [8].

Potential applications concern also wood characterization [9] that in the past was limited to frequency below 10 GHz essentially due to little advance information of material behaviour about this limit [9][10]. An attempt to use microwave imaging for diagnostics of wood was also considered in [11]. Finally, an interesting experimental set up has been developed for evaluations on reinforced concrete slabs with practical configurations and on bridges [12].

1.2 Microwave Inverse Scattering Reconstruction Techniques

A number of microwave inverse scattering reconstruction techniques has been recently proposed by the scientific community (see, for example, [13]). Microwave imaging can be defined as a technique aimed at sensing a given object or a given scene by means of interrogating microwaves. This active technique allows some parameters of the object (e.g. shape, location, complex dielectric permittivity distribution, etc.) to be reconstructed from variations in the incident field (e.g., measured scattered field). The final goal is usually to build up an image of the object. However, microwave imaging has difficulties in becoming a methodology competitive with other more classic diagnostic techniques. This is mainly due to the interaction mechanism between microwaves and materials. This mechanism is governed by the laws of the diffraction and to retrieve information from diffracted field is usually a very difficult task. The crucial points in microwave imaging approaches are essentially the following [4][14]-[18]:

• Design and realization of efficient apparatus for the illumination of the scene and the measurement of scattered-field data. These apparatus should potentially perform a large number of accurate illuminations/measurements in a very short time, with a minimum perturbation of the original field distribution. Several prototypes have been constructed to meet this task and the reader is referred to the open literature on the subject. Essentially, as far as the receiving systems are concerned, the main research areas are referring to probes that minimize the perturbation and allow for rapid measurements (e.g., modulated scattering techniques) and to imaging techniques aimed at compensating for the effects of the measurement system.

• Development of effective reconstruction procedures aimed at locating and shaping a given object and, possibly, at defining the distribution of certain electromagnetic parameters, as the permettivity distribution in imaging of dielectric objects.

The difficulties related to the development of reconstruction procedures are due to the nature of the inverse problem to be faced. The inverse problem is usually very ill-posed. The solution is not unique and it does not depend with continuity on data. In imaging of dielectric objects, this corresponds to searching for a minimum of a (usually) multimodal function.

The task of the data *inversion* is performed following essentially the following three lines:

1) use of linearized techniques, as the Born or Rytov ones, which are valid only for weakly scattering objects not too large in terms of the wavelength of the interrogating wave.

2) use of iterative procedures in which, at each step, a direct scattering problem has to be solved [19]-[22]. With a judicious choice of the starting solution, these approaches avoid to be trapped in a local minimum. Recently, very interesting reconstructions have been shown, which have been obtained by using approaches of this kind.

3) use of global optimization techniques [23][24]. Nowadays, some very efficient optimization procedures are available (e.g., simulated annealing, genetic algorithms, etc.).

The search for the global minimum of a multimodal function is performed by using stochastic concepts. Although potentially very powerful, these methods are usually very time consuming. As a consequence they do not seem to be very suitable for unconstrained imaging. Nevertheless, when a priori information on the scene is available, the efficiency of these methods can be drastically improved. In addition, these optimization methods are such that a priori information can be inserted very easily in the model. This is particularly valuable in nondestructive application, where one usually knows a lot about the "scatterer" under test. In most cases, this "scatterer" can be only a defect in an otherwise known object.

Furthermore, in order to focus the research activity toward practical applications, for several existing methodologies, the related computer codes are going to be modified in order to run on parallel computers. Stochastic optimization approaches, such as the genetic algorithm, can be naturally implemented on parallel computers, without any complex change in the problem formulation or/and in the reconstruction procedure.

• Development of efficient post-processing procedures aimed at the enhancement of the crude "images" provided by the microwave reconstruction methodologies.

The present paper is focused on the second point of the above list. In particular, we propose a stochastic optimization approach based on a genetic algorithm [25]. The first idea of this reconstruction method was presented in [26]. In that book, the present approach was developed in the framework of the first order Born approximation, which is only valid, as it is well known, when the object to be reconstructed represent only a weak discontinuity for the interrogating wave.

2. Microwave imaging in the spatial domain

Although in the past very interesting numerical and experimental results were obtained by using numerical procedures developed in the spectral domain [27][28], more recently, a large number of contributions to imaging in the space domain were presented, as, for certain applications, the main limitations of the spectral-domain methods (i.e., the low resolution and the need for low contrasts) may be too severe. This consideration seems to hold true particularly for the microwave NDE. In most cases, space-domain techniques aim to solve the equations for the inverse-scattering problem in their integral forms by using discretizations and/or matrix approaches. It is worth noting that some iterative algorithms recently developed in the spatial domain make use of parameter distributions obtained by spectral-domain techniques as starting estimates of the scene. In addition, time-domain techniques must also been mentioned.

Space-domain methods are usually based on a discretization of the continuous model represented by the following equations [29]:

$$\mathcal{L}_1\{\tau(\mathbf{r}),\mathbf{E}_i^{tot}(\mathbf{r})\} = \int_V \tau(\mathbf{u})\mathbf{E}_i^{tot}(\mathbf{u})\cdot\Gamma(\mathbf{r}/\mathbf{u})d\mathbf{u} - \mathbf{E}_i^{tot}(\mathbf{r}) = -\mathbf{E}_i^{inc}(\mathbf{r}) \qquad \mathbf{r} \in \mathrm{V} \tag{1}$$

$$\mathcal{L}_2\{\tau(\mathbf{r}),\mathbf{E}_i^{tot}(\mathbf{r})\} = \int_V \tau(\mathbf{u})\mathbf{E}_i^{tot}(\mathbf{u})\cdot\Gamma(\mathbf{r}_j/\mathbf{u})d\mathbf{u} = \mathbf{E}_{ij}^{scat}(\mathbf{r}_j) \qquad \mathbf{r}_j \in \mathrm{V}_j \ (\not\subset \mathrm{V}) \tag{2}$$

where $\mathcal{L}_1\{x\}$ and $\mathcal{L}_2\{x\}$ are nonlinear operators related to the so-called state equation and data equation, respectively. The unknown functions are $\tau(\mathbf{r}) = j(\omega\mu_o)^{-1}[k_v^2(\mathbf{r}) - k_e^2]$ ($k_v(\mathbf{r})$ and k_e being the complex wavenumbers inside and outside V) and $\mathbf{E}_i^{tot}(\mathbf{r})$ (total electric field), and $\Gamma(\mathbf{r}/\mathbf{u})$ is the Green's dyadic function for free space. Under the second order Born approximation, the total electric field inside the investigation area is approximated by the sum of the incident field and a linearized scattering term. The resulting nonlinear inverse scattering problem can be solved through the minimization of a suitable *fitness* function, which can be expressed as:

$$\mathcal{U}\{\tau(\mathbf{r})\} = \mathcal{U}\{\mathcal{L}_2\{\tau(\mathbf{r}),[\mathbf{E}_i^{inc}(\mathbf{r})+\mathcal{L}_{1B}\{\tau(\mathbf{r})\}]\} - \mathbf{E}_{ij}^{scat}(\mathbf{r}_j);\ a\ priori\ information\} \quad (3)$$

where $\mathcal{L}_{1B}\{\tau(\mathbf{r})\}$ is the operator in (1) modified by the Born linearization. The problem turns out to be nonlinear of quadratic nature and the nonlinearity concerns only the values of the dielectric parameters, which remain the only problem unknowns, resulting in a significant reduction of the computational load. A complete nonlinear version of the approach, which takes into account the state equation is currently under development.

3. Application of the genetic algorithm

The minimization of the resulting functional is performed by applying a genetic algorithm. Genetic algorithms [25][31]-[33] are global optimization procedures based on a random search of the minimum of a function by keeping the information on a fixed number of points in the solution space at each iteration. Therefore they can be naturally implemented on parallel computers.

Given the array of unknowns $T = [\tau_1\ \tau_2\ ...\tau_N]$, each τ_n value can be selected in the range $0 \le \tau_n \le \underline{\tau}_n$, $\underline{\tau}_n$ being a fixed maximum value. In the numerical results presented in the following section, we assume $\underline{\tau}_n = K\tau_{n,real}$ where $\tau_{n,real}$ is the value of the original object function in the *n*-th subdomain and K is a given constant. In several applications (as most of the microwave NDE applications), the "scatterer" under test is only a defect in a known scatterer or only certain specific permittivity values can be considered, corresponding to specific materials. This is of course a way to introduce a priori information. By means of a Q-level quantizer, a binary sequence known as a *gene* can be obtained and the set of N genes is indicated as a *chromosome.* Let π be the chromosome associated to *T*. A set of N_p chromosomes is called a *population* and it is indicated as P. At the beginning of the iterative process, a population can be randomly generated or, if available, known tentative values can be introduced as a priori information. Let us denote the population at the iteration step *k* as:

$$P^k = [\pi_1{}^k\ \pi_2{}^k\ ...\pi_{N_p}{}^k]$$

Then the population P^{k+1} can be generated from P^k by means of the following operators [25].

Selection: $S(P^k) = \pi_*{}^k$. The selection procedure known as *tournement* is applied. Two chromosomes $\pi_*{}^k$ and $\pi_{**}{}^k$ are randomly chosen in P^k and the one with the lower *fitness* function value is selected for the following operators.

Crossover: $C(\pi_*{}^k,\pi_{**}{}^k) = (\pi_*{}^{k+1},\pi_{**}{}^{k+1})$. Two selected trial chromosomes, with probability P_c, have their bits swapped from a ramdomly selected point to the end of the chromosomes themselves (see Figure 1).

Mutation: $C(\pi_*{}^k) = \pi_*{}^{k+1}$. A chromosome is chosen with probability P_m and the value of each of its bits is mutated with probability P_{mb}.

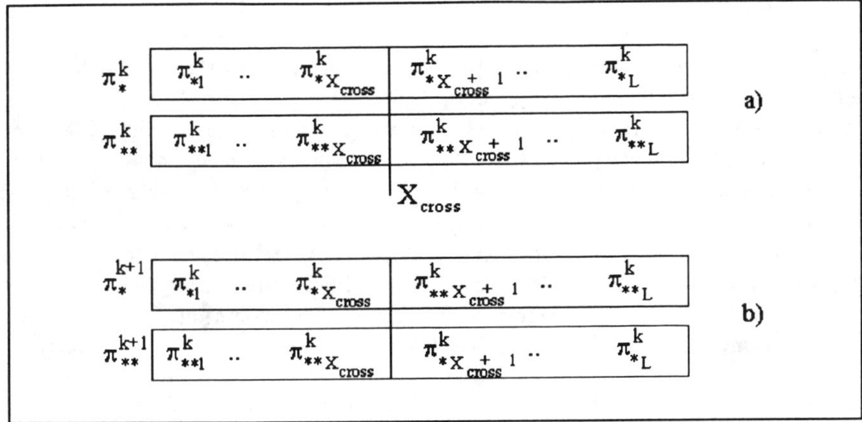

Fig. 1 Genetic algorithms: The *crossover* operator.

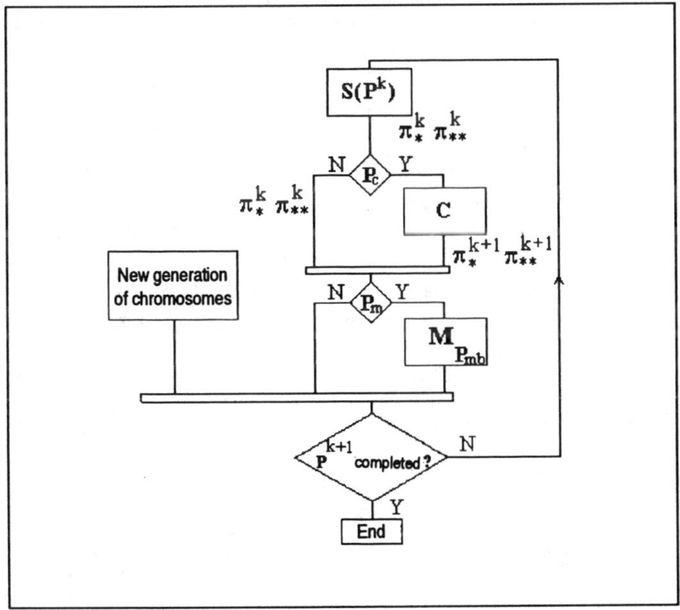

Fig. 2 Genetic algorithms: Scheme of the computational procedure.

A scheme of the implemented GA procedure is provided in Figure 2. The application of the described operators goes on until the $(k+1)$-th population is complete. The iterative process continues until a stop criterion is satisfied or a given number of iterations (generation of new populations) has been accomplished.

In the following, the genetic algorithm is applied with reference to a classic tomographic imaging configuration. A square investigation area is assumed and partitioned into square subdomains. The observation domain is made up of a set of equally-spaced measurement points located on an arc of circumference. A multiview approach is taken in which the illuminating source (unit plane wave) jointly rotates with the measurement domain.This imaging configuration is the same used in [26], where details can be found.

4. Numerical results

In the following, several numerical results are provided. In the first example the scattering object is a homogeneous circular cylinder whose diameter is $s = 0.4\lambda$, λ being the free-space wavelength corresponding to a frequency f; the coordinates of the center are $x_o = 0.363\lambda$, $y_o = 0.637\lambda$ and $\tau(x,y) = 1.0$. For this canonical scatterer the measurement data are available analytically [16]. The cross-section of the cylinder is included in a square investigation domain of side $a = \lambda$, partitioned into N = 400 subdomains. Four illuminations (V = 4) are used with uniform angular space. The equally-spaced measurement points (M = 32) are located on an arc of circumference of radius $b = 0.84\lambda$.

The data of the parameters of the genetic algorithm are the following: N_p (population dimension) = 60; P_c (probability of crossover) = 0.6; P_m (probability of mutation) = 0.5; P_{mb} (probability of bit mutation) = 0.001; Q (number of quantization levels) = 32. The choice of the above parameters for the genetic algorithm has been performed according to values suggested in the literature on the subject [19]-[25].

Figure 3 shows the pictorial representations of the reconstructed cross sections of the scatterer at various iteration steps. For the initial distribution (k = 1), completely random values are assumed. When the number of iterations increases, the background becomes empty and the location of the scatterer more accurate. At step k = 1500, it is possible to recognize the position of the object and, for k = 4000, the reconstruction process gives rather good results. As can be seen, the shape of the object is reconstructed very well. In particular, the empty part of the investigation area can be adequately individauted and very few "artefacts" (associated with false localizations) are present. However, the reconstructed cylinder exhibits a rather irregular two-dimensional profile. In several NDE applications, the shape of the object may represent sufficient information. Moreover, it is well known in microwave recontruction techniques that far more regular profiles can be obtained by adding regularizations terms in the functional to be minimized.

A version of the present approach including regularization terms in order to obtain smoother reconstructions of the homogeneous regions of the scene is still under development.

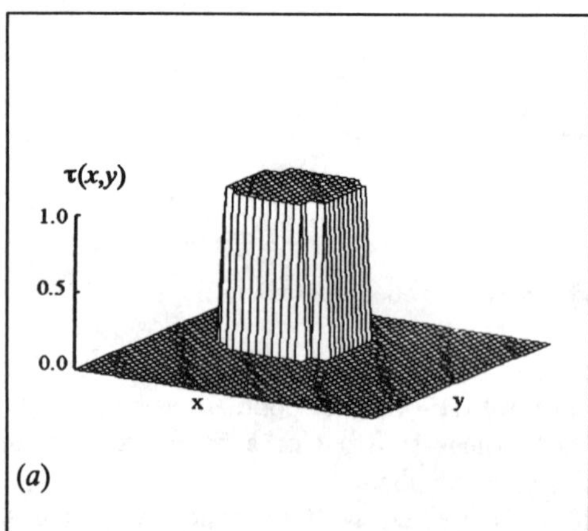

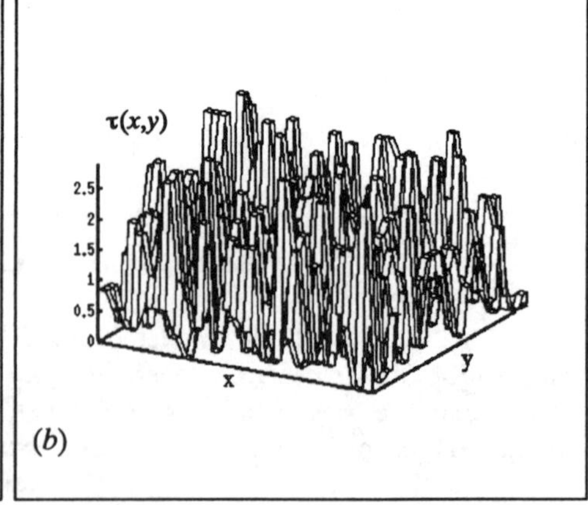

Fig. 3. Reconstruction of a circular cylinder . Images of the object function distribution reconstructed at various iteration steps. (*a*) original distribution; (*b*) *k* = 1 (initialization step); (*c*) *k* = 100; (*d*) *k* = 1000; (*e*) *k* = 1500; (*f*) *k* = 2000; (*g*) *k* = 2500; (*h*) *k* = 4000; (*i*) *k* = 6000; (*l*) *k* = 8000.

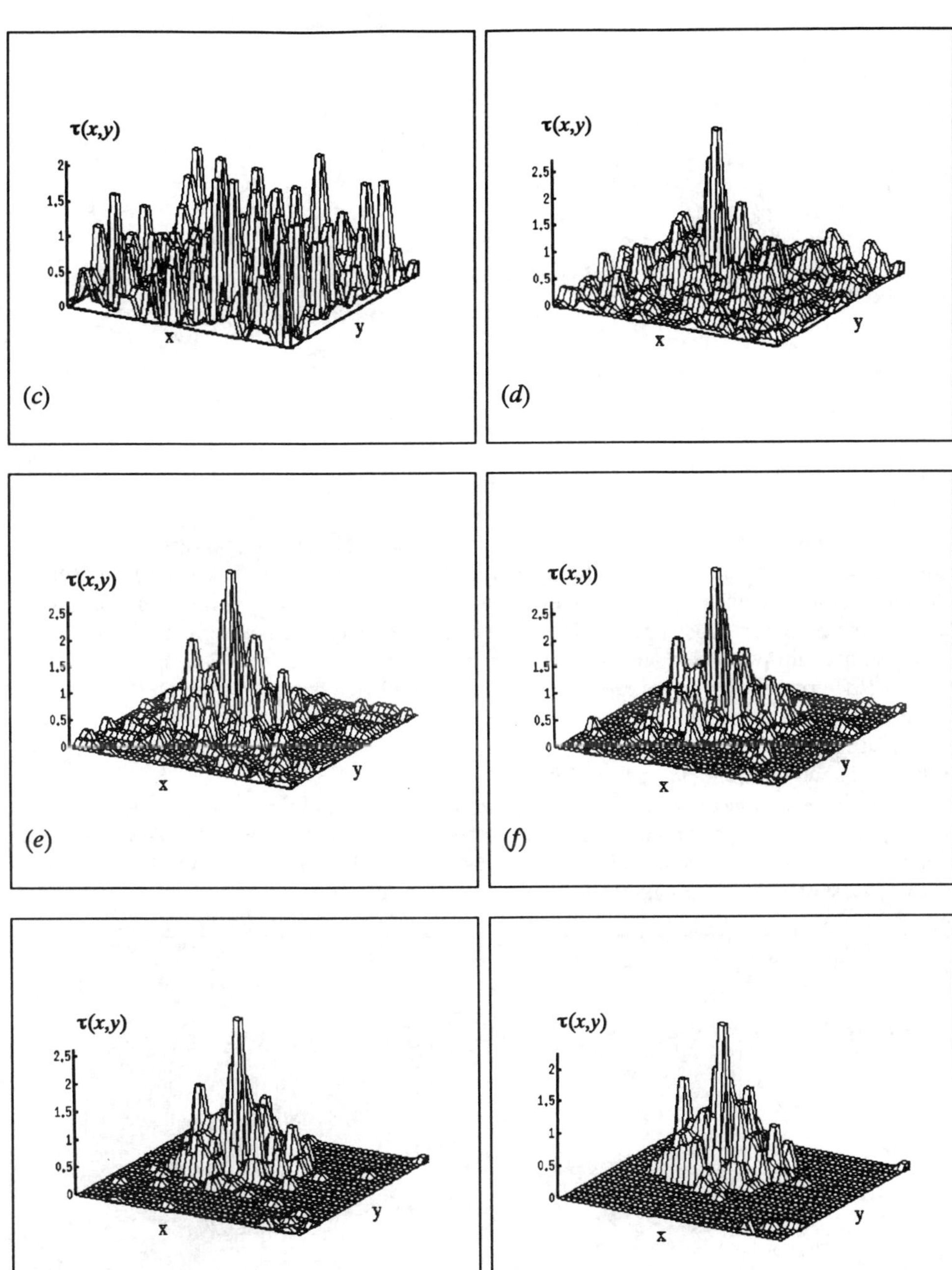

Fig. 3. *continued.*

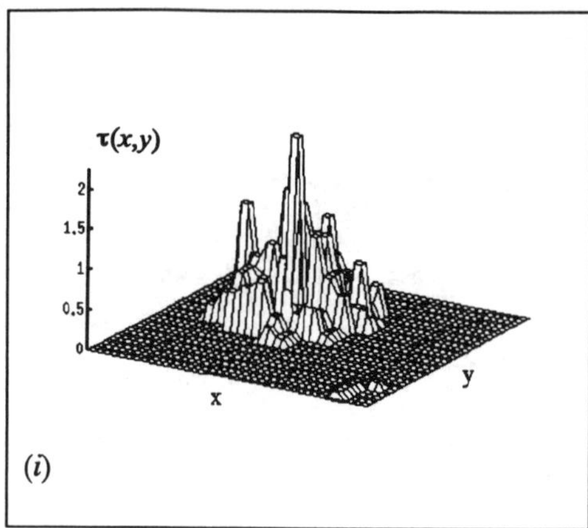

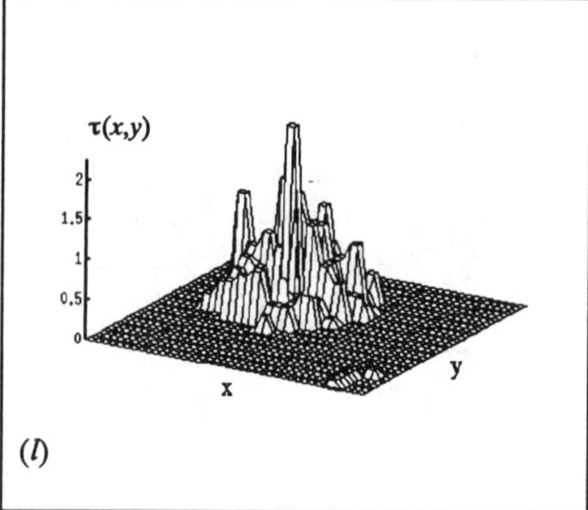

Fig. 3. *continued.*

The authors of the present paper used in the past a Markov model (associated with a simulated annealing technique) which provided very good recontructions of multilayer cylinders (see, for example, [23]). Nevertheless, when regularization terms are added to smooth the reconstruction, one needs to introduce also edge-preserving methodologies to take into account possible boundaries between different materials.

Edge preserving techniques were discussed in [34] and in [23] where a line process, similar to those applied in image processing applications [35] was used. It is worth mentioning that Markov models have been already considered in the field of NDE, for example in eddy current evaluations [36].

Other examples are devoted to comparisons between results obtained by using the first-order Born approximation and the present quadratic approach. The dielectric parameters used are the same as in the previous example, but in this case the homogeneous cylinder assumes different values of its dielectric permittivity.

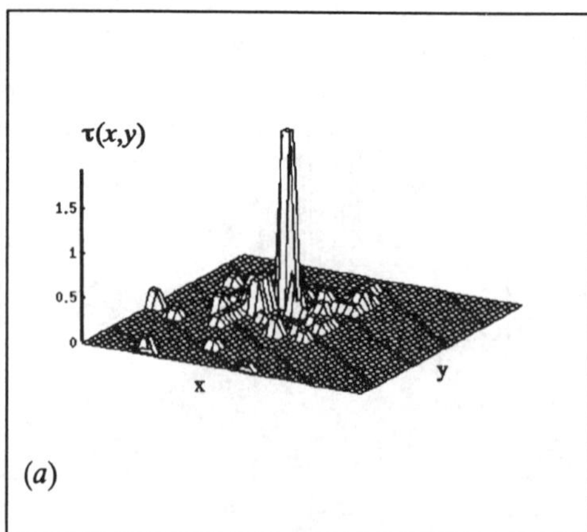

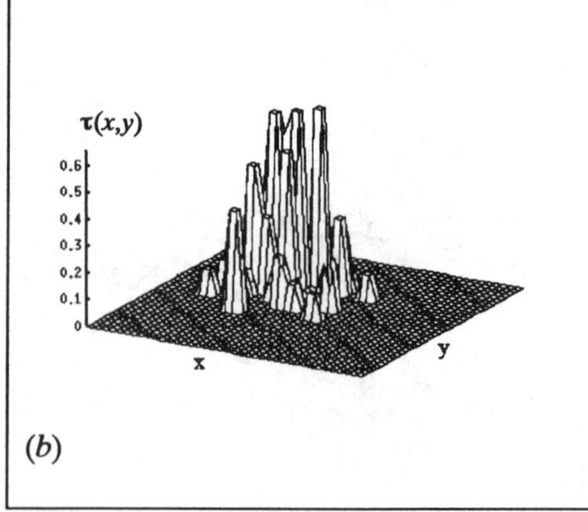

Fig. 4. Comparison between the first order Born approximation (*a*) and the quadratic model (*b*). Reconstruction of a circular cylinder. Images of the object function distribution. Original values: $\tau(x,y) = 1.2$ ($\varepsilon_r = 2.2$). $k = 4000$.

Figure 4 shows the conparison between the pictorial reconstruction of the cylinder when $\tau(x,y) = 1.2$ ($\varepsilon_r = 2.2$). As can be seen, the first order Born approach is completely unable to reconstruct the shape of the cylinder, whereas a good shaping, can be obtained by the present procedure.

When the original relative dielectric permittivity increases, the recontruction obtained by using the first order Born approximation (as expected) fails. Figure 5 refers to the case in which $\tau(x,y) = 2.4$ ($\varepsilon_r = 3.4$). Although the prospectical view in Figure 5 does not show this defect, a "hole" is present in the reconstructed cross section, with reconstructed values close to that of the background. An excellent localization and a good shaping can be obtained by the second order Born approach.

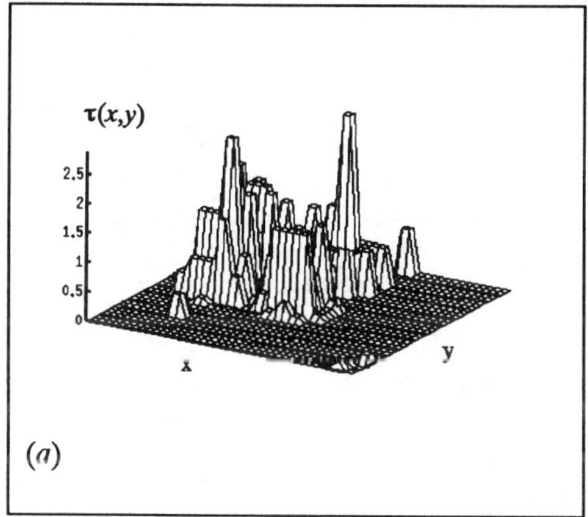

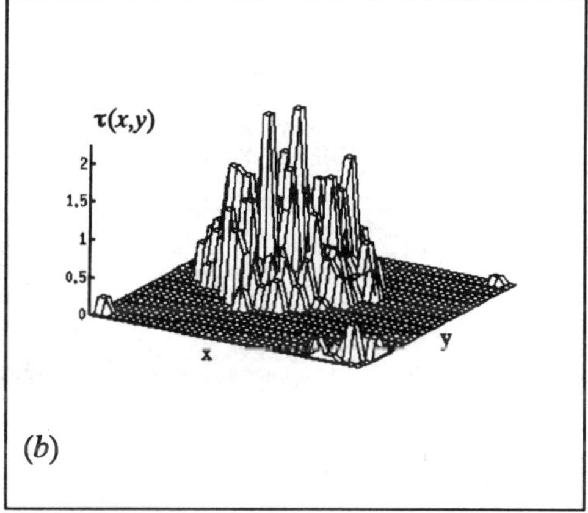

Fig. 5. Comparison between the first order Born approximation (*a*) and the quadratic model (*b*). Reconstruction of a circular cylinder. Images of the object function distribution. Original values: $\tau(x,y) = 2.4$ ($\varepsilon_r = 3.4$). $k = 4000$.

A quantitative comparison is provided in Figure 6. In particular, this figure gives the values of the following percentage mean error on the reconstruction of the scatterer (full subdomains):

$$\Psi_{\mathrm{int}} = \frac{1}{N_{\mathrm{int}}} \sum_{n=1}^{N_{\mathrm{int}}} \frac{|\tau_n - \tau_{n,rec}|}{|\tau_n|}$$

For completeness, in the case in which $\tau(x,y) = 2.4$ ($\varepsilon_r = 3.4$), Figure 7 shows: (*a*) the behaviour of the fitness function at the various iteration steps; (*b*) the behaviour of the quantitative parameter Ψ_{int}; (*c*) the behaviour of the parameter Ψ_{ext} which is an error parameter analogous to Ψ_{int} but related to the empty cells of the investigation area; and (*d*) the behaviour of the global parameter Ψ, which is defined in an analogous way and refers to the whole investigation area.

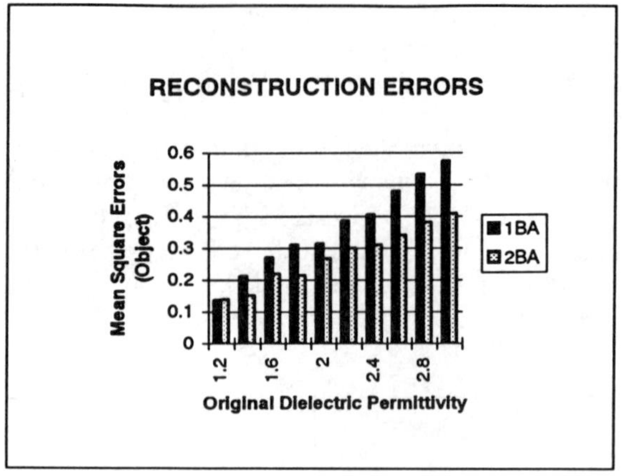

Fig. 6. Comparison between the first order Born approximation (1BA) and the quadratic model (2BA). Reconstruction errors inside the scatterer for different values of the original relative dielectric permittivity.

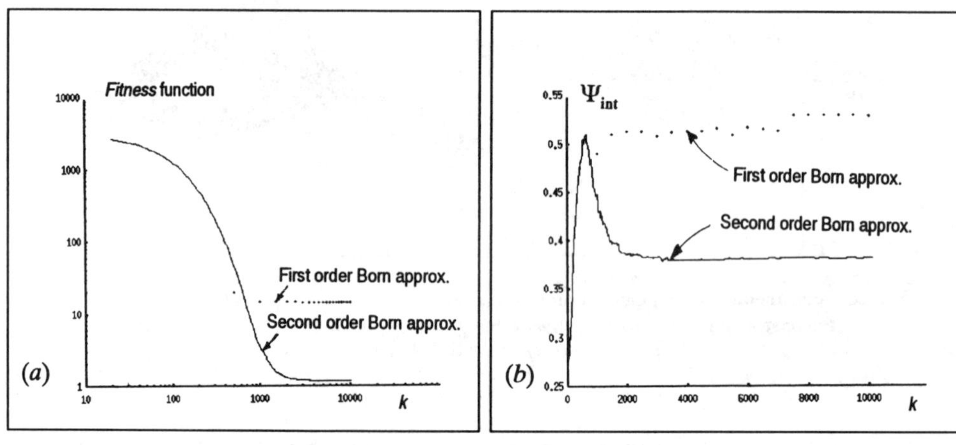

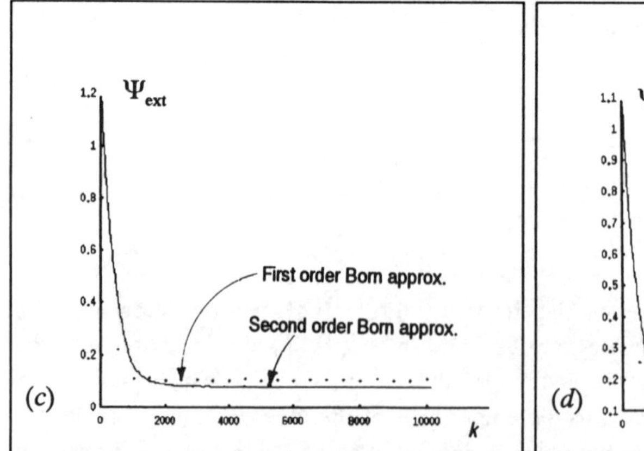

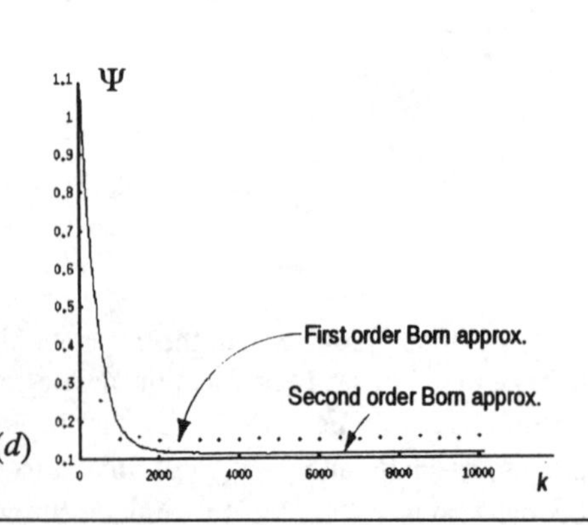

Fig. 7. Comparison between the first order Born approximation and the quadratic model. (*a*) Fitness function; (*b*)-(*d*) Behaviour of the quantitative error parameters.

5. Conclusions

A stochastic optimization approach for dielectric reconstruction in the field of microwave NDE has been proposed. Starting by the integral equations of the inverse scattering problem, a multi-illumination multiview approach has been formulated within the framework of the second order Born approximation. A genetic algorithm has been applied to minimize the resulting fitness function. Some cylindrical scatterers have been reconstructed by using analytical scattering data and comparison with linearized reconstructions (obtained by applying the first order Born approximation) have been reported. It has been shown that the range of applicability in terms of the contrast is larger for the second order Born method than for the linearized method. A version of the approach able to take into account the complete nonlinear form of the inverse scattering equations for strong scatterers is currently under development.

References

[1] R. Albanese, G. Rubinacci, T. Takagi and S. S. Udpa, Eds., *Electromagnetic Nondestructive Evaluation (II),* Studies in Applied Electromagnetics and Mechanics 14, IOS Press, 1998.

[2] S. J. Lockwood and H. Lee, "Pulse-echo microwave imaging for NDE of civil structures: Image reconstruction, enhancement, and object recognition," *International Journal of Imaging Systems and Technology*, vol. 8, pp. 407-412, 1997.

[3] M. Chang, P. Chou, and H. Lee, "Tomographic microwave imaging for nondestructive evaluation and object recognition of civil structures and materials," Proc. of ASILOMAR, vol. 29, pp. 1061-1065, 1996.

[4] J. Ch. Bolomey and N. Joachimowicz, "Dielectric metrology via microwave tomography: present and future," Mat. Res. Soc. Symp. Proc., vol. 347, pp. 259-265, 1994.

[5] K. Mayer and K. J. Langenberg, "Microwave Imaging of Defects in Solids," Proc. 21st Annual Review of Progress in Quantitative NDE, Snowmass Village, Colorado, 1994.

[6] J. C. Bolomey, "Recent European developments in active microwave imaging for industrial, scientific, and medical applications," *IEEE Trans. Microwave Theory Tech.*, vol. 37, pp. 2109-2117, Dec. 1989.

[7] N. Qaddoumi, G. Carriveau, S. Ganchev, and R. Zoughi, "Microwave imaging of thick composite panels with defects," *Materials Evaluation*, vol. 53, pp. 926-929, Aug. 1995.

[8] C. Huber et al., "Modeling of surface hairline-crack detection in metals under coatings using an open-ended rectangular waveguide," *IEEE Trans. Microwave Theory Tech.*, vol. 45, pp. 2049-2057, 1997.

[9] P. Eskelinen and P. Harju, "Characterizing wood by microwaves," *IEEE AES Systems Magazine*, pp. 34-35, Feb. 1998.

[10] Tinga et al. "Dielectric properties of materials for microwave processing - Tabulated," *J. Microwave Power*, vol. 8, 1973.

[11] J. Ch. Bolomey and Ch. Pichot, "Controle non-destructif par imagerie microonde active," Proc. JINA'88, Nice, France, pp. 485-489, 1988.

[12] K. Belkebir, Ch. Pichot, J. Ch. Bolomey, P. Berthaud, G. Gottard, X. Derobert and G. Fauchoux, "Microwave tomography system for reinforced concrete structures," Proc. 24th European Microwave Conf., Cannes, France, pp. 1209-1214, 1994.

[13] M. Pastorino, "Modern microwave inverse scattering techniques for image reconstruction," *IEEE Measurement and Instrumentation Magazine,* to appear in Dec. 1998.

[14] J. B. Detlefsen, "Industrial application of microwave imaging," Proc. 21st European Microwave Conf., Stuttgart, Germany, pp. 108-119, 1991.

[15] D. Lesselier and B. Duchene, "Wavefield inversion of objects in stratified environments: from backpropagation schemes to full solutions," *Review of Radio Sci. 1993-95,* Ed. W. R. Stone, 1996.

[18] T. C. Guo and W. W. Guo, "Computation of electromagnetic wave scattering from an arbitrary three-dimensional inhomogeneous dielectric object," *IEEE Trans. Magn.*, vol. 25, pp. 2872-2874, 1989.

[16] T. M. Habashy, M. L. Oristaglio, and A. T. de Hoop, "Simultaneous nonlinear reconstruction of two-dimensional permittivity and conductivity", *Radio Sci.*, vol. 29, pp. 1101-1118, 1994.
[17] S. Caorsi, S. Ciaramella, G. L. Gragnani, M. Pastorino, "On the use of regularization techniques in numerical inverse-scattering solutions for microwave imaging applications", *IEEE Trans. Microwave Theory Tech.*, vol. MTT-43, n. 3, pp. 632-640, 1995.
[18] S. Caorsi, E. Capra, G. L. Gragnani, A. Massa, M. Pastorino, "Microwave imaging in the spatial domain for canonical dielectric objects", *IEEE Trans. Magn.*, vol. 31, n. 3, pp. 1562-1565, 1995.
[19] W. C. Chew and Y. M. Wang, "Reconstruction of two-dimensional permittivity distribution using the distorted Born iterative method," *IEEE Trans. Med. Imaging*, vol. MI-9, pp. 218-225, 1990.
[20] A. G. Tijhuis and F. E. van Vliet, "Two-dimensional Born-type velocity inversion using multiple-frequency information," Proc. Progress in EM Research Symp., Noordwijk, The Netherlands, 1994.
[21] A. Broquetas, J. J. Mallorquì, J. M. Rius, L. Jofre, and A. Cardama, "Active microwave sensing of highly contrasted dielectric bodies," *J. Electromagnetic Wave and Appl.*, vol. 7, pp. 1439-1453, 1993.
[22] R. E. Kleinman and P. M. van den Berg, "A Modified Gradient Method for Two-Dimensional Problems in Tomography," *J. of Comput. Appl. Math.*, vol. 42, pp. 17-35, 1992.
[23] S. Caorsi, G. L. Gragnani, S. Medicina, M. Pastorino, and G. Zunino, "Microwave imaging based on a Markov random field model," *IEEE Trans. Antennas Propagat*, vol. 42, pp. 293-303, 1994.
[24] H. Carfantan and A. Mohammad-Djafari, "A Bayesian apprach for nonlinear inverse scattering tomographic imaging," Proc. ICASSP'95, 1995.
[25] D. Golberg, "Genetic and evolutionary algorithms come of age," *Comm. of the ACM*, vol. 37, pp. 113-119, 1994.
[26] S. Caorsi and M. Pastorino, "Microwave imaging using a genetic algorithm," in: R. Albanese et al. (eds.): Electromagnetic Nondestructive Evaluation (II), IOS Press, 1998 (Reference [1]), pp. 233-242.
[27] A. J. Devaney, "A computer simulation study of diffraction tomography," *IEEE Trans. Biomed. Eng.*, vol. 30, pp. 377-386, 1990.
[28] C. Pichot, "Spectral and spatial-domain technique for microwave imaging," Proc. Workshop on Inverse scattering and microwave imaging, 25th European Microwave Conf., pp. 85-90, Bologna, Italy, 1995.
[29] M. Pastorino, "Recent advances in microwave imaging", Proc. 1995 Int. Conf. Electronic Measurement and Instrum., Shangai, China, pp. 157-160, Oct. 1995.
[30] M. J. Hagmann, O. P. Gandhi, and C. H. Durney, "Upper bound on cell size for moment-method solutions," *IEEE Trans. Microwave Theory Tech.*, vol. 25, pp. 831-832, 1977.
[31] R. L. Haupt, "An introduction to genetic algorithms for electromagnetics," *IEEE Antennas Propagat. Magazine*, 37, 2, 7-15, 1995.
[32] D. S. Weile and E. Michielssen, "Genetic algorithm optimization applied to electromagnetics: a review," *IEEE Trans. Antennas Propagat*, vol. 45, pp. 343-353, March 1997.
[33] J. M. Johnson and Y. Ramat-Samii, "Genetic algorithms in engineering electromagnetics," *IEEE Antennas Propagat. Magazine*, vol. 39, pp. 7-25, 1997
[34] P. Lobel, C. Pichot, L. Blanc Feraud and M. Barlaud, "Reconstruction from experimental data using conjugate gradient and enhancement by edge-preserving regularization," *Int. J. of Imaging Systems and Technology,* vol. 8, pp. 337-342, 1998.
[35] E. Gamble and T. Poggio, "Visual integration and detection of discontinuities: the key role of intensity edges," *A.I. Memo no. 970,* MIT, 1987.
[36] L. Udpa and W. Lord, "Application of Markov models for inverse problems in eddy current nondestructive evalutations," Proc. PIERS '93, Pasadena, CA, USA, p. 13, July 1993.

Direct Modeling with Emphasis on Numerical Techniques

Electromagnetic Nondestructive Evaluation (III)
D. Lesselier and A. Razek (Eds.)
IOS Press, 1999

Superconductive and traditional electromagnetic probes in eddy current NDE for detection of deep defects

Massimo VALENTINO[1-2], Adele RUOSI[2], Giovanni P. PEPE[2] and Giuseppe PELUSO[2]
[1] *Dipartimento di Ingegneria dei Materiali per la Produzione, Università di Napoli "Federico II", Piazzale Tecchio 80, 80125 Napoli, Italy*
[2] *Istituto Nazionale per la Fisica della Materia (INFM), Unità di Napoli, Università di Napoli "Federico II", Piazzale Tecchio 80, 80125 Napoli, Italy*

Abstract. Eddy current nondestructive evaluation based on High Critical Temperature Superconductive Quantum Interference Devices (HT_C-SQUIDs) is a promising technique which is well suited for applications such as the detection of anomalies affecting conductive structures (multi-layered metallic alloys, composites, etc.) or the localization of defective components in electronic circuits. Indeed, the high sensitivity of SQUID-based magnetometers at low frequencies (typically less than 1 kHz) allows one to measure weak signals due to deeply embedded defects for which traditional eddy current probes such as induction coils and flux-gate magnetometers fail. This work is a comparative study of nondestructive testing performed by three different eddy-current measuring devices: induction coil, flux-gate and HTc-SQUID-based magnetometer. These sensors have been used for the detection of various artificial defects affecting both single-layered and multi-layered aluminum-alloy structures, without external electromagnetic shielding. Experimental data are discussed in terms of signal-to-noise ratio in the frequency range 10-1000 Hz. In particular we demonstrate that the higher sensitivity of SQUIDs in comparison to the other electromagnetic probes gives a decisive advantage when the structure under inspection is weakly conducting or multi-layered. The effect of lift-off variation and probe tilting on the output signal is also experimentally investigated.

1. Introduction

With respect to the conventional electromagnetic sensors used in Eddy-Current Nondestructive Evaluation (EC-NDE), SQUID (Superconductive Quantum Interference Device) magnetometers are attractive as multimode instruments capable of extremely high magnetic field sensitivity (detection of sub-pT signals) even in unshielded environments, high image resolution of low frequency eddy-current distributions, large bandwidth (up to MHz) and high spatial resolution [1-2]. Moreover, an optimized flux-locked-loop electronics allows this device to operate with a large dynamic range and high linearity. These characteristics are essential in order to perform sensitive measurements of magnetic field distributions produced by induced currents in test objects. Naturally, the need to operate the SQUID at cryogenic temperatures limits its use in many practical applications. The advent of High Critical Temperature Superconductors (HTc) and the development of high performance HT_C SQUIDs has renewed the interest for NDE employing superconductive sensors [3-5].

Currently, many different kinds of electromagnetic probes are used for EC-NDE measurements, the most common ones are induction coils and flux-gates. Each one has pros and cons and it is difficult to state which of them is the most effective for a particular application. An experimental characterization based on a quantitative comparison between the above-mentioned sensors is required in order to optimize their performance in relation to the problem under investigation. For aircraft industry applications such as quality tests during the fabrication process or integrity assessments of structures already in use, it is essential to have a high accuracy in the localization of both surface and deeply embedded defects (depth >10 mm) in metallic structures [6-8]. Since the skin penetration depth of the incident electromagnetic wave is given by $\delta = 1/\sqrt{(\sigma_0\mu_0\pi f)}$ (where μ_0 is the magnetic permeability of vacuum, σ_0 the material conductivity and f the incident wave frequency), probes having a high sensitivity at low frequencies are required. In this context, a comparative study of nondestructive testing performed by an induction coil, a flux-gate and a SQUID-based magnetometer is presented here. Five aluminum-alloy structures with artificial defects such as holes, slots and cracks, modeled after those commonly encountered in the aircraft industry, have been examined.

In EC-NDE measurements, noise and signal depend on the particular defect to be detected and it is difficult to measure them separately [9]. Moreover, due the intrinsically different working principle of the three probes, each is in effect measuring a different physical characteristic. Flux-gate, SQUID and induction coil detect the component of the magnetic field parallel with the core, the magnetic flux through the loop and the inductance variation, respectively. The sensor-independent parameter chosen to perform such a comparison is a type of Signal-to-Noise ratio (S/N) defined as the ratio between the maximum signal variation along the measured line-scan (S_{pp}) and the standard deviation of the N data points b_i collected along the same line scan, i.e. $\sigma = [\Sigma_i (b_i - b_m)^2/N]^{1/2}$, where b_m is the average value of N points. Since in our measurements line-scans are crossing a consistent area of the sample located far away the defect, σ can be considered as a good approximation of the noise in each measurement.

In practice, due to the spatial extent of the sensors, the value and shape of the detected signal also depend on the tilt angle between the probe axis and the direction of the sample plane. This problem together with the effect of variation in the lift-off between sensor and sample has been carefully investigated for flux-gate and SQUID.

In section 2 the three NDE systems are described. A description of the physical and geometrical characteristics of the samples is also given. Section 3 details the characterization of different sensors in terms of tilting effect, lift-off variation effect and signal-to-noise ratio. Section 4 summarizes the conclusions.

2. Experimental set-up

A schematic of the eddy-current NDE system based on HTc-SQUIDs, developed at the INFM laboratory at the University of Naples [10], is shown in Fig. 1. The operating cryogenic temperature of 77 K is reached by liquid nitrogen-bath cooling using a specially designed fiberglass cryostat. The bottom of the cryostat has a window of 12x100 mm^2 with a thickness of 1 mm in the SQUID positioning area. The measuring time is 7 hours at an evaporation rate of 0.7 l/h. Eddy currents are induced in the sample by a differential double D-shaped coil shown in Fig. 2. The geometry chosen for the coil, having a null magnetic field at its center, permits the SQUID to sense a small fraction of the unperturbed excitation field [11]. In this work two different types of coil have been used. The type *2D25* is a 25 mm diameter single-turn Cu coil patterned on a printed circuit board having a width of 1 mm, a thickness of 30 μm and 3 mm of separation between the straight lines of the double-D arrangement. Its overall thickness is 1.5 mm and the relative inductance and DC resistance are 0.5 μH and 0.2 Ω, respectively. The type *2D58* is a 58 mm diameter 53-turn coil based on a 0.10 mm thick Cu wire wound in air. Its overall thickness is 4.2 mm and the inductance and DC resistance are in this case 254 μH and 16.2 Ω respectively. A coil with a larger diameter is required when deeply embedded defects are to be detected, since the optimum diameter is proportional to $f^{-1/4}$, where f is the coil operating frequency [12]. The double-D coil is positioned beneath the stationary cryostat and parallel to the specimen at a stand-off distance z_2 = 10 mm and z_2 = 14.4 mm for the coil type *2D25* and *2D58*, respectively.

The system uses low noise iMAG-3 HTc SQUIDs [13]. Since measurements are performed in an electromagnetic unshielded environment, and in order to reject noise disturbances from distant field sources, the outputs of two SQUIDs are electronically subtracted. In this configuration, called an electronic gradiometer, it is essential to optimize the distance between SQUIDs (defined as baseline) to remove magnetic noise disturbances without subtracting a significant amount of perturbed signal. The optimized baseline used in this work is 29.5 mm.

Due to physical constraints of the probe, the minimum distance between the bottom SQUID and the inducing coil for measuring the B_z component is z_1 = 12 mm. The alignment between the axial gradiometer, located at $x = 0$, $y = 0$, $z = z_1 + z_2$, and the center of the coil, is achieved by moving the latter with a translation system placed just below the cryostat. The AC-source is a programmable low-noise HP3245A wave form synthesizer.

A newly designed non-metallic two-dimensional scanning system, with a sub-mm positioning accuracy, moves the sample beneath the stationary cryostat. The X-Y motors are located 1.5 m away from the inspected sample and are coupled to the positioning system by plexiglas rods. Both data acquisition and X-Y positioning system are computer controlled.

The flux-noise density of our system is less than 0.3 pT/√Hz. The effective field sensitivity for a signal with 6 Hz bandwidth is about 1 pT/√Hz, corresponding to a measured voltage noise density of 0.7 μV/√Hz. The SQUID output is analyzed by an EG&G 5210 dual-channel lock-in amplifier, which demodulates and filters the output with a 6 Hz bandwidth. The operating frequency of the SQUID-based magnetometer ranges from 10 Hz to 20 kHz.

A commercial low-noise three-axial flux-gate, each in an 8x8x25 mm^3 package, has been used in this work to compare the performance of different probes. The electronics of the flux-gate, operating at room temperature, is similar to the one described in Fig. 1 except for the voltage-optical converter which is not required. The magnetic field noise density of the flux-gate in the gradiometric configuration is 20 pT/√Hz, the field sensitivity is 30 pT/√Hz and the voltage noise density is 4.3 μV/√Hz in the measurement range 10 Hz-1 kHz [14]. The optimized baseline of the gradiometer is 30 mm. More details concerning the characteristics of our flux-gate probe are given elsewhere [10]. The coil and the probe can be positioned at the minimum distances z_1 = 1 mm, z_2 = 2 mm and z_2 = 9 mm for the coil type *2D25* and *2D58*, respectively.

The induction coil test was performed using the absolute probe type *PKA 22-6* [15] with a diameter of 38 mm and a thickness of 40 mm, at an operating frequency ranging from 50 Hz to 1 kHz. The maximum voltage sensitivity of this probe is 24 μV at 70 Hz. This particular kind of coil, optimized for operation at low frequency, has been chosen for a detection of deep flaws. It can be positioned on the surface of the specimen at a minimum distance $z_2 = z = 0.3$ mm.

The experimental results shown in the following have been obtained by moving the sample with a speed of 1 mm/s beneath the stationary probes.

The five aluminum-alloy specimens with various artificial defects examined in this work are sketched in Fig. 3. The orientation of the differential inducing coil during the scanning performed by SQUID and flux-gate is described for each test sample. In addition, the physical and geometrical characteristics of the samples are given in Tab. 1. In the same table, the geometrical parameters of the defects affecting those structures as thickness (t), width (w), length (L), diameter (D) and depth (d) are also reported.

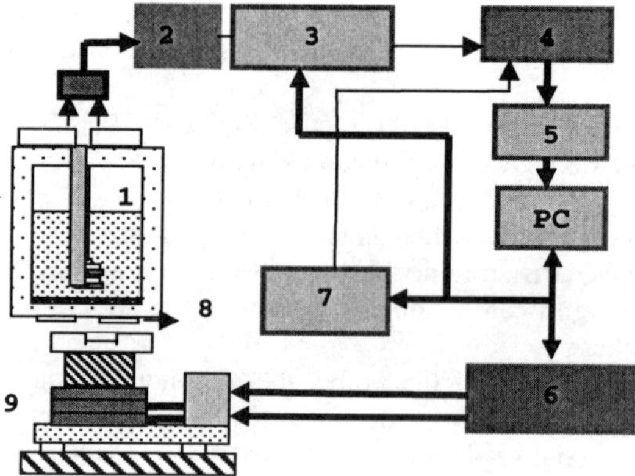

Figure 1. Scheme of the eddy-current SQUID-based NDE prototype. (1) HTc SQUIDs in fiber-glass cryostat; (2)Voltage-optical converter; (3) Optical-voltage converter and amplifier;(4) Lock-in amplifier; (5) 16 bit A/D converter; (6) X-Y controller (7) AC-current source; (8) Double–D shaped induction coil; (9) X-Y translation system.

3. Results and Discussions

3.1 Lift–off variation

The analysis of the signal attenuation due to the variation of lift-off (LO) between sample and probe, has been performed using the sample 1. In order to compare the different quantities recorded as a function of the position for each probe, the peak-to-peak signal amplitude S_{pp} detected at a distance z above the sample has been normalized to its maximum value S_{max} (S_{pp}/S_{max}) for each probe. In Fig. 4 experimental data show a sharper decrease (up to 90 %) of the induction coil signal as a function of LO for values ≥ 5 mm with respect to the curves relative to the flux-gate and SQUID. For these sensors the same signal attenuation is reached for a LO ≥ 15 mm. In case of weak signals generated by small or deep flaws, the strong dependence on the lift-off variation in induction coils can limit the accuracy and the reproducibility of the measurements. Moreover the strong signal attenuation at high lift-off limits the use of these probes in applications where, for example, the conductive layer under examination is covered by a thick electrically insulating layer.

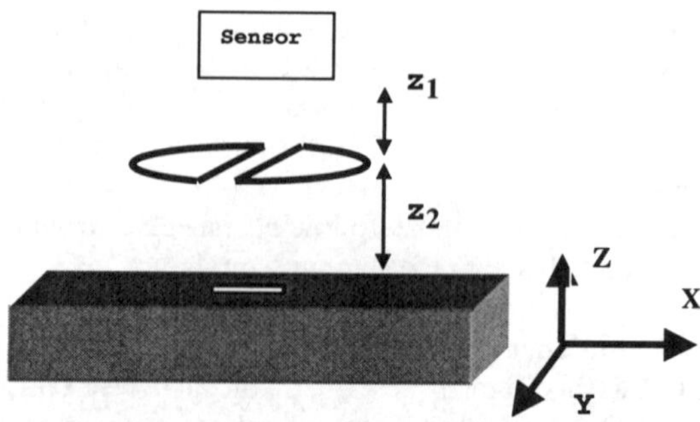

Figure 2. Scheme of the measurement technique for SQUID and flux-gate. The sensor is located at a distance z_2 above the center of the differential double-D shaped coil ($x = 0$, $y = 0$, $z = z_1 + z_2$).

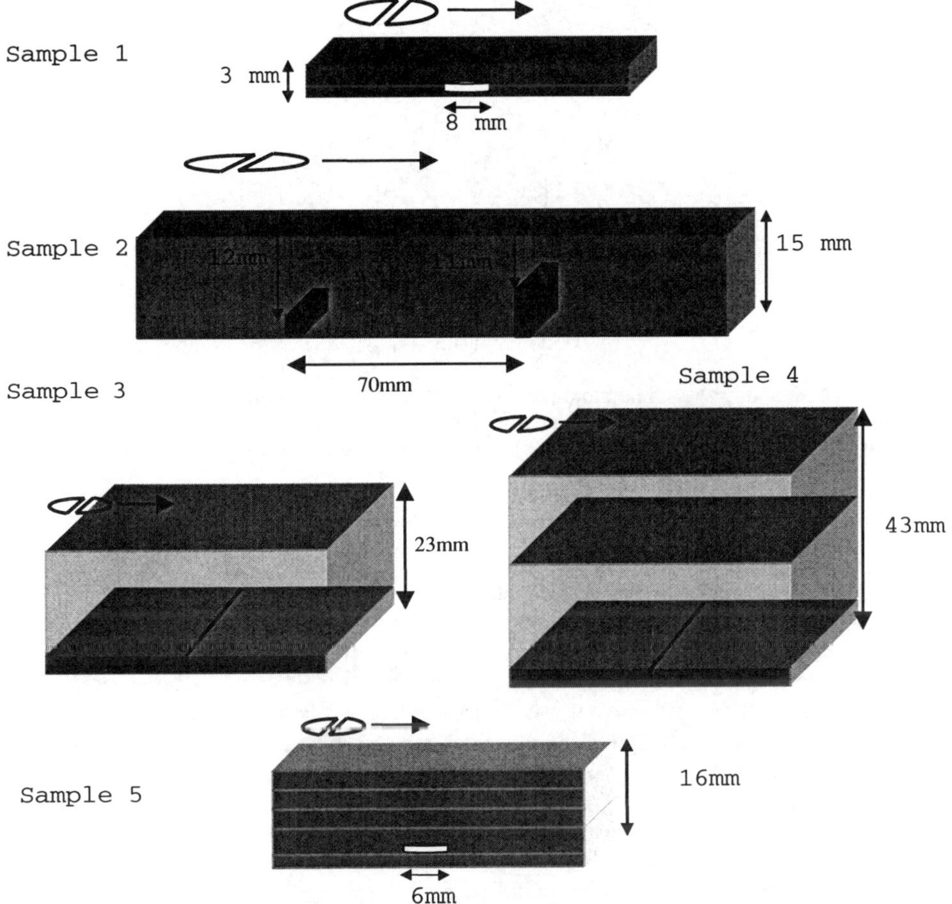

Figure 3. Sketch of the five aluminum-alloy structures with various defect configurations examined in this paper. Drawings are not to scale.

Test	Alloy	σ_0 (MS/m)	Overall Volume (mm^3)	Defect	t (mm)	w (mm)	L (mm)	D (mm)	d (mm)	Note
1	Al7071Ti651	32±20%	100x200x4	Circular hole	1			8	3	1-layer
2	Al7071Ti651	32±20%	150x1000x15	Multiple slots		0.6	40		12	1-layer
3	Anticorodal	56±20%	200x200x23	Crack	3	<0.1	400		20	2-layer
4	Anticorodal	56±20%	200x200x43	Crack	3	<0.1	400		40	3-layer
5	Al7071Ti651	32±20%	100x200x16	Circular hole	1			6	12	5-layer

Table 1. Physical and geometrical characteristics of the 5 aluminum-alloy samples which are shown in Fig. 3. σ_0, t, w, L, D, and d represent the electrical conductivity of the sample, and the thickness, width, length, diameter and depth of the particular defect, respectively.

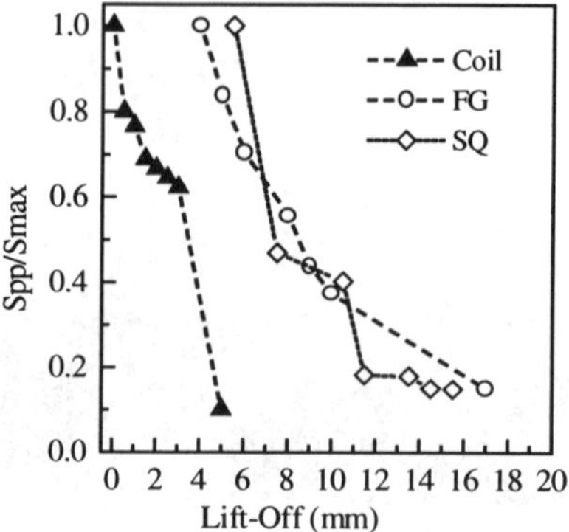

Figure 4. Peak-to peak signal amplitude S_{pp} normalized to its maximum value S_{max} as a function of lift-off between probe and sample. Triangles, circles and diamonds (connected by lines for clarity) represent the experimental data relative to the induction coil, flux-gate and SQUID, respectively.

3.2 Probe tilting

A possible source of noise in EC-NDE measurements involving magnetic sensors is due to the accidental tilting of the gradiometer axis with respect to the direction of the sample plane. In this case the contribution of other components of magnetic field can strongly influence the output signal. In order to evaluate the effect of tilting, the dipole-like magnetic field generated by the *2D25* coil has been used as a well-known electromagnetic source. The AC current feeding the coil was 1 mA at 577 Hz. The probes placed above the coil, in the absence of a sample and at the distances z_1 = 3.2 mm and z_1 = 6.8 mm for the flux-gate and the SQUID, respectively, have been tilted with an angle ranging from 0 to 4° for the first probe and from 0 to 6.5° for the second one. The axial magnetic field component B_z as a function of the line-scan coordinate has been recorded. The relevant quantities, such as the peak-to-peak amplitude S_{pp} and the spatial distance between extrema D [16], as a function of the tilt angle, are shown in Fig. 5.

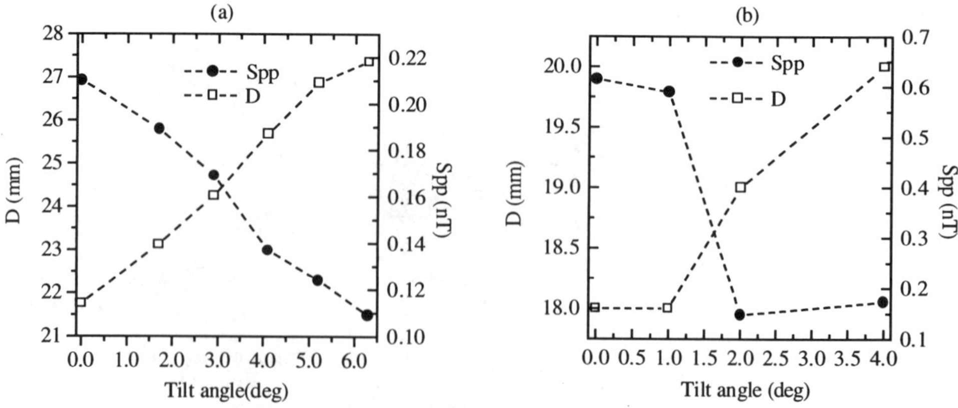

Figure 5. Peak-to-peak value (S_{pp}) of the detected axial component of the magnetic field B_z above the sample 1 (circles) and spatial distance (D) between the extrema of B_z versus line-scan coordinate (squares), as a function of the tilt angle between the probe and the perpendicular to the coil plane. The curves (a) and (b) are relative to SQUID and flux-gate, respectively.

For both sensors S_{pp} increases and D decreases with the tilt angle. The first behavior shows the presence of the non negligible contribution of the transverse components B_x and B_y to the null value B_z which should be detected when there is no misalignment. The shrinking of the extrema in the B_z curve, instead, represents a deformation of the detected magnetic field with increasing tilt angle. This problem can seriously degrade the spatial resolution of the probe. The signal amplitude S_{pp} as well as the parameter D for the flux-gate does not decrease linearly as expected for small tilt angles. This is a clear evidence of the non negligible effect of the large size of such a sensor. The almost linear dependence of both S_{pp} and D on the tilt angle for the SQUID shows that for its small area ($A_{eff} = 0.2\ mm^2$) it can be well approximated as a point sensor. The variation ΔS_{pp} due to a tilting of 1° for the SQUID is 76 pT. It is difficult to perform this experiment for our massive induction coil where a tilt angle of 1° already cancels out the amplitude of the detected signal.

3.3 Frequency dependence

In order to compare the dependence on the eddy-current frequency of the detected signals for the different probes, samples 1 and 2 were chosen. The frequency dependence of the peak-to-peak signal amplitude normalized to its maximum value is shown in Fig. 6 (a) for sample 1. The large output signals of the probes at 600 Hz correspond to a penetration depth of the incident wave of 3.6 mm for the alloy Al7071T6. It is related to the 1 mm defect located at a depth of 3 mm in the plate. The better sensitivity of the SQUID below 600 Hz, with respect to the induction coil, demonstrates its superior sensitivity to probe for defects at larger depths in the aluminum-alloy structures. The coil used in this measurement for both magnetic probes was the type *2D25*. The AC current feeding the coil was 60 mA.

A more detailed comparison of the SQUID and induction coil in the detection of deeper defects as a function of frequency, was performed using sample 2 with 11 mm and 12 mm deep slots. In this case the frequency dependence of the Signal-to-Noise ratio (S/N) is measured in order to take in account the contribution of noise coming from the different probes. The coil used for the magnetic probes was type *2D25*. The operating frequency was 77 Hz and the current amplitude 120 mA for the SQUID and 180 mA for the flux-gate. The frequency used for the induction coil was 70 Hz. In Fig. 6 (b) the S/N of the induction coil and SQUID as a function of the frequency is shown for the two slots.

The curves exhibit a maximum value of S/N at 80 Hz and 110 Hz for induction coil and SQUID respectively. These values correspond to penetration depth $\delta = 8.5$ mm and 9.9 mm in this aluminum-alloy. The disagreement between these values and the defects depths at 11 mm and 12 mm is due to the approximation made in calculating the penetration depth from the skin effect relation in which an incident planar wave is considered. This effect is stronger for deeper defects. In this case a correction factor to δ, dependent on the geometrical parameters of the inducing coil, taking into account a different wave propagation is required. These experimental data also show significantly larger S/N values for SQUID as compared to the induction coil for the frequency range 60 Hz-200 Hz. This demonstrates the superior performance of the SQUID in detecting defects at depth higher than 10 mm in a single-layer structure. Moreover, the measurements performed on the sample 1 and 2 show that the frequency response for the SQUID, in the bandwidth 60 Hz-1 kHz and 60 Hz-200 Hz, is flatter than the one of the other probes. This is essential in order to get a more uniform response in the detection of both surface and deep embedded flaws. It is worthwhile noticing that, at low frequency, the SQUID sensitivity and S/N is superior to that of the flux-gate and the induction coil, in spite of the larger lift-off ($z = 22$ mm for the SQUID and 3 mm and 0.3 mm for the flux-gate and the coil, respectively).

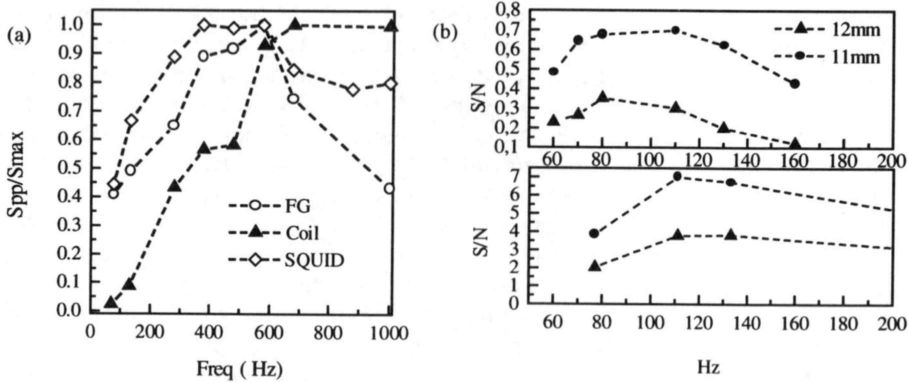

Figure 6 (a)-(b). Dependence of the peak-to-peak signal amplitude S_{pp} on the eddy current frequency, normalized to its maximum value, for sample 1 (a). The circles, triangles and squares (connected by lines) correspond to measurements performed by the flux-gate, induction coil and SQUID, respectively. Dependence on frequency of the signal-to-noise ratio (S/N) (b) of the induction coil (above) and SQUID (below) for sample 2. Triangles, and circles are relative to the 12 mm and 11 mm deep slot, respectively.

3.4 Detection of deep flaws

To compare the probe sensitivity to deep flaws, output signals as a function of the line-scan coordinate were recorded for samples 2-4. Each of the structures used in this study and described in Fig. 1 was chosen to test the sensitivity of the sensors in different circumstances. Sample 2 gives information on the detection of multiple slots at depths greater than 10 mm in a single-layer structure. Measurements on samples 3 and 4 give information on the detection of a long crack in a three-layer structure at even larger depths (20 mm and 40 mm), which is severely testing the sensitivity of the probes, as shown below. Finally, sample 5 is used for the detection of a millimeter-sized defect in a 5-layer structure at depths larger than 10 mm. The presence of many layers and of insulating gaps between them results both in a significant reduction of the conductivity of the whole structure and in complex wavefields reflected at the interfaces. The attenuation of the output signal can be significant in this case.

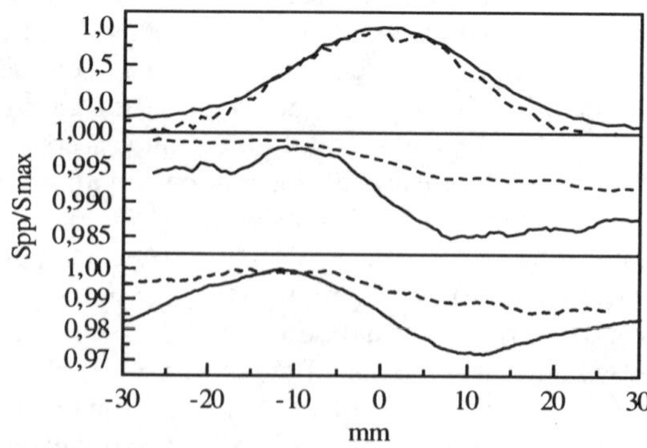

Figure 7. Peak-to peak signal amplitude S_{pp} normalized to its maximum value S_{max} as a function of the line-scan coordinate for sample 2. Measurements of the 12 mm deep (dashed line) and 11 mm deep (full line) slots are shown for induction coil (above), flux-gate (middle) and SQUID (below).

All the S/N data discussed are summarized in Tab. 2. In figure 7 (see above) one compares the sensitivity of the three probes to slots 11 mm and 12 mm deep in sample 2. The inducing coil used here for the magnetic probes was the type *2D25* at an operating frequency of 77 Hz, and a current amplitude of 120 mA (SQUID) and 180 mA (flux-gate). The inductive coil frequency was 70 Hz. The S/N ratios of the measurements of the two slots were 4.20 and 1.95 for the SQUID, 2.46 and 0.64 for the flux-gate, 0.65 and 0.25 for the induction coil. This result exemplifies the higher sensitivity of the SQUID when compared to the other probes, in agreement with the discussion of the previous section on the frequency-dependent measurements of this sample.

Figure 8 shows the dependence of the magnetic field B_z on the line-scan coordinate measured by the flux-gate with the inducing coil *2D25* (Fig. 8(a)) and *2D58* (Fig. 8 (b)) at 133 Hz and 180 mA for sample 3. The larger magnetic field generated by the larger diameter coil results in a S/N = 0.92 which should be compared with S/N = 2.07 for the smaller diameter coil. In addition, the spread of the magnetic field in the area of the sample around the defect (located at 0 mm) is greater in the first measurement than in the second one. At a higher frequency (1 kHz), when the coil type *2D25* is used, the different eddy-current distribution in the sample results in extremely large peaks relative to the edge-effect of the plate in the measurement. This effect is complicating the detection of defects.

The signal outputs of the induction probe at 70 Hz, measured along both a line-scan across the defect in sample 3 and across a free-defect area of an equivalent sample, are shown in Fig. 8 (c). In this case S/N = 3.49. The measurement of B_z performed on sample 3 by the SQUID (Fig. 8 (d)) with an inducing coil type *2D25* at 111 Hz and 120 mA, results in a S/N = 2.54. These results suggest that an inducing coil of large diameter increases the sensitivity, which is what is aimed at for an accurate detection of deep flaws. Adversely, this results in a noticeable reduction of the spatial resolution of the measurements. In short, this highlights the trade-off between sensitivity and spatial resolution which applies generally for all probes. The detection of the very deep crack in sample 4 was performed by the flux-gate and SQUID with a coil type *2D58* at 13 Hz and 180 mA. The relative values of the signal-to-noise ratio are S/N = 0.56 and S/N = 2.78. The induction coil gave no measurable signal for this sample.

The measurements on the last structure studied (sample 5, with insulating gaps between the stacked layers) are summarized in Fig. 9 (a)-(d).The experimental results for the flux-gate with the inducing coil *2D25* at 133 Hz and 77 Hz and 180 mA are shown in Fig. 9 (a). The output signal in this case did not reveal the defect even after lowering the frequency. After removing the upper layer, the probe detected a signal with a S/N = 1.85 at 133 Hz and 180 mA (Fig. 9 (b)). The induction coil operating at 70 Hz detected the hole in sample 5 with a S/N = 0.16. The response improved upon removal of one and two upper layers, as shown in Fig. 9 (c). Finally, Fig. 9 (d) shows the SQUID response with the inducing coil *2D25* at 120 mA for three different frequencies (133 Hz, 177 Hz and 577 Hz). The highest signal-to-noise ratio S/N = 3.573 is obtained at 133 Hz. It decreased slightly at 177 Hz and, at 577 Hz, no clear signal is observed due to the different eddy-current distribution in the plates. In this challenging structure the superior performance of SQUID is clearly demonstrated.

Figure 10 summarizes the S/N ratios as function of defects depth for the above described measurements.

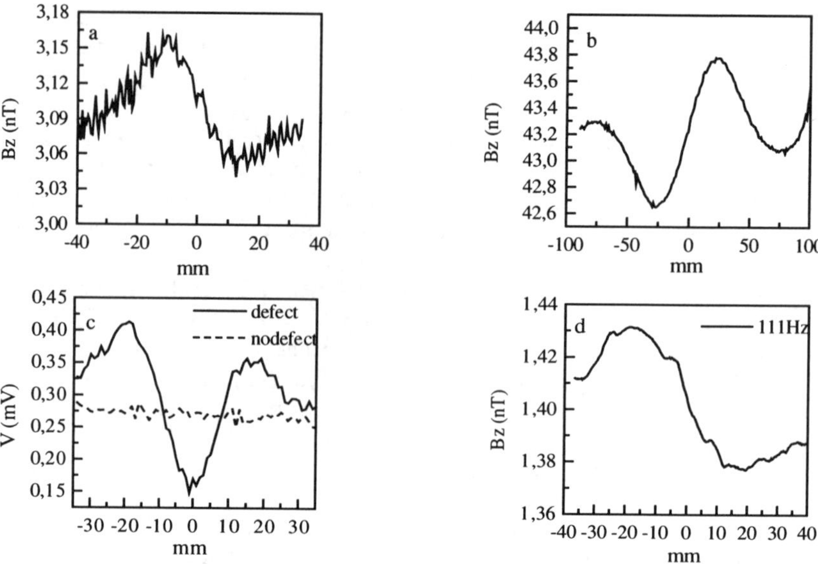

Figure 8 (a)-(d). Dependence on the line-scan coordinate of the output signals for flux-gate with the inducing coil *2D25* (a) and *2D58* (b) at 133 Hz and 180 mA; inductive probe at 70 Hz (c) and SQUID with the coil *2D25* at 111 Hz and 120 mA (d).

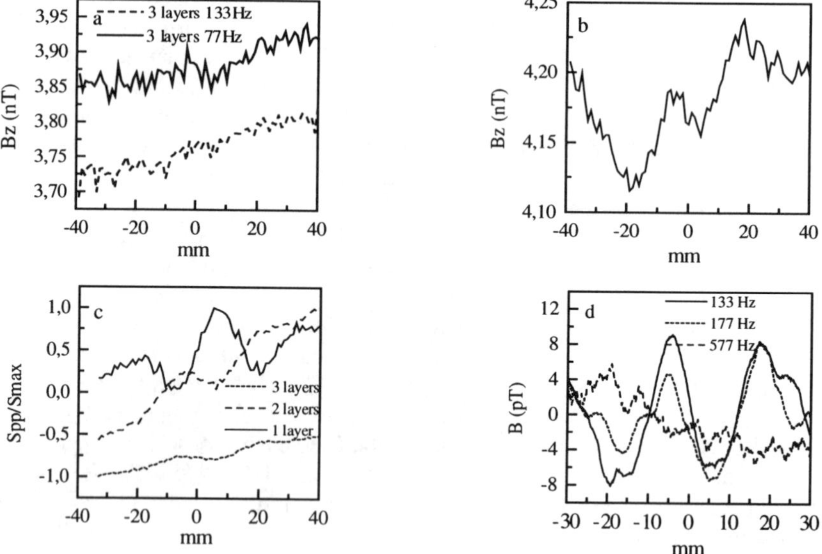

Figure 9 (a)-(d). Experimental results for the flux-gate with the inducing coil *2D25* at 133 Hz and 77 Hz and 180 mA in sample 5 (a) and after removing the upper layer at 133 Hz (b); induction coil at 70 Hz in sample 5 (3-layer) and after removing one and two upper layers (c); SQUID response with the inducing coil *2D25* at 120 mA for three frequencies (133 Hz, 177 Hz, and 577 Hz) (d).

Table 2. a summary of the S/N ratios for the different probes in the examined samples

Sample	1	2	3	4	5
Coil S/N	2,58-1kHz 2.38-580 Hz 2.34 -70 Hz	0.65 -70Hz (11mm) 0.25 (12 mm)	0.92 70Hz	-	0.164 70Hz
FG S/N	9.16 -1kHz 6.97-580Hz 2.84 -77Hz	2.46 -77Hz (11mm) 0.64 (12 mm)	3.59 133Hz* 2.07 133Hz	0.56 13Hz *	-
SQUID S/N	2.621-1kHz 5.21-580Hz 2.19-77Hz	4.20 -77Hz (11mm) 1.95 (12 mm)	2.54 133Hz	2.78 13Hz *	3.573 133Hz

(*) coil type *2D58*

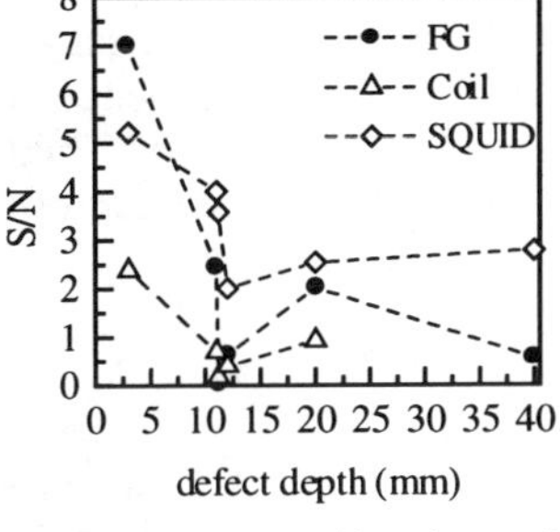

Figure 10. Signal-to-noise ratio S/N versus the defects depth.

4. Conclusion

An experimental characterization based on a quantitative comparison between different eddy-current probes has been carried out. In particular, the performance of an induction coil, a flux-gate and a HTc-SQUID-based magnetometer has been investigated on the basis of comparative experiments on five Al-alloy test samples with artificial defects such as holes, slots and cracks.

The effect of larger probe lift-off distances (representing, e.g., an insulating layer at the surface of the structure) is less negative in the case of the SQUID and the flux-gate. Studies of the effect of a probe tilt angle reveal that the SQUID can be considered, to a very good extent, as a point sensor, which is simplifying both the measurements and the numerical simulations.

Measurements of the frequency dependence of the S/N response demonstrate the superior performance of the SQUID and the flux-gate in detecting subsurface and deep defects at depths greater than 10 mm in a single-layer structure. Moreover, the flatter response up to 1 kHz for the SQUID is desirable for the detection of both surface and deeply embedded flaws.

The overall superiority of the SQUID sensor can be appreciated from the summarizing Tab. 2 where it is seen that, while it always equally or more sensitive than the other sensors in each individual case, it is the only sensor which succeeds in identifying the defects in all structures examined here. In the presence of voids and insulating layers in multi-layered samples, the conductivity is significantly reduced and wavefield reflections at the interface become quite complex. As demonstrated with sample 5, the SQUID is the only sensor capable of detecting these anomalies with sufficient sensitivity.

Emphasize finally that this study has compared the performance of intrinsically different probes in their optimized operating conditions. This has required optimization of the relevant parameters, for example, z_1, z_2, f and I, so as to maximize the S/N response for each probe rather than rigorously keeping the same values for all three probes.

Acknowledgements

We are grateful to D. Lesselier and V. Monebhurrun for many helpful discussions, and to M. Prencipe for useful suggestions on the induction coil measurements. A. Maggio and S. Marrazzo are acknowledged for their technical support. This work has been carried out in the frame of the project: "Eddy Current Non-destructive Evaluation Using Superconducting Devices" with the support of the National Institute for Matter Physics (INFM).

References

[1] H. Weinstock and M. Nisenoff, "Non destructive evaluation of metallic structures using a SQUID gradiometer," in H. D. Halbohm and H. Lubbig (eds), *SQUID'85*, de Gruyter, (1985) 853-858 .

[2] S. Pagano *et al.*, "HTc SQUID for non destructive evaluation," in *Electromagnetic Nondestructive Evaluation (II)*, R. Albanese *et al.* (eds.), IOS Press, Amsterdam, (1998) 206-214.

[3] A. Cochran *et al.*, "SQUID systems for non destructive testing by AC field mapping," *IEEE Trans. Appl. Superconductivity* **3** (1993) 1926-1929.

[4] Y. Tavrin *et al.*, "A second order SQUID gradiometer operating at 77K," *Supercond. Sci. Technol.* **7** (1994) 265-268.

[5] M. Mück *et al.*, "Eddy current non destructive evaluation based on HTS SQUIDs," *Physica C* **282-287** (1997) 407-410.

[6] Y. Pei Ma and J. P. Wikswo Jr., "Depth-selective SQUID eddy current techniques for second layer flaw detection," in *Review of Progress in Quantitative Nondestructive*, D. O. Thompson and D. E. Chimenti (eds.), 401, Plenum Press, New York, (1996).

[7] A. Ruosi *et al.*, "Experimental and numerical results of electromagnetic non-destructive testing with HTc SQUIDs," submitted to ASC Conference 1998, to *appear IEEE Trans. Appl. Sup erconductivity* (1998).

[8] V. Monebhurrun *et al.*, "Eddy current non-destructive evaluation using SQUIDs," to appear in *Electromagnetic Nondestructive Evaluation (III), D. Lesselier et A. Razek* (eds.), IOS Press, Amsterdam, (1999).

[9] J. Pávó *et al.*, "Proposal for benchmark problem qualifying some aspects of the performance of ECT probes," in *Electromagnetic Nondestructive Evaluation (II)*, R. Albanese *et al.* (eds.), IOS Press, Amsterdam, (1998) 337-342.

[10] A. Ruosi *et al.*, "Eddy-current non-destructive measurements using SQUIDs: preliminary results," in *Electromagnetic Nondestructive Evaluation (II)*, R. Albanese *et al.* (eds.), IOS Press, Amsterdam, (1998) 215-224.

[11] M. Valentino. *et al.*, "Eddy-current non-destructive measurements with different HTc-SQUID spatial orientations," *Journal de Physique IV* (1998) 249-252.

[12] P. Burrascano *et al.*, "Physical modelling applied to eddy current non-destructive testing," in *Electromagnetic Nondestructive Evaluation (II)*, R. Albanese *et al.* (eds.), IOS Press, Amsterdam, (1998) 24-30.

[13] Conductus Inc., Sunnyvale, California 94086, USA. Low-noise HTc-SQUID, IMAG-3.

[14] Bartington Instrument Inc., 087XGE Oxford England. Low-noise three-axis flux-gate MAG-3.

[15] Rohmann, Germany. System ELOTEST-B1-SDM; probe PKA 22-6

[16] G. Pepe *et al.*, "SQUID based NDE: a comparison between experimental data and numerical FEM modeling," to appear in *Review of Progress in QNDE* Vol. 17,. D. O. Thompson and D. E. Chimenti (eds.), Plenum Press, New York, (1998) in press.

Electromagnetic Nondestructive Evaluation (III)
D. Lesselier and A. Razek (Eds.)
IOS Press, 1999

Eddy current non-destructive evaluation using SQUIDs

V. Monebhurrun, D. Lesselier and B. Duchêne

Département de Recherche en Électromagnétisme
Laboratoire des Signaux et Systèmes, CNRS-SUPÉLEC
Plateau de Moulon, 91192 Gif-sur-Yvette Cedex - France

A. Ruosi†, M. Valentino‡, G. P. Pepe† and G. Peluso†

†Dipartimento di Scienze Fisiche
‡Dipartimento di Ingegneria dei Materiali per la Produzione,
Istituto Nazionale per la Fisica della Materia (INFM), Unita' di Napoli
Università di Napoli "Federico II", Piazzale Tecchio 80, 80125 Napoli - Italy

Abstract. Eddy current non destructive evaluation of planar stratified structures such as those encountered in the aircraft industry is investigated. A numerical model is proposed for the development of a SQUID-based measuring system. Here, we focus onto the forward modelling of the interaction of a probing signal with a 3-D bounded defect in a homogeneous, non-magnetic and isotropic metallic slab; only non-magnetic sources are considered (typically, a double-D shaped coil) and the SQUIDs are assumed to be point sensors. Anomalous fields in the presence of the defect are cast within a vector wavefield integral equation framework. Appropriate dyadic Green's functions of the planar structure are defined. A Method of Moments is applied and the forward problem is tackled by a Conjugate-Gradient Fast Fourier Transform (CG-FFT)-based algorithm. To deal with our specific configuration where the probe and the sensor are displaced simultaneously, several approximations are investigated, e.g., the Localized NonLinear Approximation. Numerical simulation results are compared with results obtained with a finite-element model and with experimental data.

1 Introduction

Recent advances on SQUIDs (Superconducting Quantum Interference Devices) show that measuring systems based on such devices are potentially attractive for non-destructive evaluation applications [1]. Compared to traditional eddy current measuring devices, SQUID-based sensors offer higher spatial resolution and greater sensitivity which make them particularly suited for the detection of surface as well as sub-surface defects. Although rather bulky for certain specific applications (the inspection of thin steam generator tubes, for example [2]), a portable measuring system for the detection of defects affecting planar structures is a practical application of SQUIDs.

For the configuration considered herein (see figure 1), the SQUID is mounted on the top of a double-D shaped exciting coil and positioned at a given lift-off just above the centre of the coil where, in the absence of defects, the vertical component of the magnetic field is cancelled. Obviously, a high sensitivity is required from the measuring system since for extreme defect configurations, the eddy current perturbation can be quite low (imperceptible with a fluxgate magnetometer, for example). In a typical measurement set-up, both coil and sensor are displaced within a plane parallel to the structure, resulting in a surface scan of the magnetic field which is characteristic of the defect.

A numerical model is essential to fully understand the electromagnetic phenomena, after which optimisation of the pertinent parameters of the measuring system becomes possible. The simultaneous displacement of the double-D coil and of the SQUID entails a high computational cost and hinders a finite-element analysis whereby a new mesh is usually required for each position of the measuring system. But the vector wavefield integral formulation which we adopt, proves to be appropriate for such a configuration. Although we restrict our study to planar isotropic structures, for which adequate dyadic Green's functions have to be derived, the same analysis can be easily extended to handle anisotropic structures.

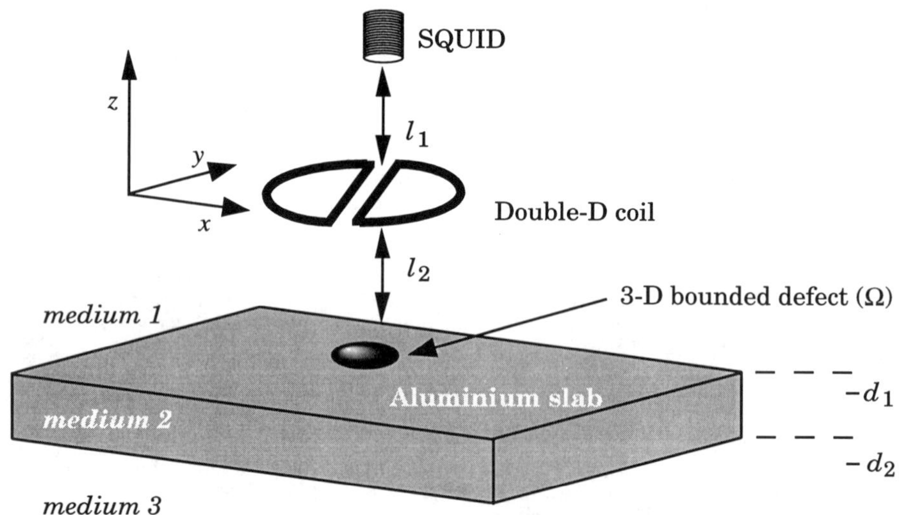

Figure 1. Configuration under study.

The wavefield integral modelling implies two Fredholm's vector integral equations [3]: the first is a coupling equation that links, via a dyadic Green function [4], the induced current density to an unknown fictitious current density (Huygens-type sources) in the flawed region. The second is an observation equation which, once the fictitious current density above is known, gives the anomalous magnetic field via an other dyadic Green function. Clearly, the derivation of the Green functions is a crucial step in the wavefield formulation; to derive these functions for the planarly-layered structure, we choose a vector wave (or eigenfunction) expansion method [4-5] and develop the electromagnetic field into uncoupled transverse electric (*TE*) and transverse magnetic (*TM*) fields after taking the two-dimensional spatial Fourier transform of the in-plane $((x, y))$ components of the fields (i.e., a spectral decomposition of the fields).

After discretisation of the integral equations using a method of moments [6] with a pulse-basis point-matching scheme, we employ a conjugate gradient algorithm [7] to solve for the fictitious current density. 2-D Fast Fourier Transforms [8] are applied during the computation since the kernels of the integral equations exhibit a convolutional nature. The use of Fourier transforms also brings considerable simplification when accounting for the simultaneous displacement of the SQUID and the source. But obviously, some appropriate approximation has to be devised in view of solving the inverse problem, because for a typical surface scan, the computational cost is still important.

The working frequency (typically, a few hundred Hertz) suggests the use of the Born approximation which yields a straightforward solution for the anomalous magnetic field. However a better estimate is obtained if the quasi-static limit is assumed whereby the normal component of the induced current density is neglected [9] and the forward problem is solved only for the corresponding in-plane components (i.e., the depolarisation of the field is restricted to the plane of the structure only). As for the so-called localized nonlinear approximation [10-11], which is similar to the Born approximation in the sense that the solution of the coupling equation via the conjugate gradient scheme is by-passed, it is expected to provide satisfactory results for the configuration considered herein.

The paper is organised as follows. In section 2 we present the wavefield integral formulation. The expressions for the dyadic Green functions are derived in section 3 (developments leading to the final expressions are provided in appendix). The solution of the forward problem using a conjugate gradient scheme is discussed in section 4 and the different approximations mentioned above are introduced in section 5. Comparisons of numerical simulation results with experimental data follow in section 6.

2 The vector volume integral formulation

We consider a planar stratified structure, as shown in figure 1, where media 1 and 3 consist of air (with electromagnetic parameters defined as (ε_0, μ_0)) while medium 2 is a highly conductive (18.1 *MS/m*), isotropic, linear, homogeneous and non-magnetic metal (with electromagnetic parameters defined as (σ_0, ε_0, μ_0)). A defect in this metal occupies a volume Ω and is characterised by electromagnetic parameters (σ, ε_0, μ_0). Under harmonic excitation of frequency ω, the source (a double-D coil in practice) generates eddy currents in the metallic structure. In the presence of a defect, the eddy current pattern is perturbed and anomalous fields arise, which are measured with a SQUID-based magnetic sensor.

The derivation of the Fredholm second-kind vector integral equation is straightforward after application of Green's theorem to Maxwell's equations. Assuming a time-harmonic excitation $\exp(-j\omega t)$, this integral equation, known as the coupling equation, reads:

$$\mathbf{E}_2^{inc}(\mathbf{r}) \;=\; \mathbf{E}_2(\mathbf{r}) \;-\; j\omega\mu_0 \int_\Omega \overline{\mathbf{G}}_{22}^{ee}(\mathbf{r},\, \mathbf{r}')\; \mathbf{J}_2(\mathbf{r}')d\mathbf{r}' \tag{1}$$

where the fictitious current density $\mathbf{J}_2(\mathbf{r})$ is equal to $\sigma_0\chi(\mathbf{r})\;\mathbf{E}_2(\mathbf{r})$, letting $\chi(\mathbf{r})$ be the conductivity contrast function $[\sigma(\mathbf{r})/\sigma_0 - 1]$ (for the particular case of void-type defects, it reduces to -1 inside the volume of the defect and to 0 outside). $\mathbf{E}_2^{inc}$ and $\mathbf{E}_2$ are respectively the incident and the total electric fields in medium 2. $\overline{\mathbf{G}}_{22}^{ee}$ is the electric-electric dyadic Green function which transforms a point source of electric current in region 2 into the electric field it generates in the same region.

More generally, the dyadic Green function $\overline{\mathbf{G}}_{lm}^{ee}(\mathbf{r},\, \mathbf{r}')$, $(l = 1,\, 2,\, 3;\; m = 1,\, 2,\, 3)$ is the electric field observed at point $\mathbf{r}$ in medium l due to a point source of electric current at point $\mathbf{r}'$ in medium m. It obeys the Helmholtz equation:

$$\nabla \times \nabla \times \overline{\mathbf{G}}_{lm}^{ee}(\mathbf{r},\, \mathbf{r}') \; - \; k_l^2(\mathbf{r})\, \overline{\mathbf{G}}_{lm}^{ee}(\mathbf{r},\, \mathbf{r}') \; = \; \overline{\mathbf{I}}\, \delta(\mathbf{r}-\mathbf{r}')\, \delta_{lm} \tag{2}$$

where $\overline{\mathbf{I}}$ is the unit dyad, $\delta(\mathbf{r})$ is the Dirac delta distribution, δ_{lm} is the Kronecker symbol and k_l is the wave propagation constant in medium l defined as $k_l^2(\mathbf{r}) = \omega^2 \mu_0 \varepsilon_l$, where $\varepsilon_1 = \varepsilon_3 = \varepsilon_0$ and $\varepsilon_2 = \varepsilon_0 + j\sigma_0/\omega$. The incident electric field in region 2 is obtained by applying the dyadic Green function $\overline{\mathbf{G}}_{21}^{ee}$ to the current density $\mathbf{J}_{probe}$ carried by the exciting coil:

$$\mathbf{E}_2^{inc}(\mathbf{r}) \; = \; j\omega\mu_0 \int_{probe} \overline{\mathbf{G}}_{21}^{ee}(\mathbf{r},\, \mathbf{r}')\, \mathbf{J}_{probe}(\mathbf{r}')\, d\mathbf{r}' \tag{3}$$

Using the same procedure as above, we derive an equation similar to Equ. (1) for the magnetic flux density observed in medium 1, the so-called observation equation:

$$\mathbf{B}_1^{obs}(\mathbf{r}) \; = \; \mathbf{B}_1^{inc}(\mathbf{r}) \; + \; \mu_0 \int_{\Omega} \overline{\mathbf{G}}_{12}^{me}(\mathbf{r},\, \mathbf{r}')\, \mathbf{J}_2(\mathbf{r}')\, d\mathbf{r}' \tag{4}$$

where $\overline{\mathbf{G}}_{12}^{me}$ is the magnetic-electric dyadic Green function which transforms a point source of electric current in region 2 into its associated magnetic field observed in region 1. Again, the incident magnetic flux density is obtained by applying the dyadic Green function $\overline{\mathbf{G}}_{11}^{me}$ to the current density carried by the probe:

$$\mathbf{B}_1^{inc}(\mathbf{r}) \; = \; \mu_0 \int_{probe} \overline{\mathbf{G}}_{11}^{me}(\mathbf{r},\, \mathbf{r}')\, \mathbf{J}_{probe}(\mathbf{r}') d\mathbf{r}' \tag{5}$$

Expressions for $\overline{\mathbf{G}}_{11}^{me}$, $\overline{\mathbf{G}}_{12}^{me}$, $\overline{\mathbf{G}}_{21}^{ee}$ and $\overline{\mathbf{G}}_{22}^{ee}$ are derived below.

3 The dyadic Green function

Owing to the inherent singularity (which occurs when $\mathbf{r} = \mathbf{r}'$) of the Green function [12 - 13], it is often desirable to split this function into a singular term $\delta_{lm}\, \overline{\mathbf{G}}_{lm}^{(s)}$ and a reflection term $\overline{\mathbf{G}}_{lm}^{(r)}$. For most problems of practical interest, analytical expressions for the reflection term exist only in the spectral domain (associated to one or more variables of the spatial coordinates) while expressions for the singular term can be derived indifferently in the spatial and in the spectral domain. However to avoid numerical convergence problems, the singular term, which corresponds to the electric (or magnetic) field response of a point source of electric current in a homogeneous medium, is better described in the spatial domain and its expression is easily derived from classical textbooks. Here we only consider the reflection term $\overline{\mathbf{G}}_{lm}^{(r)}$.

The analysis which follows is inspired from [4-5] and is based on the vector wave function or eigenfunction expansion of the dyadic Green function. It is beyond the scope of this paper to dwell on it and the reader should refer to these two references for further details. In simple terms, the electromagnetic field is developed into two uncoupled transverse electric (*TE*) and transverse magnetic (*TM*) vector wave functions denoted as $\mathbf{\Pi}$ and $\mathbf{\Gamma}$ respectively. The basis sets formed by each of these two functions are then employed to expand the solution of the Helmholtz equation, i.e. the dyadic Green function $\overline{\mathbf{G}}_{lm}^{ee(r)}$ or $\overline{\mathbf{G}}_{lm}^{me(r)}$ is expressed as a plane wave expansion (or spectral decomposition) of the electromagnetic field:

$$\overline{\mathbf{G}}_{lm}^{ee/me(r)}(\mathbf{r},\, \mathbf{r}') \; = \; \frac{1}{(2\pi)^2} \int_{-\infty}^{\infty} \int_{-\infty}^{\infty} d\mathbf{k}_s \left[\overline{\mathbf{\Gamma}}_{lm}^{ee/me}(\mathbf{k}_s,\, \mathbf{r},\, \mathbf{r}') + \overline{\mathbf{\Pi}}_{lm}^{ee/me}(\mathbf{k}_s,\, \mathbf{r},\, \mathbf{r}') \right] \tag{6}$$

where $d\mathbf{k}_s = dk_x dk_y$, and the $\mathbf{\Pi}$ and $\mathbf{\Gamma}$ vectors (each representing a three-component electric or magnetic field for the *TE* and *TM* modes) have been transformed into their associated $\overline{\mathbf{\Pi}}$ and $\overline{\mathbf{\Gamma}}$ tensors, respectively, to account for the nine components of the dyadic Green function (each column of either tensor represents a three-component electric or magnetic field for one of the three possible independent polarizations of the source). These tensors read:

$$\overline{\mathbf{\Gamma}}^{ee}_{lm}\,(\mathbf{k}_s,\ \mathbf{r},\ \mathbf{r}') \ = \ \frac{j}{2\ k_{mz}\ k_s^2}\ (\nabla\times\mathbf{z})(\nabla'\times\mathbf{z})\ \mathrm{e}^{j\mathbf{k}_s\cdot(\mathbf{r}_s-\mathbf{r}'_s)}\ \Phi^{TE}_{lm}(z,\ z') \tag{7}$$

$$\overline{\mathbf{\Pi}}^{ee}_{lm}\,(\mathbf{k}_s,\ \mathbf{r},\ \mathbf{r}') \ = \ \frac{j}{2\ k_{mz}\ k_s^2}\ \frac{(\nabla\times\nabla\times\mathbf{z})(\nabla'\times\nabla'\times\mathbf{z})}{\omega^2\ \varepsilon_l\ \mu_0}\ \mathrm{e}^{j\mathbf{k}_s\cdot(\mathbf{r}_s-\mathbf{r}'_s)}\ \Phi^{TM}_{lm}(z,\ z') \tag{8}$$

with $\mathbf{r}_s = (x,\ y)$, ${k_s}^2 = {k_x}^2 + {k_y}^2$, ${k_{mz}}^2 = {k_m}^2 - {k_s}^2$ and ${k_{lz}}^2 = {k_l}^2 - {k_s}^2$. The magnetic counterparts $\overline{\mathbf{\Gamma}}^{me}_{lm}$ and $\overline{\mathbf{\Pi}}^{me}_{lm}$ are deduced from $\overline{\mathbf{\Gamma}}^{ee}_{lm}$ and $\overline{\mathbf{\Pi}}^{ee}_{lm}$, respectively, by applying the curl operator. The roots k_{mz} and k_{lz} are taken with positive imaginary parts to satisfy the radiation conditions.

The scalar wave functions $\Phi^{TE/TM}_{lm}$ describe, for each mode, the well-known plane wave propagation along the vertical axis (z) in a three-layer medium and they account for the electromagnetic phenomena which occur at the different interfaces of the structure. For the configuration under study, these wave functions are as follows (the superscript *TE* or *TM* is dropped for clarity but it is understood that the wave function depends on the mode of propagation):

$$\Phi_{11}(z,\ z') \ = \ \left[\mathrm{e}^{\ jk_{1z}(z'+2d_1)}\ \tilde{R}_{12}\right]\ \mathrm{e}^{\ jk_{1z}z} \tag{9}$$

where $\tilde{R}_{12}$ is known as a generalised reflection coefficient and is easily deduced from the local (Fresnel) reflection and transmission coefficients. It is given by:

$$\tilde{R}_{12} \ = \ R_{12} \ + \ T_{12}\ R_{23}\ T_{21}\ \mathrm{e}^{\ 2jk_{2z}(d_2-d_1)}\ \tilde{M}_2 \tag{10}$$

where $\tilde{M}_2$ accounts for multiple reflection in medium 2 and is given by:

$$\tilde{M}_2 \ = \ \frac{1}{[1\ -\ R_{21}R_{23}\mathrm{e}^{\ 2jk_{2z}(d_2-d_1)}]} \tag{11}$$

and where

$$R^{TE}_{lm} = \frac{k_{lz}-k_{mz}}{k_{lz}+k_{mz}},\ R^{TM}_{lm} = \frac{\varepsilon_m k_{lz}-\varepsilon_l k_{mz}}{\varepsilon_l k_{mz}+\varepsilon_m k_{lz}},\ T^{TE}_{lm} = \frac{2\ k_{lz}}{k_{lz}+k_{mz}},\ T^{TM}_{lm} = \frac{2\ \varepsilon_m k_{lz}}{\varepsilon_l k_{mz}+\varepsilon_m k_{lz}}.$$

As for the three other functions, they read:

$$\Phi_{21}(z,\ z') \ = \ \left[\mathrm{e}^{-\ jk_{2z}z} \ + \ \mathrm{e}^{\ jk_{2z}(z+2d_2)}\ R_{23}\right]\ \mathrm{e}^{-\ jk_{2z}d_1}\ T_{12}\ \mathrm{e}^{\ jk_{1z}d_1}\ \mathrm{e}^{\ jk_{1z}z'}\ \tilde{M}_2 \tag{12}$$

$$\Phi_{12}(z,\ z') \ = \ \mathrm{e}^{\ jk_{1z}(z+d_1)}\ T_{21}\ \mathrm{e}^{\ -\ jk_{2z}d_1}\ \left[\mathrm{e}^{-\ jk_{2z}z'} \ + \ \mathrm{e}^{\ jk_{2z}(z'+2d_2)}\ R_{23}\right]\ \tilde{M}_2 \tag{13}$$

and

$$\begin{aligned}\Phi_{22}(z,\ z') \ = \ & \Big[R_{21}\ \mathrm{e}^{\ jk_{2z}(-z-z'-2d_1)} \ + \ R_{23}\ \mathrm{e}^{\ jk_{2z}(z+z'+2d_2)} \\ & +\ \mathrm{e}^{\ 2jk_{2z}(d_2-d_1)}\ R_{21}\ R_{23}\ \left(\mathrm{e}^{\ jk_{2z}(z-z')} \ + \ \mathrm{e}^{\ jk_{2z}(-z+z')}\right)\Big]\ \tilde{M}_2\end{aligned} \tag{14}$$

4 The forward problem

For the numerical processing of the wavefield integral equations, we adopt a method of moments [6] to obtain their discrete counterparts. A simple pulse basis point-matching scheme is employed, which means that the electromagnetic fields, the current densities and the flaw parameters are supposed to be constant valued over an elementary voxel of size $\Delta x \times \Delta y \times \Delta z$ chosen such that each side of the voxel has a length of about $\delta/7$ (where δ is the skin depth), a size considered as optimal for our purpose here [2].

Analytical expressions for the triple integration implied by the kernel of the integral equations exist only for the reflection term of the Green function. A Gaussian quadratic numerical scheme [14] is therefore employed for the evaluation of the triple integration in the case of the singular term. Regarding the reflection term, special care is to be taken when handling the spectra of the different Green functions. Indeed, one is either faced with slow convergence ($\overline{\mathbf{G}}_{22}$) or fast decay ($\overline{\mathbf{G}}_{21}$) of the spectrum. The use of classical Fast Fourier Transforms (FFTs) [8] (where the bandwidth of the spectrum is in effect taken to be inversely proportional to the step of discretisation) to evaluate the 2-D inverse Fourier transforms implied by the Green functions would either require a fine discretisation and/or a considerable amount of sampling points.

Nevertheless, these inverse Fourier transforms can be elegantly calculated by applying the algorithms proposed in [15] and developed to handle FFTs for non-equispaced data, and a Hanning filter employed for smoothing, for example. Furthermore, by multiplying both sides of the coupling equation (Equ. 1) by $\sigma_0 \chi(\mathbf{r})$, we restrict the search for the unknown fictitious current density $\mathbf{J}_2$ to the region effectively occupied by the flaw:

$$\mathbf{J}_2^{inc}(\mathbf{r}) = \mathbf{J}_2(\mathbf{r}) - j\omega\mu_0\sigma_0\chi(\mathbf{r}) \int_{\Omega} \overline{\mathbf{G}}_{22}^{ee}(\mathbf{r},\, \mathbf{r}')\, \mathbf{J}_2(\mathbf{r}')\, d\mathbf{r}' \qquad (15)$$

Discretisation of the above then results in an equation of the type $Y = AX$ with X (a vector) as unknown; such an equation is solved by using a conjugate gradient algorithm, since the number of unknowns for a three-dimensional vector problem can reach a very high value.

Summations implied by the coupling and the observation equations are efficiently carried out by using 2-D FFTs where we take full advantage of the convolutional nature of the kernels (the 2-D forward Fourier transform of the current density $\mathbf{J}_{probe}$ needed for this purpose is preferably carried out numerically by using its expression in the spatial domain); furthermore, the use of Fourier analysis greatly simplifies the algorithm which accounts for the simultaneous displacement of the SQUID and the double-D coil.

As a first validation of our model, the numerical simulation results are cross-checked with data obtained with a finite-element model.

We consider a 2 mm deep circular hole of diameter 6 mm drilled from the top surface of a 4 mm thick metallic slab; the other two dimensions of the slab are considered long enough to neglect edge effects. The lift-off between the SQUID and the double-D coil is 25 mm while the lift-off between the coil and the test sample is 14 mm. The double-D consists of 60 turns of a very thin wire which carries a 1 mA (peak to peak) current. The working frequency is 277 Hz and we calculate the normal component of the anomalous magnetic flux density observed when the measuring system is displaced along a line passing through the centre of the defect and perpendicular to the straight parts of the double-D. Figure 2 shows excellent agreement between the two models.

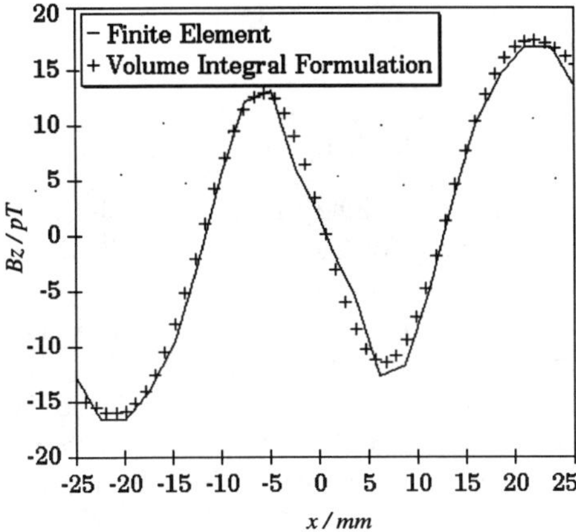

Figure 2. Cross-check of our model with a finite-element model.

5 Some approximations of interest

For the particular scanning configuration considered above, involving 100 points of vector field computation, the run time on a Cray C94 was about 7-8 minutes. The discretisation adopted (1 *mm* along each direction of the co-ordinate system) implied $96\times96\times2$ voxels ($x\times y\times z$) to correctly describe the flawed region. Indeed, zero-padding of the current density carried by the probe (having a diameter of 25 *mm*) implies a total of 50 points along the x and y directions but better results are obtained with at least 64 points along these directions. In addition, we need to include more sampling points to account for the displacement of the measuring system. That is, we need to solve for $3 \times 96 \times 96 \times 2$ ($\approx$ 60 000) discrete values of the unknown vector current density for each position of the double-D coil. Clearly, for a typical surface scan consisting of 40×40 probe positions, the run time can easily reach about 2 hours.

To reduce the run time, we may, as a first approximation, neglect the normal component of the induced current density (the z-component) and only consider the in-plane components when solving the state equation, i.e., the quasi-static limit is assumed [9]. Consequently, for the configuration considered above, the number of unknown values for the current density drops to $2 \times 96 \times 96 \times 2$ ($\approx$ 40 000). Furthermore, only 4 out of the 9 components of the dyadic Green function are now needed.

But a much faster algorithm results if we employ the Born approximation whereby the induced current density is supposed to undergo very little perturbation. This assumption is usually valid for a small defect and at very low frequencies, as is to some extent the case here. In the frame of the Born approximation one by-passes the state equation since the anomalous magnetic flux density follows from the observation equation after substituting the incident current density.

A better approximation, yet resulting in as fast an algorithm as the one within the Born approximation framework, is possible if the Localized NonLinear Approximation (LNLA) is applied [2, 10-11]. The state equation is employed to first deduce an estimate of the fictitious current density after making an assumption on the singular behaviour of the Green function; in short, the singularity of the Green function is such that a point source of electric current is seen as influencing only neighbouring voxels in such a way that the diagonal terms of the Green dyad suffice to describe the electromagnetic phenomena. Therefore, the coupling equation can be written as:

$$\mathbf{J}_2(\mathbf{r}) = \Xi(\mathbf{r})\, \mathbf{J}_2^{inc}(\mathbf{r}) \tag{16}$$

where

$$\overline{\overline{\Xi}}(\mathbf{r}) = \left[\overline{\overline{\mathbf{I}}} - j\,\omega\,\mu_0 \int_{\Omega} \overline{\overline{\mathbf{G}}}_{22}^{ee(diag)}(\mathbf{r},\, \mathbf{r}')\, \chi(\mathbf{r}')\, d\mathbf{r}'\right]^{-1} \tag{17}$$

In the above, $\overline{\overline{\mathbf{G}}}_{22}^{ee(diag)}$ consists of the diagonal components of the dyadic Green function. Substitution into the observation equation yields the desired anomalous magnetic flux density.

Figure 3 shows a comparison of the exact solution with numerical results obtained by using the different approximations presented above. We observe that only the Born approximation does not provide accurate results although the shape of the curve remains consistent. A comparison of the different run times clearly exemplifies that the localized nonlinear approximation offers the best compromise both for the forward and inverse problems [16].

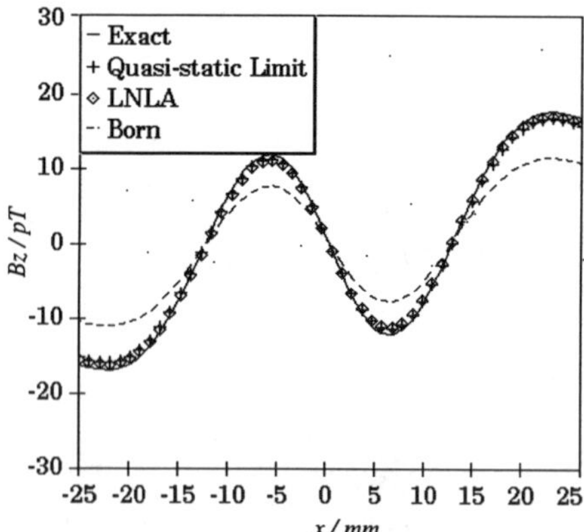

ALGORITHM EMPLOYED	CPU (CRAY C94) / min. (100 scanning points)
Exact Formulation	7.5
Quasi-static Limit	3.8
Localized NonLinear Approximation	1.1
Born Approximation	1.0

Figure 3. Comparison of the exact solution with the different approximations.

6 Experimental validation

A first comparison of numerical simulation results with experimental data for the 2 *mm* deep circular hole considered previously is shown in figure 4. The lift-off between the SQUID and the double-D coil is now 12 *mm* while the lift-off between the coil and the test sample is 11 *mm* (the two experimental lift-offs are known with a maximal error of 1 *mm*). The working frequency is 1 *kHz* and the injected current is 1 *mA*.

Other defect configurations are considered in figures 5-7. The lift-off between the SQUID and the double-D is 10 *mm* and the lift-off between the coil and the test sample is 13 *mm*. All defects are 1 *mm* deep rectangular slots cut from the top surface of the metallic slab. In the case of figure 5, the slot is 20 *mm* long (the length is taken to be along the scanning direction, i.e., along x) and 2 *mm* wide; in the case of figure 6, a 4 *mm* long and 2 *mm* wide slot is considered, while for figure 7 the slot is 4 *mm* long and 0.5 *mm* wide. Clearly, the last configuration is critical since the perturbation is rather small (a few pico-Teslas) but, although quite noisy, the eddy current response still enables the detection of the defect.

7 Conclusion

A three-dimensional model, based on the volume integral formulation, for the eddy current non-destructive evaluation of planar isotropic structures has been investigated and a numerical code developed for simulation. The measurement configuration adopted requires the simultaneous displacement of the SQUID-based sensor and of the exciting coil, and our modelling proves to be adequate for such a configuration. Good agreement is observed when comparing numerical simulation results with experimental data and with results obtained with a finite-element model. The computational cost can be reduced if the quasi-static limit is assumed. Yet, a much faster algorithm results if the localized nonlinear approximation is employed. Its efficiency for the configuration at hand makes it the ideal candidate for inversion.

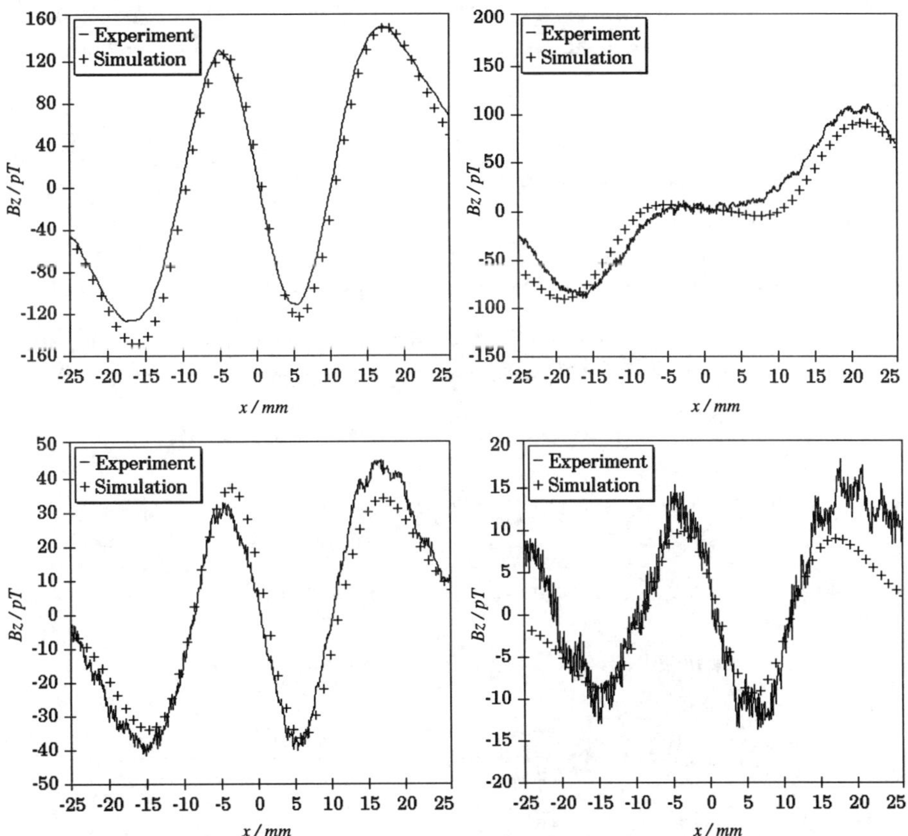

Figures 4-7. The defect configurations are as follows: figure 4 (top left): 2 *mm* deep circular hole of diameter 6 *mm*; figure 5 (top right): 1 *mm* deep, 2 *mm* wide and 20 *mm* long slot; figure 6 (bottom left): 1 *mm* deep, 2 *mm* wide and 4 *mm* long slot; figure 7 (bottom right): 1 *mm* deep, 0.5 *mm* wide and 4 *mm* long slot.

Acknowledgements

We are grateful to Dr. D. Tescione and co-workers from CIRA for providing us with simulation data obtained with their FEM code. The 2-D FFTs have been developed from 1-D FFTs for non-equispaced data kindly provided by V. Rokhlin. The computations have been carried out with a Cray C94 at IDRIS - CNRS - Orsay.

Appendix. The dyadic Green functions

Depending on the sign of z ($\pm$) and z' ($\pm$) of the exponential term of the wave function $\Phi_{lm}^{TE/TM}$ and due to the specific Fourier dependence $e^{j\mathbf{k}_s.(\mathbf{r}_s - \mathbf{r}'_s)}$ of the terms involved therein, we can write $\nabla\times = j\,\mathbf{k}_{l\,\pm}$ and $\nabla'\times = -\,j\,\mathbf{k}_{m\,\mp}$ with letting $\mathbf{k}_{l\,\pm} = \mathbf{k}_s \pm k_{lz}\mathbf{z}$ and $\mathbf{k}_{m\mp} = \mathbf{k}_s \mp k_{mz}\mathbf{z}$. In general, in function of the signs of z and z', distinct tensors $\overline{\mathbf{\Gamma}}_{lm}$ and $\overline{\mathbf{\Pi}}_{lm}$ are required. However one may observe that $(\nabla\times\mathbf{z})(\nabla'\times\mathbf{z}) = (\mathbf{k}_s\times\mathbf{z})(\mathbf{k}_s\times\mathbf{z})$, which means that $\overline{\mathbf{\Gamma}}_{lm}^{ee}$ is independent of the sign of z and z', while $\overline{\mathbf{\Gamma}}_{lm}^{me}$ depends on the sign of z only.

So, if we first write:

$$\mathbf{k}_{l\alpha} = \mathbf{k}_s + \alpha\, k_{lz}\,\mathbf{z}, \text{ where } \begin{cases} \alpha = 1 \text{ if } sign(z) > 0 \\ \alpha = -1 \text{ if } sign(z) < 0 \end{cases}$$

and

$$\mathbf{k}_{l\beta} = \mathbf{k}_s + \beta\, k_{lz}\,\mathbf{z}, \text{ where } \begin{cases} \beta = -1 \text{ if } sign(z') > 0 \\ \beta = 1 \text{ if } sign(z') < 0 \end{cases}$$

where $sign(z)$ and $sign(z')$ correspond to the sign which shows up in front of z and z' in the exponential term of $\Phi_{lm}^{TE/TM}$, we are able to express the $\overline{\mathbf{\Gamma}}_{lm}$ and $\overline{\mathbf{\Pi}}_{lm}$ tensors as:

$$\overline{\mathbf{\Gamma}}_{lm}^{ee/me}(\mathbf{k}_s,\,\mathbf{r},\,\mathbf{r}') = \overline{\mathbf{M}}_{lm}^{ee/me(\alpha)}\, e^{j\mathbf{k}_s.(\mathbf{r}_s-\mathbf{r}'_s)}\, \Phi_{lm}^{TE}(z,\,z')$$

$$\overline{\mathbf{\Pi}}_{lm}^{ee/me}(\mathbf{k}_s,\,\mathbf{r},\,\mathbf{r}') = \overline{\mathbf{N}}_{lm}^{ee(\alpha\beta)/me(\alpha\beta)}\, e^{j\mathbf{k}_s.(\mathbf{r}_s-\mathbf{r}'_s)}\, \Phi_{lm}^{TM}(z,\,z')$$

where

$$\overline{\mathbf{M}}_{lm}^{ee}(\mathbf{k}_s,\,\mathbf{r},\,\mathbf{r}') = \frac{j}{2\,k_{mz}\,k_s^2}\,(\mathbf{k}_s\times\mathbf{z})(\mathbf{k}_s\times\mathbf{z})$$

$$\overline{\mathbf{M}}_{lm}^{me(\alpha)}(\mathbf{k}_s,\,\mathbf{r},\,\mathbf{r}') = -\,\frac{\mathbf{k}_{l\alpha}}{2}\times\overline{\mathbf{M}}_{lm}^{ee}(\mathbf{k}_s,\,\mathbf{r},\,\mathbf{r}')$$

$$\overline{\mathbf{N}}_{lm}^{ee(\alpha\beta)}(\mathbf{k}_s,\,\mathbf{r},\,\mathbf{r}') = \frac{j}{2\,k_{mz}\,k_s^2}\,\frac{(\mathbf{k}_{l\alpha}\times\mathbf{k}_s\times\mathbf{z})(\mathbf{k}_{m\beta}\times\mathbf{k}_s\times\mathbf{z})}{\omega^2\,\varepsilon_l\,\mu_m}$$

$$\overline{\mathbf{N}}_{lm}^{me(\alpha\beta)}(\mathbf{k}_s,\,\mathbf{r},\,\mathbf{r}') = -\,\frac{\mathbf{k}_{l\alpha}}{2}\times\overline{\mathbf{N}}_{lm}^{ee(\alpha\beta)}(\mathbf{k}_s,\,\mathbf{r},\,\mathbf{r}')$$

The derivation of the $\overline{\mathbf{M}}_{lm}$ and $\overline{\mathbf{N}}_{lm}$ tensors is straightforward from the above but the limits of these tensors as k_x and/or k_y tend to zero have to be carefully defined before proceeding to any numerical computation.

References

[1] A. Ruosi *et al.*, "Eddy-current non-destructive measurements using SQUIDs: preliminary results," in *Electromagnetic non-destructive Evaluation (II)*, 215-224, R. Albanese *et al.* (eds.), IOS Press, Amsterdam, 1997.

[2] V. Monebhurrun *et al.*, "Evaluation of a 3-D bounded defect in the wall of a metal tube at eddy current frequencies: the direct problem," *J. Electromagn. Waves Applic.* **12** (1998) 315-347.

[3] J. R. Bowler *et al.*, "A theoretical and computational model of eddy-current probes incorporating volume integral and conjugate gradient methods," *IEEE Trans. Mag.* **MAG-25** (1989) 2650-2664.

[4] Chen-To Tai, *Dyadic Green's Functions in Electromagnetic Theory*, IEEE Press, New York, 1994.

[5] W. C. Chew, *Waves and Fields in Inhomogeneous Media*, IEEE Press, New York, 1995.

[6] R. F. Harrington, *Field Computation by Moment Methods*, Macmillan, New York, 1968.

[7] T. K. Sarkar *et al.*, "A limited survey of various conjugate gradient methods for solving complex matrix equations arising in electromagnetic wave interactions," *Wave Motion* **10** (1988) 527-546.

[8] E. O. Brigham, *The Fast Fourier Transform*, Prentice-Hall, Englewood Cliffs, 1974.

[9] J. R. Bowler, "Eddy current calculations using half-space Green's functions," *J. Appl. Phys.* **61** (1987) 833-839.

[10] T. Habashy *et al.*, "Beyond the Born and Rytov approximations: A nonlinear approach to electromagnetic scattering," *J. Geophys. Res.* **98** (1993) 1759-1775.

[11] C. Torres-Verdin and T. Habashy, "Rapid 2.5-dimensional forward modeling and inversion via a new nonlinear scattering approximation," *Radio Sci.* **29** (1994) 1051-1079.

[12] Ching-Chuan Su, "A simple evaluation of some principal value integrals for dyadic Green's function using symmetry property," *IEEE Trans. Antennas Propagat.* **AP-35** (1987)1306-1307.

[13] W. C. Chew, "Some observations on the spatial and eigenfunction representations of dyadic Green's functions," *IEEE Trans. Antennas Propagat.* **AP-37** (1989) 1322-1327.

[14] M. Abramovitz and I. A. Stegun, *Handbook of Mathematical Functions*, Dover, New York, 1970.

[15] A. Dutt and V. Rokhlin, "Fast Fourier Transforms for nonequispaced data," *SIAM J. Sci. Comput.* **14** (1993) 1368-1393.

[16] V. Monebhurrun *et al.*, "3-D inversion of eddy current data for Non-Destructive Evaluation of steam generator tubes," *Inverse Problems* **14** (1998) 707-724.

Electromagnetic Nondestructive Evaluation (III)
D. Lesselier and A. Razek (Eds.)
IOS Press, 1999

Some Finite Element Models of the FLUXSET Sensor

D.Rodger and P.K.Vong
University of Bath, Claverton Down, Bath, BA2 7AY, UK

Abstract. Some finite element models of a FLUXSET sensor are described. These are of varying degrees of complexity and can take into account magnet non-linearities and eddy currents in the amorphous ribbon material. Results from volume and thin sheet methods are compared.

1 Introduction

The FLUXSET sensor consists of a strip of amorphous alloy ribbon with a set of excitation coils and pick up coils wrapped around it. A typical arrangement is shown in Figure 1. In this contribution various finite element techniques for modelling the FLUXSET are compared.

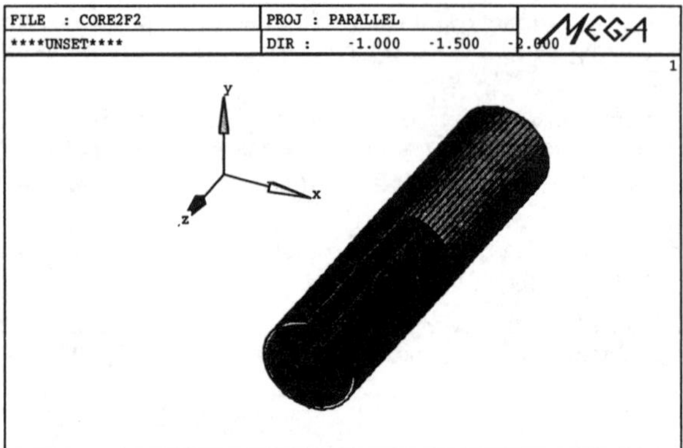

Figure 1: A FLUXSET sensor.

2 Finite Element Formulations

Nodal variable finite elements are described here, non-conductors are modelled in terms of magnetic scalar potentials and conductors are modelled using magnetic vector potentials.

2.1 Non-conductors

Non-conducting regions may be modelled economically in 3D using magnetic scalar potentials, either the total scalar ψ ,defined as $\mathbf{H}_T = -\nabla\psi$, or the reduced scalar ϕ, defined as $\mathbf{H}_T = -\nabla\phi + \mathbf{H}_S$. Here $\mathbf{H}_T$ is the total magnetic field intensity and $\mathbf{H}_S$ is the field defined as $\nabla\times\mathbf{H}_S = \mathbf{J}_S$, where $\mathbf{J}_S$ is the source current density. The basic total and reduced scalar method can be extended to allow multiply connected conditions [1], and to produce cuts for solving multiply connected problems. Both scalars give rise to a Laplacian type equation which has to be solved:

$$\nabla\cdot\mu\nabla\psi = 0 \tag{1}$$

2.2 Thin Non-conducting Regions

The ribbon is thin compared with the other dimensions (20μ m compared with a length of 9mm), it is therefore useful to consider using a thin sheet formulation for the ribbon. The formulation described here has been shown to compare favourably with measurements around thin iron sheets, of thickness comparable to an electrical machine lamination [2], here we show that it is applicable to amorphous material. If the conductivity of the ribbon may be assumed to be zero as was done in [3],[4], all of the regions may be modelled using magnetic scalar potentials, either total or reduced. Thin elements are required in the ribbon. This may give rise to numerical ill-conditioning or an excessively fine mesh. In order to avoid this, the thin high permeability ribbon can be modelled as a sheet, as follows: The Galerkin method applied to the Laplace equation yields,

$$\int_{V_e} \nabla N_i \cdot \mu\nabla\psi \, \mathrm{d}V = \int_{\Omega_e} \mu N_i \frac{\partial\psi}{\partial n} dS \tag{2}$$

In a thin region, the volume integral in (2) may be replaced by

$$\int_{\Omega_e} h \left[\frac{\partial N_i}{\partial s}\mu\frac{\partial\psi}{\partial s} + \frac{\partial N_i}{\partial t}\mu\frac{\partial\psi}{\partial t} \right] dsdt \tag{3}$$

In (3) s and t are two tangential directions on the sheet surface and h is the thickness of the sheet. This implies that the field is constant across the thickness of the sheet. The boundary conditions for ψ on the sheet surface are exactly the same as those which would be applied to a volume region under the same conditions.

2.3 Conducting Regions

Fields in conductors can be modelled using $\mathbf{A}$, the magnetic vector potential, and V, the electric scalar potential. In regions of constant conductivity, surrounded by magnetic scalar potential, it is sometimes possible to avoid using V. In this case, using $\mathbf{B} = \nabla\times\mathbf{A}$ and $\mathbf{E} = -\frac{\partial\mathbf{A}}{\partial t}$, we obtain

$$\nabla\times\frac{1}{\mu}\nabla\times\mathbf{A} = -\sigma\frac{\partial\mathbf{A}}{\partial t} \tag{4}$$

Table 1: FLUXSET dimensions

drive coil i.d. m	2 mm
drive coil length	9 mm
drive coil turns	470
pick up coil i.d.	2 mm
pick up coil length	5 mm
pick up coil turns	100
ribbon length	9.5 mm
ribbon thickness	20μm
ribbon width	0.6 mm
ribbon conductivity	1×10^5 S/m
ribbon ribbon relative permeability magnitude $B = 0 - 0.65T$	85000
ribbon ribbon relative permeability magnitude $B > 0.65T$	1

2.4 Time Transient Conditions

Application of the Galerkin technique and 3D finite elements to (1) and (4) results in a system of equations which has the following standard form,

$$[\mathbf{K_a}]\boldsymbol{x_a} + [\mathbf{C_a}]\dot{\boldsymbol{x_a}} = \boldsymbol{f_a} \quad (5)$$

in which $\boldsymbol{x_a}$ is a vector of the unknown ψ and $\mathbf{A}$ variables.This is solved as follows: Calculate $\boldsymbol{\alpha_n}$ from

$$\boldsymbol{\alpha_n} = ([\mathbf{C_a}] + \Theta\Delta t[\mathbf{K_a}])^{-1}\{\boldsymbol{f_{an}} - [\mathbf{K_a}]\boldsymbol{x_{an}}\} \quad (6)$$

Then calculate the new $\boldsymbol{x_a}$

$$\boldsymbol{x_{an+1}} = \boldsymbol{x_{an}} + \boldsymbol{\alpha_n}\Delta t \quad (7)$$

When non-linear materials are involved, (5) becomes non-linear and is solved using standard iteration techniques, such as Newton Raphson [5] or simple iteration.

3 Results

A device with the same dimensions as that given in [3] and[4] was modelled using both a conventional volume 3D finite element scheme and also the thin sheet method. Main dimensions are given in Table 1.

Figs 2 and 3 show the normal and axial components of **B** along the centre line of the ribbon, plotted from the centre to one end. Both volume and sheet finite elements were used for the calculations. Here the conductivity of the ribbon was neglected. A constant unsaturated ribbon permeability was used. Exploiting symmetry, only one eighth of the device had to be modelled. The results show that both methods give similar results.

4 Response to External Fields

In this section it is assumed that the conductivity of the ribbon may be taken to be zero. The magnetic non-linearity is taken into account. The FLUXSET is intended to be used to measure magnetic fields and is intended to respond only to the axial component of external field. Here we confirm that this is likely to be the case.

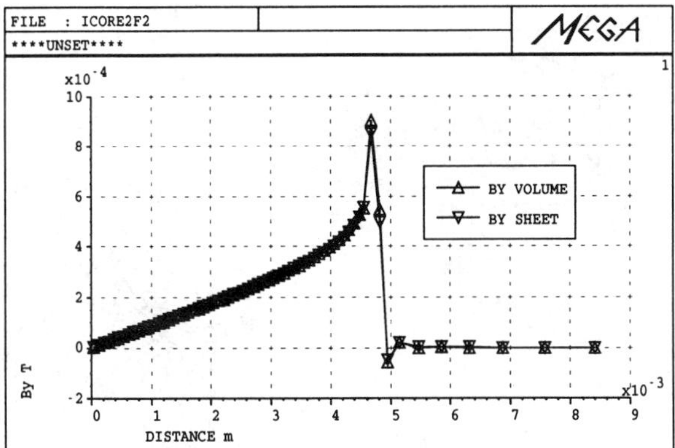

Figure 2: B_y just above the ribbon.

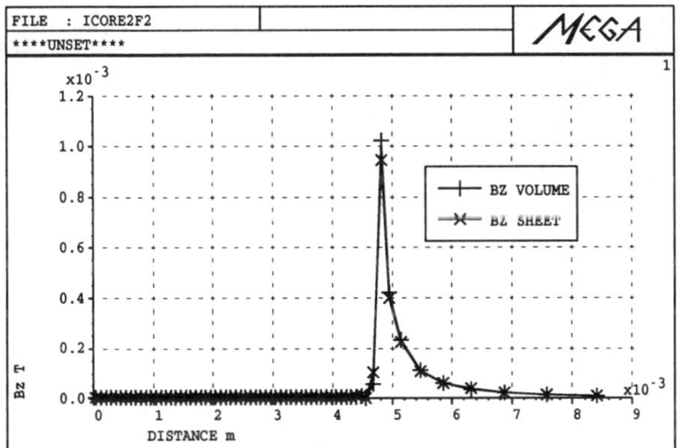

Figure 3: B_z just above the ribbon.

4.1 Axial External Fields

The FLUXSET was modelled immersed in various axial external fields. Fig 4 shows the average magnetic flux in the pickup coil for external fields of 0 and 200 μT respectively. Multiplication of this graph by the pickup coil turns and differentiation with respect to time gives the pickup coil voltage, which is the desired output of the device. Also shown on Fig 4 is the triangular drive coil current (peak 12mA). The resultant pickup coil voltage is shown on Fig 5.

4.2 Non-axial External Fields

In order to find out whether the device will measure only the axial component of field, or whether it would be affected by transverse components of field, the device was modelled while placed in an external magnetic field at various angles to the axis of the device. Fig 6 shows results for a 200 and 20 μT external field at an angle of 45° to the ribbon axis. Also shown are results for two uniform axial fields of magnitude 200 $\cos(\pi/4)$ and 20 $\cos(\pi/4)$. The results are identical, which would confirm that the device is unaffected by transverse fields. The numerical test carried out here is more stringent than required, considering what will happen in reality, since the permeability

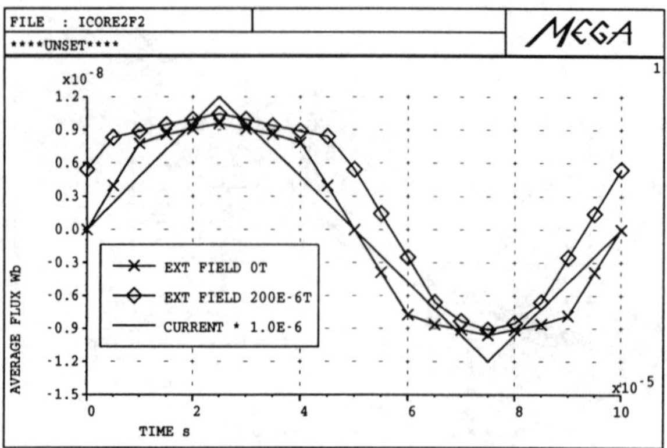

Figure 4: Average flux in the pick up coil.

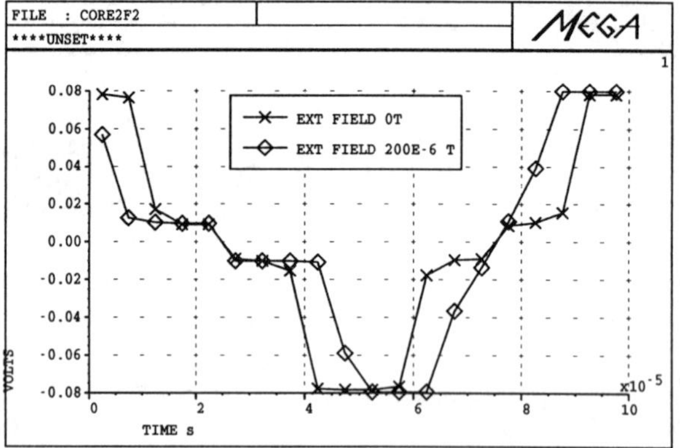

Figure 5: Volts in the pick up coil.

in the transverse direction of the real ribbon material is much poorer than in the axial direction. In the numerical test, the same permeability was used in both directions.

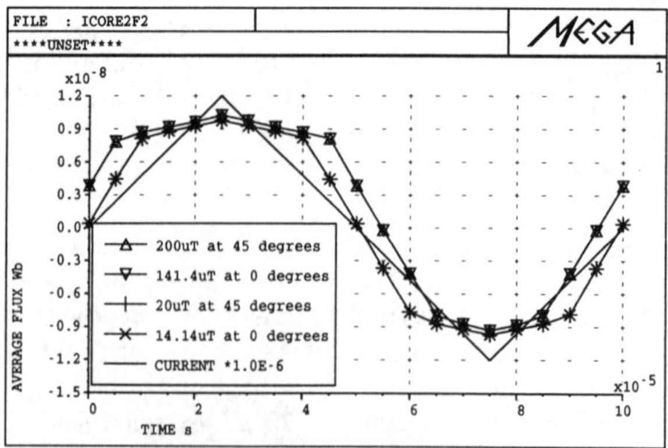

Figure 6: Average flux for various transverse fields.

5 Eddy Current Simulation

Consideration of the classical eddy current skin depth, $\delta = \sqrt{2/(\omega\sigma\mu)}$ implies that at 100kHz δ is around 17μm, comparable with the thickness of the ribbon. A time transient simulation was therefore carried out, allowing eddy currents to flow in the ribbon but allowing the permeability to remain constant, to save computer time. The drive coil current was as already described. Fig 7 shows that there is negligible eddy current effect at this frequency. The average magnetic flux in the pick up coil is shown for both conducting and non-conducting ribbons; these are substantially identical.

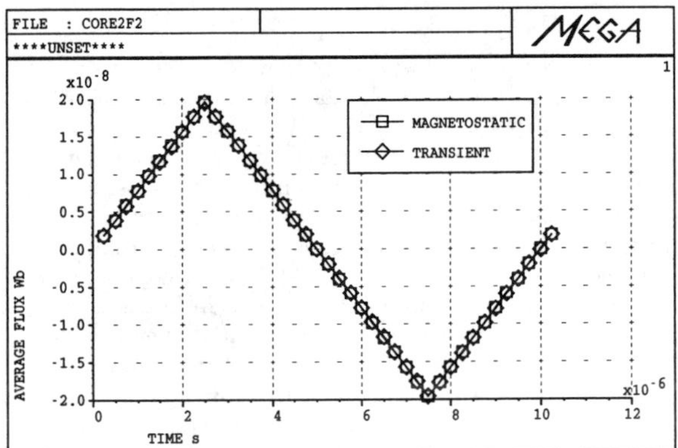

Figure 7: Time transient solution at 100kHz.

6 Conclusions

It has been shown that the thin sheet method can be useful for modelling amorphous ribbon material in that results are similar to those of volume elements and computer times are faster. Preliminary work appears to indicate that transverse fields do not adversely affect the described FLUXSET sensor. Since the device is 'long and thin', it behaves in a similar way to the classical long magnetic rod in a uniform field.

Acknowledgement Work supported by EU contract IC15-CT96-0703.

References

[1] P.J.Leonard and D.Rodger. "Modelling voltage forced coils using the reduced scalar potential method". *IEEE Trans. Magn.*, MAG 28(2):1615–1617, March 1992.

[2] D.Rodger, P.J.Leonard, and H.C.Lai. "Surface elements for modelling 3D fields around thin iron sheets". *IEEE Trans. Magn.*, 29(2):1483–1486, 1993.

[3] D.Ioan, M.Rebican, G.Ciuprina, and P.Leonard. "3D FEM model of a fluxset sensor". *E'NDE 97 Proc, Reggio Calabria ITALY*, September 1997.

[4] K.Preis, I.Bardi, J.Pavo, A.Gasparics, I.Ticar, O.Biro, and K.Richter. "Numerical simulation and design of a fluxset sensor by finite element method". *Conference record COMPUMAG 97*, pages 99–100, November 1997.

[5] D.Rodger, N.Allen, H.C.Lai, and P.J.Leonard. " Calculation of transient 3D eddy currents in non linear media - Verification using a rotational test rig". *IEEE Trans. Magn.*, 30(5):2988–2991, September 1994.

Electromagnetic Nondestructive Evaluation (III)
D. Lesselier and A. Razek (Eds.)
IOS Press, 1999

Prediction of eddy-current signals due to cracks concealed by surface deposits

Nicola HARFIELD and John R. BOWLER
Department of Physics, University of Surrey, Guildford, Surrey GU2 5XH, England

Abstract. Many components which are routinely tested using eddy-current methods suffer from the build-up of surface deposits. It is important to assess the effect of surface deposits because the defect signal in an eddy-current probe may be reduced, even to the point where defect detection is impossible. Here, the impedance change in a probe due to a surface crack in a half-space conductor is calculated taking into account a layer of conducting and/or permeable material on the conductor surface. The calculation is performed in the thin-skin regime, in which the electromagnetic skin-depth in the conductor is assumed to be much smaller than the dimensions of the crack. In this regime, an analytical approach can be adopted in which a scalar potential obeys Laplace's equation on the planar domain of the crack. The effect of the surface layer enters the formulation in the boundary condition on the potential along the line of the crack mouth. This boundary condition is written as an integral equation using Green's theorem, and the Green's function kernel takes account of the surface layer. Example calculations are performed for a long crack in a conductor with a surface layer of various thickness, conductivity and permeability.

1. Introduction

Eddy-current non-destructive evaluation is commonly performed in environments where, over a period of time, surface deposits build up on the structures to be tested, making the task of inspection more difficult. These deposits are typically metal oxides which are electrically conducting and may be magnetic. They therefore provide an alternative path for induced eddy currents, weakening the signal from cracks in the metal. Assessment of the magnitude of this effect is important since some cracks may become undetectable due to surface deposits. An example is given in Figure 1, in which the effect of a layer on the probe signal from a long, surface crack is shown. The change in probe impedance due to a crack in mild steel is plotted as a function of layer thickness, for a layer with conductivity one thousandth and permeability one tenth of the values in the steel. The parameters used in the calculation are given in Table 1. Clearly, the probe signal is significantly reduced by the presence of the layer. The dominant process in reducing the signal is the skin effect in the layer, although some of the signal reduction is due to lift-off.

Oxides may form inside a crack, as well as on the surface of the conductor. The effect of oxide within a crack has been analysed elsewhere [1]. Here, we consider a surface layer of arbitrary conductivity and permeability deposited on a cracked metal

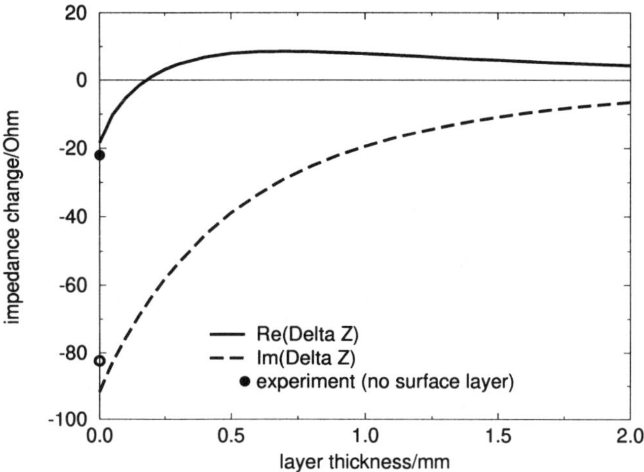

Figure 1: Probe impedance change due to a crack, versus layer thickness.

substrate. In the theoretical development, it is assumed that the electromagnetic skin depth in the metal is much smaller than the dimensions of the crack. The eddy currents then flow in a thin skin at the surface of the conductor and at the crack faces. This occurs practically in many eddy-current inspections and, in the theory, an analytical approach can be taken which reduces the computational cost of the calculations. We present calculations of eddy-current probe signals for a long surface crack in a conductor with a surface layer.

2. Scalar Formulation

2.1 Surface Potential

Consider a planar crack in a conductor with a surface layer, shown in Figure 2. It is assumed that the electromagnetic field varies as the real part of $\exp(-i\omega t)$, that the material properties are linear, that the displacement current is negligible and that the conductor is sufficiently thick to behave as a half-space. For the purpose of calculating the fields, it is also assumed that the crack is ideal, having negligible opening but forming a perfect barrier to the flow of current. The calculation of the impedance change due to the defect (Section 4.2) does, however, allow for the possibility of finite gape.

The field calculation proceeds by decomposing the electric and magnetic fields in the conductor into transverse electric (TE) and transverse magnetic (TM) components [2]:

$$\mathbf{E}(\mathbf{r}) = i\omega\mu\left[\nabla \times \hat{x}\psi'(\mathbf{r}) - \nabla \times \nabla \times \hat{x}\psi''(\mathbf{r})\right], \qquad z \leq 0, \tag{1}$$

$$\mathbf{H}(\mathbf{r}) = \nabla \times \nabla \times \hat{x}\psi'(\mathbf{r}) - k^2\nabla \times \hat{x}\psi''(\mathbf{r}), \qquad z \leq 0, \tag{2}$$

where the preferred direction, $\hat{x}$, is normal to the crack plane, ψ' is the TE potential, ψ'' is the TM potential and $k^2 = i\omega\mu\sigma$ with μ and σ being values of permeability and conductivity in the conductor. (The subscript 3 shown in Figure 2 is not used in this

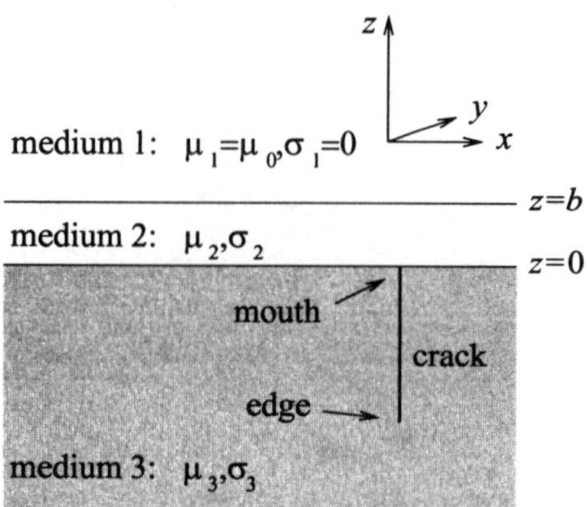

Figure 2: Cracked conductor with a surface layer.

section.) Assuming that electric field components in the same plane as the crack do not interact with it, the TE interaction with the crack is negligible and the problem can be formulated in terms of the TM potential alone. The assumption of negligible TE interaction is exact for cracks in non-magnetic materials, and is a good approximation for cracks in magnetic materials provided that the crack opening is not too large. This formulation has been given in detail elsewhere [3], and only the key stages are given here.

Assuming that no current flows across the crack, $\mathbf{E}\cdot\hat{x} = 0$ at the crack surface and, from (1),

$$\nabla_x^2 \Psi(\mathbf{r}) = 0, \tag{3}$$

where $\nabla_x^2 = \partial^2/\partial y^2 + \partial^2/\partial z^2$ and $\Psi = k^2\psi''$, for notational convenience. The TM potential can therefore be found by solving the Laplace equation according to appropriate boundary conditions at the crack perimeter.

2.2 Boundary Conditions

While Ψ satisfies the surface Laplace equation given in (3) at arbitrary frequency, the boundary conditions at the crack edge and mouth given in this section are valid only in the thin-skin regime. The crack mouth is defined as the line of intersection between the surface of the crack and the conductor surface. The crack edge is defined as the line bounding the crack surface in the body of the conductor. In the thin-skin regime, the component of the magnetic flux density normal to the crack edge is zero [4] and it can be deduced that

$$\Psi(\mathbf{r}_e) = 0, \tag{4}$$

where $\mathbf{r}_e$ denotes the coordinate of a point at the crack edge [3].

At the crack mouth, a boundary condition is derived from an integral equation which determines Ψ. It is possible to use this equation as the basis for computing numerical solutions, but here it will be used for the more limited purpose of defining the crack mouth boundary condition.

The integral equation is formed by applying Green's theorem on a surface, S, which encloses the crack. Since Ψ obeys the Helmholtz equation, integrating over the volume enclosed by S yields, using Green's theorem,

$$\int_S [G(\mathbf{r},\mathbf{r}')\nabla'\Psi(\mathbf{r}') - \Psi(\mathbf{r}')\nabla' G(\mathbf{r},\mathbf{r}')] \cdot dS' = \Psi(\mathbf{r}). \tag{5}$$

If S is collapsed onto the crack faces [5], equation (5) can be simplified using the properties of Ψ across the crack. If $\mathbf{r}_+$ and $\mathbf{r}_-$ denote field points on opposite faces, then continuity of the normal component of $\mathbf{H}$ across the crack implies that

$$\Psi(\mathbf{r}_+) - \Psi(\mathbf{r}_-) = 0. \tag{6}$$

Further, in the thin-skin regime, the fields exhibit exponential decay normal to the crack faces. This means that the x-dependence of Ψ close to the crack is approximately $\exp(ik|x|)$ and

$$\left.\frac{\partial\Psi(\mathbf{r})}{\partial x}\right|_{\mathbf{r}_+} - \left.\frac{\partial\Psi(\mathbf{r})}{\partial x}\right|_{\mathbf{r}_-} \approx 2ik\Psi(\mathbf{r}_+). \tag{7}$$

Using (6) and (7) in (5) yields

$$\Psi(\mathbf{r}) = \Psi^{(i)}(\mathbf{r}) + 2ik \int_{S_c} G(\mathbf{r},\mathbf{r}')\Psi(\mathbf{r}')dS' \tag{8}$$

where S_c is the open surface defining the crack, and the superscript (i) denotes the incident, or unperturbed, field.

Integral equation (8) will be used to give a boundary condition for Ψ at the crack mouth, but first it is helpful to consider the form of the Green's function. As shown in [3], a two-dimensional Fourier transform method can be used to obtain the following form for a half-space Green's function:

$$G(\mathbf{r},\mathbf{r}') = \frac{1}{(2\pi)^2}\int_{-\infty}^{\infty}\int_{-\infty}^{\infty}\frac{1}{2\gamma}\left[e^{-\gamma|z-z'|} + e^{\gamma(z+z')} + (\Gamma_{22}-1)\,e^{\gamma(z+z')}\right]e^{iu(x-x')+iv(y-y')}dudv, \tag{9}$$

where

$$\tilde{\tilde{F}}(u,v,z) = \int_{-\infty}^{\infty}\int_{-\infty}^{\infty} F(\mathbf{r})e^{-i(ux+vy)}\,dxdy \tag{10}$$

defines the two-dimensional Fourier transform with u and v the Fourier space coordinates, and $\gamma = (u^2+v^2-k^2)^{1/2}$. The coefficient Γ_{22} is a reflection coefficient which embodies the coupling of the fields at the interface between the conductor and air. When a surface layer is present, these equations still apply but Γ_{22} is modified to take account of the layer, as shown in the next section. Equation (9) can be written in the form

$$G(\mathbf{r},\mathbf{r}') = G_0(\mathbf{r},\mathbf{r}') + G_i(\mathbf{r},\mathbf{r}'') + \frac{1}{k^2}\frac{\partial^2}{\partial y^2}U(\mathbf{r},\mathbf{r}''), \tag{11}$$

where G_0 is the free space scalar Green's function given by

$$G_0(\mathbf{r},\mathbf{r}') = \frac{e^{ik|\mathbf{r}-\mathbf{r}'|}}{4\pi|\mathbf{r}-\mathbf{r}'|}, \tag{12}$$

G_i represents an image source with $\mathbf{r}'' = \mathbf{r}' - 2\hat{z}z'$ and

$$U(\mathbf{r},\mathbf{r}'') = -\frac{k^2}{(2\pi)^2}\int_{-\infty}^{\infty}\int_{-\infty}^{\infty}\frac{(\Gamma_{22}-1)}{v^2}e^{iu(x-x')+iv(y-y')+\gamma(z+z')}dudv. \tag{13}$$

Although (8) is valid only in the thin-skin regime, this Green's function can be applied at arbitrary frequency.

The mouth boundary condition is obtained from (8) by differentiating with respect to z and restricting $\mathbf{r}$ to points at the crack mouth, $\mathbf{r}_m$:

$$\left.\frac{\partial \Psi(\mathbf{r})}{\partial z}\right|_{\mathbf{r}=\mathbf{r}_m} = \left.\frac{\partial \Psi^{(i)}(\mathbf{r})}{\partial z}\right|_{\mathbf{r}=\mathbf{r}_m} + \frac{2}{ik}\int_{-c/2}^{c/2} \frac{\partial^2}{\partial y^2} U(0, y-y', 0)\Psi(0, y', 0)dy'. \quad (14)$$

The crack length is c. Integration with respect to z' has been carried out by assuming that $\Psi(0, y', z')$ varies slowly with z' and is therefore roughly constant over the effective range of the kernel in the z-direction. Equations (4) and (14) are the boundary conditions which must be applied to solve (3) for Ψ.

3. Layered Half-Space Field Analysis by Scalar Decomposition

In this section, the reflection coefficient Γ_{22} which appears in the expression for the boundary condition at the crack mouth will be determined. In the case of a half-space conductor, Γ_{22} can be written

$$\Gamma_{22} = 1 + \frac{2\mu_r v^2 k^2}{\kappa(\gamma + \mu_r \kappa)(u^2 - k^2)}, \quad (15)$$

where $\kappa^2 = u^2 + v^2$. For a half-space conductor with a surface layer, the situation is more complicated and a different expression for Γ_{22} is required.

Consider the electromagnetic field in a domain divided into three regions as shown in Figure 2. The upper half-space is non-conducting with free-space permeability. The layer and base conductor have uniform conductivities σ_2 and σ_3 and permeabilities μ_2 and μ_3 respectively. The field is to be expressed in terms of TE and TM potentials defined with respect to different preferred directions in these regions. In the upper half-space and the layer, regions 1 and 2, the preferred direction is normal to the interface between them, the z-direction. In the host material, region 3, the preferred direction is tangential to the interface and chosen to be the x-direction.

In homogeneous, isotropic regions, the TE and TM potentials give rise to independent solutions. Here, because different preferred directions are used in regions 2 and 3, cross-coupling occurs between the modes at the interface between the regions. The effect of this cross-coupling is written in a 2×2 matrix of interface transmission and reflection coefficients which allow, for example, a TE mode to be reflected by an incident TM mode and vice versa. The mode coupling is built into a half-space scalar Green's function following preliminary analysis which determines the transmission and reflection coefficients. The coefficients are derived for an arbitrary source in the conductor using a two-dimensional Fourier transform of the field.

3.1 Scalar Decomposition

Assuming time dependence $\exp(-i\omega t)$, the electric and magnetic fields in the three regions can be written in terms of TE and TM potentials (ψ' and ψ'' respectively) as follows [2]:

$$\mathbf{E}_1(\mathbf{r}) = i\omega\mu_0[\nabla \times \hat{z}\psi_1'(\mathbf{r}) - \nabla \times \nabla \times \hat{z}\psi_1''(\mathbf{r})], \quad z \geq b, \quad (16)$$
$$\mathbf{E}_2(\mathbf{r}) = i\omega\mu_2[\nabla \times \hat{z}\psi_2'(\mathbf{r}) - \nabla \times \nabla \times \hat{z}\psi_2''(\mathbf{r})], \quad 0 \leq z \leq b, \quad (17)$$
$$\mathbf{E}_3(\mathbf{r}) = i\omega\mu_3[\nabla \times \hat{x}\psi_3'(\mathbf{r}) - \nabla \times \nabla \times \hat{x}\psi_3''(\mathbf{r})], \quad z \leq 0, \quad (18)$$

$$\mathbf{H}_1(\mathbf{r}) = \nabla \times \nabla \times \hat{z}\psi_1'(\mathbf{r}), \qquad z \geq b, \tag{19}$$
$$\mathbf{H}_2(\mathbf{r}) = \nabla \times \nabla \times \hat{z}\psi_2'(\mathbf{r}) - k_2^2 \nabla \times \hat{z}\psi_2''(\mathbf{r}), \quad 0 \leq z \leq b, \tag{20}$$
$$\mathbf{H}_3(\mathbf{r}) = \nabla \times \nabla \times \hat{x}\psi_3'(\mathbf{r}) - k_3^2 \nabla \times \hat{x}\psi_3''(\mathbf{r}), \quad z \leq 0, \tag{21}$$

where $\hat{x}$ and $\hat{z}$ are unit vectors and $k_i^2 = i\omega\mu_i\sigma_i$ with subscript $i = 2, 3$ denoting the region.

Using Maxwell's equations it can be shown that

$$\nabla^2 \nabla_z^2 \psi_1(\mathbf{r}) = 0, \qquad z \geq b, \tag{22}$$
$$(\nabla^2 + k_2^2)\nabla_z^2 \psi_2(\mathbf{r}) = 0, \qquad 0 \leq z \leq b, \tag{23}$$
$$(\nabla^2 + k_3^2)\nabla_x^2 \psi_3(\mathbf{r}) = 0, \qquad z \leq 0, \tag{24}$$

where the ψ_i carry either a single or double prime depending on whether they represent the TE or TM potential and

$$\nabla_x = \nabla - \hat{x}\frac{\partial}{\partial x}, \qquad \nabla_z = \nabla - \hat{z}\frac{\partial}{\partial z}. \tag{25}$$

Expressions for the transmission and reflection coefficients are found through consideration of the field continuity conditions at the interfaces;

$$\hat{z} \times \mathbf{E}_1 = \hat{z} \times \mathbf{E}_2, \qquad \hat{z} \times \mathbf{H}_1 = \hat{z} \times \mathbf{H}_2, \qquad z = b \tag{26}$$

and

$$\hat{z} \times \mathbf{E}_2 = \hat{z} \times \mathbf{E}_3, \qquad \hat{z} \times \mathbf{H}_2 = \hat{z} \times \mathbf{H}_3, \qquad z = 0. \tag{27}$$

These equations express the continuity of tangential electric and magnetic field components across the boundaries at $z = b$ and $z = 0$. The transmission and reflection coefficients are most conveniently found by using the two-dimensional Fourier transform with respect to the x- and y-coordinates defined in (10). Transforming (22), (23) and (24) using (10) reveals that ψ_1, ψ_2 and ψ_3 obey the following second-order differential equations in z:

$$\left(\frac{\partial^2}{\partial z^2} - \kappa^2\right)\widetilde{\widetilde{\psi}}_1(u, v, z) = 0, \qquad z \geq b, \tag{28}$$
$$\left(\frac{\partial^2}{\partial z^2} - \gamma_2^2\right)\widetilde{\widetilde{\psi}}_2(u, v, z) = 0, \qquad 0 \leq z \leq b, \tag{29}$$
$$\left(\frac{\partial^2}{\partial z^2} - \gamma_3^2\right)\widetilde{\widetilde{\psi}}_3(u, v, z) = 0, \qquad z \leq 0, \tag{30}$$

where $\kappa = (u^2 + v^2)^{1/2}$, $\gamma_i = (u^2 + v^2 - k_i^2)^{1/2}$ and roots with positive real parts are taken.

3.2 Internal Source

Explicit expressions for the transmission and reflection coefficients at the interfaces will now be established for a source within the conductor. The solution of (28), vanishing as $z \to \infty$, may be written

$$\begin{bmatrix} \widetilde{\widetilde{\psi}}_1'(u, v, z) \\ \widetilde{\widetilde{\psi}}_1''(u, v, z) \end{bmatrix} = \begin{bmatrix} e' \\ e'' \end{bmatrix} e^{-\kappa z}. \tag{31}$$

The solution of (29) has the form

$$\begin{bmatrix} \widetilde{\widetilde{\psi}}_2'(u,v,z) \\ \widetilde{\widetilde{\psi}}_2''(u,v,z) \end{bmatrix} = \begin{bmatrix} c' \\ c'' \end{bmatrix} e^{-\gamma_2 z} + \begin{bmatrix} d' \\ d'' \end{bmatrix} e^{\gamma_2 z} \tag{32}$$

and the solution of (30) in the region above the source but below the surface of the conductor has the form

$$\begin{bmatrix} \widetilde{\widetilde{\psi}}_3'(u,v,z) \\ \widetilde{\widetilde{\psi}}_3''(u,v,z) \end{bmatrix} = \begin{bmatrix} a' \\ a'' \end{bmatrix} e^{-\gamma_3 z} + \begin{bmatrix} b' \\ b'' \end{bmatrix} e^{\gamma_3 z}. \tag{33}$$

The first term on the right hand side of (33) represents the free-space solution in the conductor. The a^i are source-dependent coefficients with $i = {}'$ or $''$. The b^i and c^i are related to the a^i via reflection and transmission coefficients which are found by using the field continuity conditions at the interface between the conductor and the surface layer, equations (27). Similarly, the d^i and e^i are related to the c^i by using the field continuity conditions at the interface between the surface layer and air, equations (26). In the latter case there is no cross-coupling between TM and TE potentials since the preferred direction in each region is the same. This means that the following relationships can be defined:

$$\begin{bmatrix} d' \\ d'' \end{bmatrix} = \begin{bmatrix} \Gamma' & 0 \\ 0 & \Gamma'' \end{bmatrix} \begin{bmatrix} c' \\ c'' \end{bmatrix} \tag{34}$$

$$\begin{bmatrix} e' \\ e'' \end{bmatrix} = \begin{bmatrix} T' & 0 \\ 0 & T'' \end{bmatrix} \begin{bmatrix} c' \\ c'' \end{bmatrix}, \tag{35}$$

where the Γ^i are reflection coefficients and the T^i are transmission coefficients. At the interface between the conductor and the surface layer, however, the possibility of cross-coupling must be allowed for. This allowance can be made by defining the following:

$$\begin{bmatrix} b' \\ b'' \end{bmatrix} = \begin{bmatrix} \Gamma_{11} & \Gamma_{12} \\ \Gamma_{21} & \Gamma_{22} \end{bmatrix} \begin{bmatrix} a' \\ a'' \end{bmatrix} \tag{36}$$

$$\begin{bmatrix} c' \\ c'' \end{bmatrix} = \begin{bmatrix} T_{11} & T_{12} \\ T_{21} & T_{22} \end{bmatrix} \begin{bmatrix} a' \\ a'' \end{bmatrix}. \tag{37}$$

The Γ_{ij} $(i, j = 1, 2)$ are reflection coefficients and the T_{ij} are transmission coefficients. Since the Green's function we require uses only Γ_{22}, see equation (9), the T^i will not be stated explicitly. First, the reflection coefficients in (34) are determined by applying (26) to give

$$\begin{aligned}
e^{-\kappa b}(iue' - iv\kappa e'') &= \frac{\mu_2}{\mu_0}[iu(c'e^{-\gamma_2 b} + d'e^{\gamma_2 b}) - iv\gamma_2(c''e^{-\gamma_2 b} - d''e^{\gamma_2 b})] \\
e^{-\kappa b}(ive' + iu\kappa e'') &= \frac{\mu_2}{\mu_0}[iv(c'e^{-\gamma_2 b} + d'e^{\gamma_2 b}) + iu\gamma_2(c''e^{-\gamma_2 b} - d''e^{\gamma_2 b})] \\
e^{-\kappa b} iv\kappa e' &= iv\gamma_2(c'e^{-\gamma_2 b} - d'e^{\gamma_2 b}) - iuk_2^2(c''e^{-\gamma_2 b} + d''e^{\gamma_2 b}) \\
e^{-\kappa b} iu\kappa e' &= iu\gamma_2(c'e^{-\gamma_2 b} - d'e^{\gamma_2 b}) + ivk_2^2(c''e^{-\gamma_2 b} + d''e^{\gamma_2 b}).
\end{aligned}$$

Expressing the solution of these equations in the form of (34) gives

$$\Gamma' = \frac{\gamma_2 - \frac{\mu_2}{\mu_0}\kappa}{\gamma_2 + \frac{\mu_2}{\mu_0}\kappa} e^{-2\gamma_2 b}, \qquad \Gamma'' = -e^{-2\gamma_2 b}. \tag{38}$$

Using (27) and expressing d' and d'' in terms of c' and c'' it is found that

$$\begin{aligned}
-iu(1+\Gamma')c' + iv\gamma_2(1-\Gamma'')c'' &= \frac{\mu_3}{\mu_2}[\gamma_3(a'-b') - uv(a''+b'')] \\
-iv(1+\Gamma')c' - iu\gamma_2(1-\Gamma'')c'' &= \frac{\mu_3}{\mu_2}(u^2-k_3^2)(a''+b'') \\
-iv\gamma_2(1-\Gamma')c' + iuk_2^2(1+\Gamma'')c'' &= uv(a'+b') - \gamma_3 k_3^2(a''-b'') \\
iu\gamma_2(1-\Gamma')c' + ivk_2^2(1+\Gamma'')c'' &= -(u^2-k_3^2)(a'+b').
\end{aligned}$$

The solution of these equations expressed in the form of (36) and (37) shows that

$$\begin{aligned}
D\Gamma_{11} &= \gamma_2\gamma_3\kappa^2(u^2-k_3^2)\left[\left(\frac{\mu_3}{\mu_2}\right)^2 k_2^2(1-\Gamma')(1+\Gamma'') - k_3^2(1+\Gamma')(1-\Gamma'')\right] \\
&\quad -\frac{\mu_3}{\mu_2}(\gamma_3^2u^2+k_3^2v^2)\left[\gamma_3^2k_2^2(1+\Gamma')(1+\Gamma'') - \gamma_2^2k_3^2(1-\Gamma')(1-\Gamma'')\right] \qquad (39)\\
D\Gamma_{12} &= 2\frac{\mu_3}{\mu_2}\gamma_3k_3^2uv[\gamma_3^2k_2^2(1+\Gamma')(1+\Gamma'') - \gamma_2^2k_3^2(1-\Gamma')(1-\Gamma'')] \qquad (40)\\
D\Gamma_{21} &= 2\frac{\mu_3}{\mu_2}\gamma_3uv[\gamma_3^2k_2^2(1+\Gamma')(1+\Gamma'') - \gamma_2^2k_3^2(1-\Gamma')(1-\Gamma'')] \qquad (41)\\
D\Gamma_{22} &= -\gamma_2\gamma_3\kappa^2(u^2-k_3^2)\left[\left(\frac{\mu_3}{\mu_3}\right)^2 k_2^2(1-\Gamma')(1+\Gamma'') - k_3^2(1+\Gamma')(1-\Gamma'')\right] \\
&\quad -\frac{\mu_3}{\mu_2}(\gamma_3^2u^2+k_3^2v^2)\left[\gamma_3^2k_2^2(1+\Gamma')(1+\Gamma'') - \gamma_2^2k_3^2(1-\Gamma')(1-\Gamma'')\right] \qquad (42)
\end{aligned}$$

where

$$D = \kappa^2(u^2-k_3^2)\left[\frac{\mu_3}{\mu_2}\gamma_2(1-\Gamma') + \gamma_3(1+\Gamma')\right]\left[\frac{\mu_3}{\mu_2}\gamma_3k_2^2(1+\Gamma'') + k_3^2\gamma_2(1-\Gamma'')\right]. \qquad (43)$$

The reflection coefficient required in (9) is given by (42) with (43). Note that the analysis in this section is not restricted to the thin-skin regime but applies for arbitrary skin depth.

4. Long Crack

4.1 Field Calculation

Consider a crack of gape a (in x) and depth d (in z) which is infinite in y ($c \to \infty$). The field solution is then best found by Fourier transforming (3) in y. If $\tilde{\Psi}(v,z)$ is the Fourier transform of $\Psi(y,z)$, then

$$\tilde{\Psi}(v,z) = A(v)\sinh[v(z+d)], \qquad (44)$$

which vanishes at the defect edge ($z=-d$) as required by (4). The coefficient $A(v)$ is to be determined from the boundary condition at the defect mouth.

Fourier transforming (14) in y gives

$$2\frac{v^2}{ik}\tilde{U}(v)\tilde{\Psi}(v,0) + \left.\frac{\partial\tilde{\Psi}(v,z)}{\partial z}\right|_{z=0} = \left.\frac{\partial\tilde{\Psi}^{(i)}(v,z)}{\partial z}\right|_{z=0}, \qquad (45)$$

where $k^2 = i\omega\mu\sigma$ again refers to values in the conductor and $\widetilde{U}(v)$ can be deduced from (13), (42) and (43). From (2) write

$$\mathbf{H}_{TM}(\mathbf{r}) = -\nabla \times \hat{x}\Psi(\mathbf{r}). \tag{46}$$

Then

$$\widetilde{H}^{(i)}_{yTM}(v,0) = -\left.\frac{\partial\widetilde{\Psi}^{(i)}(v,z)}{\partial z}\right|_{z=0}. \tag{47}$$

Substituting for the incident field in (45) using the above relation and for $\widetilde{\Psi}$ using (44) allows the determination of $A(v)$. We find

$$A(v) = -\frac{\widetilde{H}^{(i)}_{yTM}(v,0)}{v[\cosh(vd) + 2(v/ik)\widetilde{U}(v)\sinh(vd)]}. \tag{48}$$

4.2 Coil Impedance

The probe impedance change due to a crack, ΔZ, can be calculated by integrating the product of the TM potential and $H^{(i)}_y$ along the line of the crack mouth. The following expression can be obtained, with some manipulation, from a knowledge of the electromagnetic fields in the vicinity of the crack by using the principle of reciprocity [3]. In the thin-skin regime,

$$\begin{aligned}\Delta Z &= \frac{1}{\sigma_0 I^2}\left\{\left(k^2a + 2ik\right)\int_{-\infty}^{\infty} H^{(i)}_y(y,0)\Psi(y,0)\,dy\right.\\ &\qquad \left. + ika\int_{-\infty}^{\infty} H^{(i)}_y(y,0)[H_y(y,0) - H_{yTM}(y,-d)]\,dy\right\},\end{aligned} \tag{49}$$

where I is the current flowing in the coil. The first term in (49) represents contributions from the crack volume (the term of order k^2) and faces (the term of order k), and the second term accounts for the loss of conductor surface at the crack mouth and creation of surface at the crack base. The total magnetic field at the crack mouth, $H_y(y,0)$, can be calculated from an integral equation in a manner similar to the calculation of Ψ [3], and $H_{yTM}(y,-d)$ can be deduced from Ψ using (46). For a normal, air-cored coil, the incident field can be obtained from existing closed-form expressions [6].

We now present some calculations of the impedance change in an eddy-current coil due to a surface crack concealed by a layer. In these calculations, the layer is assumed to have lower values of conductivity and permeability than the host material, but the formulation is not restricted to these cases. The theory is valid for arbitrary values of conductivity and permeability in the layer and in the host. Since this is a thin-skin formulation, however, the electromagnetic skin-depth in the host must be somewhat smaller than the dimensions of the defect.

In Figure 3, the impedance change in an eddy-current coil due to a long slot in an aluminium test piece is shown as a function of the ratio of the layer conductivity to the host conductivity, for layers of various thickness. For low values of layer conductivity, the reduction in the magnitude of the impedance change with increasing layer thickness is a consequence of the fact that the probe is separated from the metal substrate. This is a lift-off effect; as the layer thickness increases, the probe becomes more distant from the crack and there is a reduction in the signal. For this particular coil and crack

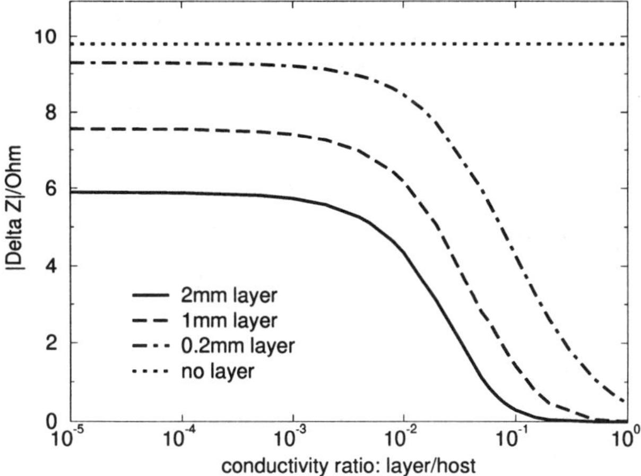

Figure 3: Probe impedance as a function of layer conductivity.

Table 1: Parameters.

Figure		3	1, 4
material	σ/Ω^{-1}m	1.667×10^7	5.587×10^6
	μ_r	1	87 ± 5
EDM slot	depth/mm	2.96	2.95 ± 0.05
	gape/mm	0.42 ± 0.06	0.33 ± 0.03
coil	inner radius/mm	9.34	3.015
	outer radius/mm	18.40	5.46
	stand-off/mm	2.05	2.94
	length/mm	8.8	1.29
	no. of turns	408	900
	frequency/kHz	100	100

configuration, whose parameters are given in the first column of Table 1, the theory predicts a dramatic reduction in the measured impedance change for layer conductivities greater than one hundredth of the host conductivity. For example, for a layer 0.2mm thick with $\sigma_2/\sigma_3 = 0.1$, $|\Delta Z|$ is about 50% of the value obtained with $\sigma_2/\sigma_3 = 10^{-5}$. For a layer 2mm thick, the crack signal practically vanishes for $\sigma_2/\sigma_3 = 0.1$, dropping to 5% of the value obtained for a non-conducting layer. When the conductivity of the layer matches that of the host, the magnitude of the signal also practically vanishes for layers 1 and 2mm thick. There is still a small signal from the crack buried beneath a 0.2mm layer since the skin depth in this case is about 0.4mm and some induced current penetrates through the layer into the host and interacts with the crack. In Figure 4, the effect of variations in the permeability of a conducting layer deposited on a mild steel substrate is examined. Predictions of coil impedance are plotted for several values of layer thickness. The coil and material parameters are given in the second column of Table 1. In this calculation, the layer is assigned conductivity one thousandth of that

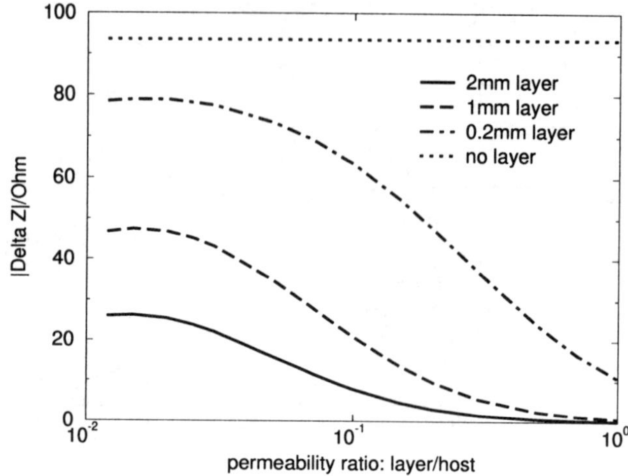

Figure 4: Probe impedance as a function of layer permeability.

of the host material since this value is thought to be typical of oxide deposits. Again, there is a reduction in probe signal magnitude as the value of the layer permeability increases towards that of the host. It is clear that the presence of a conducting and/or permeable layer may significantly reduce and even obscure the signal from a crack in the metal below.

5. Conclusion

A method has been presented for calculating the impedance change in an eddy-current probe due to a crack in a conductor with a surface layer of a different material. The theory is based on a thin-skin model of a crack in a half-space conductor, in which a single scalar potential is used to represent the electromagnetic fields in the conductor. The scalar potential obeys Laplace's equation on the domain of the crack, subject to certain boundary conditions along the line of the crack mouth and edge. The effect of a layer on the surface of the conductor is to modify the boundary condition at the crack mouth.

The impedance change in an eddy-current probe has been calculated for a long crack concealed by layers of various thickness, conductivity and permeability. Even relatively thin layers with typical values of conductivity and permeability significantly reduce the signal magnitude. In practice, deposited material often fills a crack, providing a conducting pathway between its faces [1], as well as forming a surface layer. We conclude that the combination of these two effects can easily reduce the crack signal to a point where some cracks will not be detected.

Acknowledgements

This work was conducted for Magnox Electric Plc and Nuclear Electric Ltd. The authors gratefully acknowledge the provision of experimental data by S. K. Burke.

References

[1] N. Harfield and J. R. Bowler, A Thin-Skin Theory of Eddy-Current Leakage across Surface Cracks, In: R. Albanese *et. al.* (eds.), Studies in Applied Electromagnetics and Mechanics 14, IOS Press, Amsterdam, 1998, 1–8.

[2] L. B. Felsen and N. Marcuvitz, Radiation and Scattering of Waves. Prentice-Hall, Englewood Cliffs, 1973.

[3] N. Harfield and J. R. Bowler, Theory of Thin-Skin Eddy-Current Interaction with Surface Cracks, *J. Appl. Phys.*, **82** (1997) 4590–4603.

[4] A. M. Lewis, A Theoretical Model of the Response of an Eddy-Current Probe to a Surface-Breaking Metal Fatigue Crack in a Flat Test-Piece, *J. Phys. D: Appl. Phys.* **25** (1992) 319-326.

[5] J. R. Bowler, Eddy-Current Interaction with an Ideal Crack. I. The Forward Problem, *J. Appl. Phys.* **75** (1994) 8128–8137.

[6] C. V. Dodd and W. E. Deeds, Analytical Solutions to Eddy-Current Probe-Coil Problems, *J. Appl. Phys.* **39** (1968) 2829–2838.

Electromagnetic Nondestructive Evaluation (III)
D. Lesselier and A. Razek (Eds.)
IOS Press, 1999

Characterisation of the Electromagnetic Field Generated by Eddy Current Probes With Measurement and Modelling

Stéphanie MASTORCHIO
NonDestructive Testing Group, EDF/Research and Development Division,
6, quai Watier - 78401 CHATOU (FRANCE)

Françoise THEVENOT
Advanced Technologies Development Center, Dassault Aviation
Z.A. Louis Bréguet, BP 12, 78141 VÉLIZY-VILLACOUBLAY Cedex, France

Abstract. In order to ensure the quality of Non Destructive Testing measurements, eddy current probes need to be characterised for their configuration of inspection. This involves not only an improved design of the emitters for a good interaction with the defects, but also an optimisation of the sensitivity of the receivers. In this paper means to characterise eddy current probes according to the electromagnetic field that are based on both experimentation and modelling are discussed.

1 Introduction

Eddy currents are commonly used for the inspection of metallic components in nuclear power plants and in the aeronautic field. The design of an Eddy Current (EC) probe leads to the optimisation of both receiver and emitter components to detect cracks with a good sensitivity. Current researches tend to focus on high-sensitivity receivers such as SQUID, magneto-resistance, and Hall effect probes. But these sensitive components require higher precision when setting up the inspection process, and strict calibration procedures.

Dassault Aviation owns an electromagnetic field acquisition bench employing magnetoresistor receivers for a large range of frequencies. EDF/DER uses the 3D modelling code TRIFOU in order to simulate and analyse EC inspections. In this paper, a quantitative comparison of electromagnetic field calculations with modelling and acquisition is carried out for an EC probe.

2 Modelling the right configuration with TRIFOU

TRIFOU is a combined Finite Elements - Boundary Integral Formulation code for solving Maxwell's equations. It has been developed by EDF/R&D since ten years (project initiated by A. Bossavit and J-C. Verite, EDF/DER, France). TRIFOU is capable to treat different electromagnetic problems such as magnetostatic, magnetodynamic (transient or harmonic regime), electrocinetic, electrostatic, and microwave ones.

The aim of every modelling code is to get as close as possible to the experimental configuration. This means to fully take into account the complexity of the geometry, of the material, and of the inductors.

2-1 CAD input

CAD input data can be read by TRIFOU for a simple format including nodes position and connection between the nodes, group property preparing material and symmetry attribution. The current format is the one generated by the CAD software I-deas, though connection is possible with other CAD software. In order to draw simple geometry such as probes, plates and tubes with artificial defects, a library of simple geometrical shapes is available. In figures 1 and 2 we show how the map of a complex component (here a ferrite pot) is represented as a CAD file and modelled with TRIFOU.

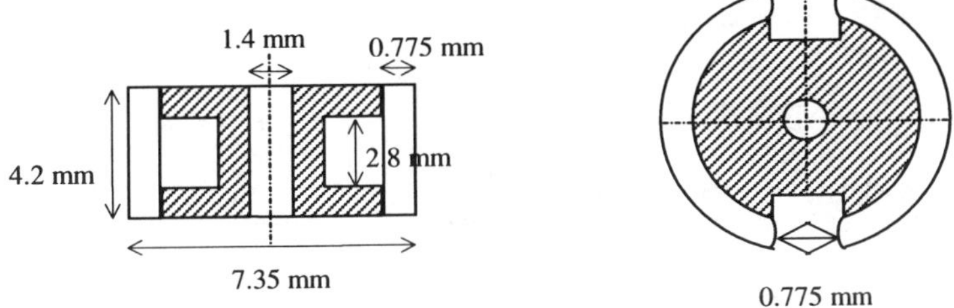

Fig 1 : ferrite description.
2D view of the ferrite pot description (the bobbin is placed just inside the pot).

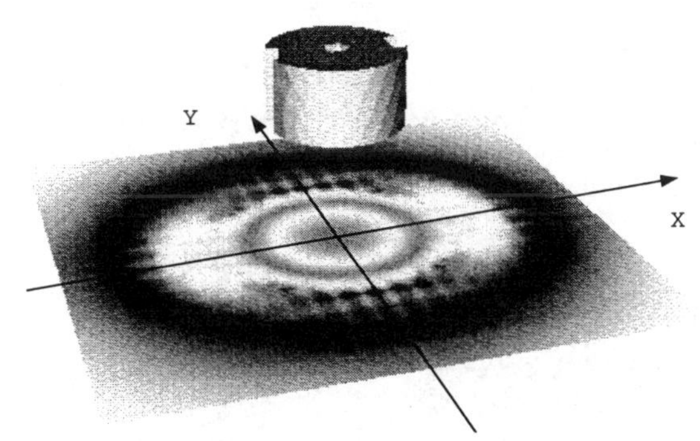

Fig 2 : TRIFOU modelling of the ferrite.
3D view with the value of the amplitude of the magnetic field in a plane.

2-2 Material input

TRIFOU can take into account any kind of linear isotropic material for both transient and harmonic testing. A non-linear model is used in the static regime and is being extended to periodic and transient cases. This parameter is important for ferrite and ferromagnetic material such as steel. But, the non-linear curve is not always available, which is why an approximation of linear behaviour is useful to simulate the inspection and can be validated.

The material representation of the defect is certainly as important as its shape representation. A defect can be modelled in TRIFOU as a meshed voluminous defect, or as a surface defect with certain conductivity properties. Changing this value of conductivity, to go from an impenetrable defect to the case without defect (the conductivity becomes equal to the one of the block) has a great impact on the signal. The example below (figure 3) shows the change in amplitude and in phase for an impedance probe when the resistivity is varied from infinite (case 1) to the one of an Inconel plate (case 7).

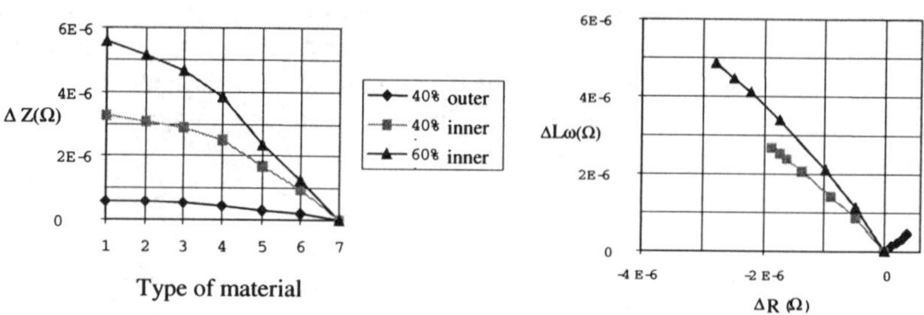

Fig 3 : TRIFOU modelling of impedance probe responses on an Inconel test block with 3 different defects. Influence of the resistivity of the artificial rectangular defect at 220 kHz.

2-3 Probe characteristic input

In order to improve NDT inspection, many configurations of probes are tested and optimised so as to fit to the defect to be detected. TRIFOU offers us a library of ready-to-use probes, which is taking into account the most common configurations encountered in NDT : coil bobbin, D-probe, square probe, two parallel-axis coil bobbins, two axisymetric coil bobbins, two crossed bobbins, etc. (figure 4). Few parameters are to be fixed such as centres distance, outer and inner radii, and of course power supply. In case of test block symmetry, the calculation can be simplified with probe symmetry. This facilitates the construction of multi-element probes with several bobbins that are emitting simultaneously.

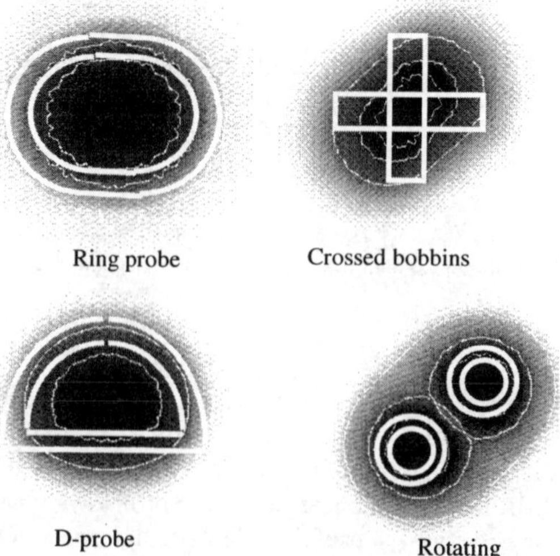

Fig 4 : TRIFOU modelling ring probe, crossed bobbin probe, D-probe and rotating probe (two parallel coils). Amplitude of the magnetic field in the air 0.5 mm below the probes.

3 Measuring the magnetic field with the Dassault Aviation bench

The detection and the quantification (sizing) of defects in metallic materials with Eddy Current techniques are based on the performances of the EC equipment and probes. The EC probes, which get all the information from the materials to be inspected, need to be well characterised so as to ensure the quality of the Eddy Current inspection. This characterisation has to take into account their geometry, their electrical and electromagnetic characteristics as well as their behaviour in relation to materials and defects.

The knowledge of this electromagnetic field is very important for a better understanding of the field distribution and its influence on defects but also to compare probes between one another and to follow their evolution. So, an equipment has been developed in order to carry out the electromagnetic characterisation of eddy current probes (figure 5). This equipment can also be used to validate the electromagnetic model and to design EC probes.

3-1 Principle of measurement

Electromagnetic field measurement consists in the determination, at any point of a prescribed plane or a volume, of the module and of the direction of the magnetic field generated by an Eddy Current probe. The module and the direction of this field, H, is calculated from the measurement of its three components H_x, H_y and H_z in an orthogonal reference system $(Oxyz)$.

The prototype developed so far enables us to realise a 2D map of the electromagnetic field generated by Eddy Current probes. Two types of maps can be obtained: transverse cross-section maps, realised in a section perpendicular to the Eddy Current probe axis, and longitudinal cross-section maps, realised in a section parallel to the Eddy Current probe axis. These measurements are performed with magnetoresistors. These magnetic sensors are quite appropriate for the characterisation of the electromagnetic field generated by Eddy Current probes. In particular, they operate in a large frequency range from DC up to several MHz.

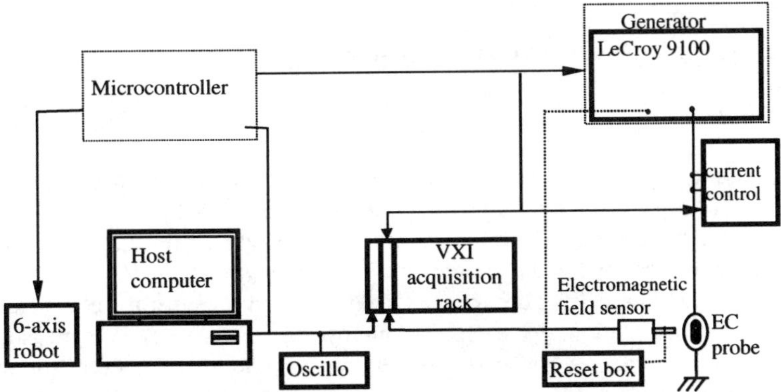

Fig 5. Electromagnetic field acquisition bench.

3-2 Electromagnetic field acquisition bench

Motorised displacement equipment, consisting of a 6-axis robot with numerical control connected to a remote host computer, enables the mechanical positioning and displacement of the EC probes. A Lecroy 9100 generator supplies the exciting signal to the EC probe. An amplifier has been developed, which allows the amplification of the signal output from the magnetoresistor. A digital acquisition system, based on VXI standard, collects and digitises the information from the amplifier. A HP 9000 computer drives all the installation and allows the automation of the data acquisition and treatment. Because of the magnetoresistor flipping, a reset box has been developed to ensure the initialisation of the magnetoresistive sensor. Transfer function of the measuring system has been determined, allowing us to define the calibration procedures of the system.

The electromagnetic field measuring system is balanced in air. The magnetoresistor is placed in front of the EC probe. The correct position of the magnetoresistor is obtained by the determination of the energy maximum along each axis of the mechanical scanner. The map is realised by scanning a prescribed cross-section in space. The scanning step can be chosen ranging from 0.01 mm up to 5 mm. The frequency bandwidth of the electromagnetic field measuring system goes from 100 Hz up to 1.5 MHz, the field magnitude ranging from 0 A/m up to 500 A/m.

The acquisition toolbox was developed in C language on the HP 9000 computer. The magnetoresistor response is tone burst. This tone burst (with a fixed length) is recorded at each point of the examined cross-section. The analysis tools box was developed with the PV-WAVE software on the same computer. It allows to visualise the acquired electromagnetic signal (tone burst) at each point of the examined cross-section, and the 2D electromagnetic field reconstruction (2D image, surface plot, contour and H profile line). Signal processing is also possible with this toolbox, such as spectral analysis, signal deconvolution, analysis of the module and phase of the measured signal.

The different results presented herein has been obtained with a 1-axis measuring head, i.e., we measure a single component only of the electromagnetic field. But we also developed and validated a 3-axis measuring head. This particular head allows the measurement of the three components H_x, H_y and H_z of the electromagnetic field, and the reconstruction of the module and phase of H can be done by software.

4 Comparing experimental and modelled data

4-1 Configurations

The chosen configuration consists of an impedance probe with a complex ferrite pot placed over an aluminium plate (5 mm depth) with an artificial crack (20% OD). This defect is considered as an impenetrable, infinite defect (same length as the plate. The ferrite pot geometry has been described in the paragraph *2.1 CAD input*.

Acquisition and modelling are realised and compared for several frequencies (10 kHz and 20 kHz) and alimentation. For each configuration the magnetic field is measured below the aluminium test-block in a plane situated at 1 mm below the surface of the test block. The acquisition is made with Dassault Aviation bench with a step of 0.5 mm over a 30 mm x 30 mm zone in the plane of measurement centred in the middle of the aluminium plate. This provides us with a 60 x 60 point image representing the magnetic amplitude for the normal component of the field (figures 6 and 7).

Table 1 Parameters for field calculation and measurement

Probe type	Probe SNEC AF1T100 (impedance probe)
Probe dimensions	2 mm (inner diameter) x 3 mm (outer diameter) x 2 mm (high)
Ferrite pot	$\rho = 0.2\ \Omega.m$ and $\mu r = 2280$
Test block	Square plate
material	Aluminium alloy
conductivity	$\sigma = 1.66 \times 10^6$ S/m
dimensions	99 mm x 98 mm x 5 mm (depth)
Defect	artificial rectangular 20% defect
dimensions	99 mm (length) x 0.5 mm (width)

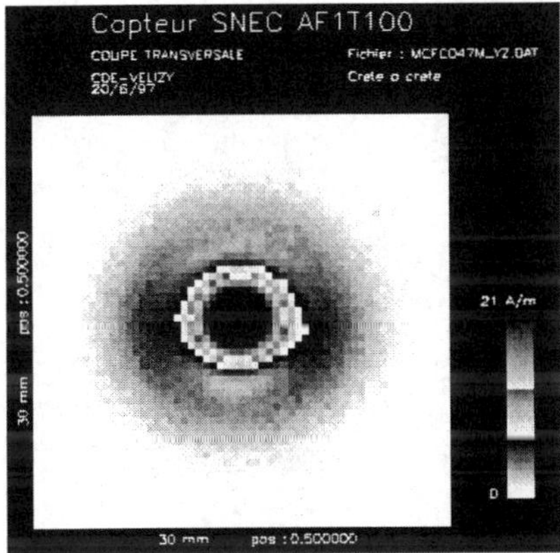

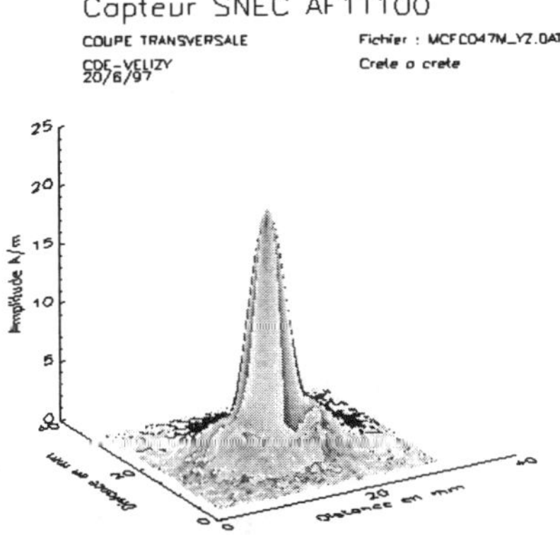

Fig 6 : Dassault acquisitions with case 1 (10 kHz, 200 mA).
2D and 3D views of the amplitude of the normal component of the magnetic field.

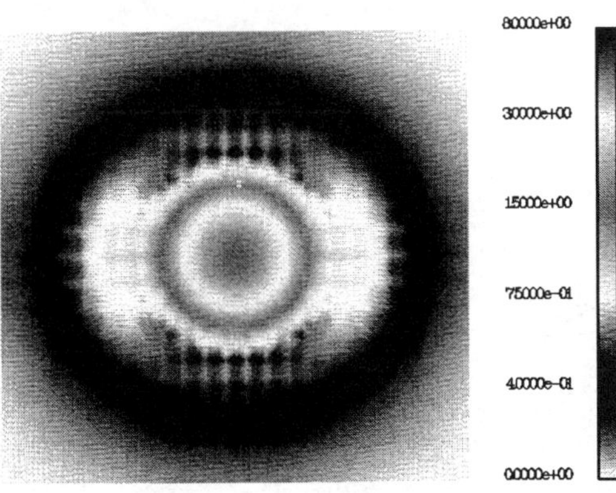

Fig 7 : TRIFOU modelling with case 1 (10 kHz, 200 mA).
2D view of the amplitude of the normal component of the magnetic field (A/m).

Table 2 Configuration for comparison

Case 1	
Frequency	10 kHz
Alimentation	200 mA
Tension	5.4 Volt
Case 2	
Frequency	20 kHz
Alimentation	100 mA
Tension	8.6 Volt

4-2 Quantitative comparison

The images of modelling and acquisition show the similitude between the two methods (figure 8 and figure 9). In order to proceed with a quantitative comparison, we extract two lines corresponding to the X axis and the Y axis represented in figure 2, the two lines crossing the maximum of intensity of the magnetic field just below the probe centre.

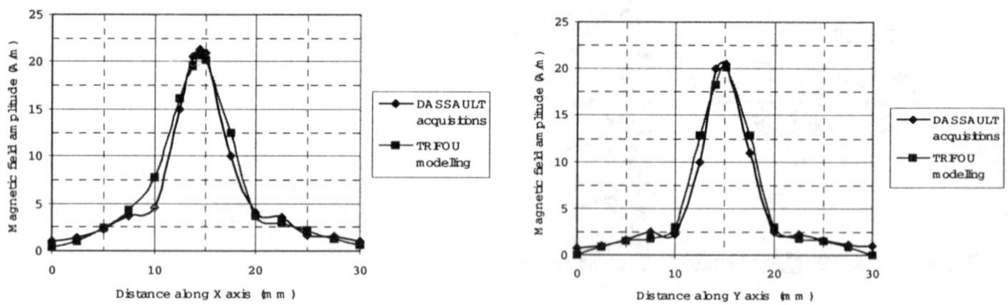

Fig 8 : Comparison between modelling and acquisitions.
Case 1 (10 kHz / 200 mA) along the axes X and Y in the zone of measurement.

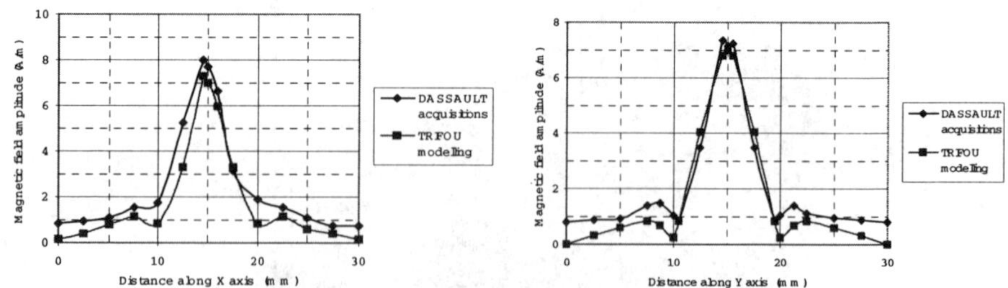

Fig 9 : Comparison between modelling and acquisitions.
Case 2 (20 kHz / 100 mA) along the axes X and Y in the zone of measurement.

The comparison shows good results for the two cases (figures 8 and 9). Some parameters have to be taken into account with modelling such as wire diameter and the number of turns. The approximation made for the ferrite pot also seems to be quite good for the two given frequencies.

5 Conclusion

The knowledge of the electromagnetic field is very important if one wishes to better understand the field distribution and its influence on the defects, but also to compare probes between them and to follow their evolution. The choice of the technology adapted to a given problem depends on the amplitude of the signal due to the defect to be detected. The type of receiver (coil, thin film coil, magnetoresistor, Hall effect probe, SQUID) depends on the amplitude of the signal and on the range of variation. For example, electromagnetic field can be useful when positioning the receiver component of a probe. This position can be optimised by calculating the normal difference field as :

$$Hdiff = \frac{H_{with\_defect} - H_{without\_defect}}{H_{without\_defect}}$$

Both modelling and experimental equipment are available in order to carry out this electromagnetic characterisation of Eddy Current probes. We presented in this paper the capabilities of both products. Comparisons between modelling and experimental results showed good agreement. So they appear complementary for the characterisation of EC probes.

References

[1] P.O. Gros, "Numerical modelling of eddy current steam generator inspection : Comparison with experimental data," *Review of Progress in Quantitative Nondestructive Evaluation*, 257-261, D. O. Thompson and D. E. Chimenti (eds), Plenum Press, New York, 1996.

[2] S. Mastorchio and P.O. Gros, "TRIFOU : 3D Modelling for Eddy Current testing in the steam generator tubes," *Electromagnetic Nondestructive Evaluation*, 137-145, R. Albanese *et al.* (eds), IOS Press, Amsterdam, 1997.

[3] S. Mastorchio and N. Harfield, Modelling of Eddy-Current interaction with cracks in the thin-skin regime: two approaches, *Proceedings of the 7th European Conference on Non-Destructive Testing*, 2398-2405, Copenhagen, 1998.

[4] F. Thevenot, M. Dessendre, and H. Trétout, "Characterisation of the electromagnetic field generated by Eddy Current probes," *Review of Progress in Quantitative Nondestructive Evaluation*, 981-987, D. O. Thompson and D. E. Chimenti (eds), Plenum Press, New York, 1996.

[5] S. Mastorchio, F. Thevenot, and M. Dessendre, "A modelling approach for non destructive testing with eddy current," *Contrôle Industriel,* N° 210, pp 62-63 (March 1998).

Electromagnetic Nondestructive Evaluation (III)
D. Lesselier and A. Razek (Eds.)
IOS Press, 1999

Simulation of the behavior of Cecco probes to lift-off and tilt effects during tube inspection

Michel MAYOS, Stéphanie MASTORCHIO, Laurence AUBRY
EDF/DER, 6 Quai Watier, 78401 Chatou Cedex, France
Kenzo MIYA
NERL, The University of Tokyo, Tokai-Mura, Ibaraki 319-11, Japan

Abstract. The sensitivity of Cecco probes to geometrical lift-off and tilt effects has been simulated using 3D modelling codes (EDF's TRIFOU and the University of Tokyo's FEM-BEM code), in order to evaluate some of the functional characteristics of these configurations in a tube-testing situation. Both the signal created in the absence of defect (noise evaluation) and the influence of the geometry effect on the signal response have been observed. It is shown how numerical simulation can be used to simplify to a significant extent characterization experiments and parametric studies, as well as to foresee the behavior of the probe in the actual inspection situation.

1. Introduction

At the present time, the eddy current non destructive inspection of steam generator tubes in French nuclear power plants is performed through the use of bobbin coils and rotating probes, which give satisfactory performance. Nonetheless, the anticipation of future needs leads to an expression of interest towards multi-element probes, among which the Cecco probes constitute the only commercially available ones [1]. These probes are in theory less sensitive to distance effects such as tilt and lift-off, due to their separate transmit-receive configuration. Nonetheless, their ability has to be demonstrated in tube testing situations, where the inspection of expansion zones represents a critical issue, in which one faces the problem of detecting crack-like defects in a varying geometry. In this regard, it appears very challenging to use simulation tools to reduce the volume and the cost of characterization experiments and mock-up design.

The present paper shows how the numerical modelling approach can be used to simulate the evaluation of the geometrical lift-off and tilt effects, taking into account their two aspects :

- the signal created by lift-off and tilt alone (no-defect situation), leading to noise evaluation ;
- the effect on defect detection (defect situation).

In order to simulate the inspection of steam generator tubes, all the conductors modelled in this study are in Inconel 600.

2. Lift-off and tilt effects in the absence of defect

The simulation has been achieved using TRIFOU, which is a 3D EDF-developed FEM-BEM code [2]. In this part of the study, the modelled probe configuration is an absolute one, unlike the commercial Cecco probes, in order to have an existing lift-off effect. The geometry considered is the one represented on figure 1.

The analyzed parameter, in a rather classical way for no-defect configurations, has been the impedance variation : $\Delta Z = Z - Z_{air}$,
where Z and Z_{air} are the probe impedances in the current testing situation and in the air.

2.1. Plane geometry

In a first step, calculation has been achieved for a plane geometry both for lift-off and tilt variations. Lift-off is taken as the distance from the probe lowest point to the plane and tilt as the angle between the probe plane and the inspected plane, as shown in figure 2. Lift-off was given values from 0 to 1 mm and tilt from 0 to 15°.

Figures 3 and 4 display the effect of each geometrical parameter (i.e. lift-off variation for a constant tilt and vice-versa) on the impedance plane trajectories. It is observed that both effects are rather similar in phase.

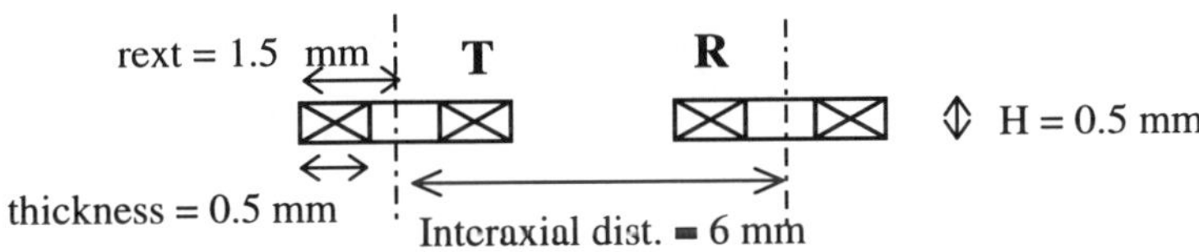

Figure 1 : Probe geometry for the no-defect situation

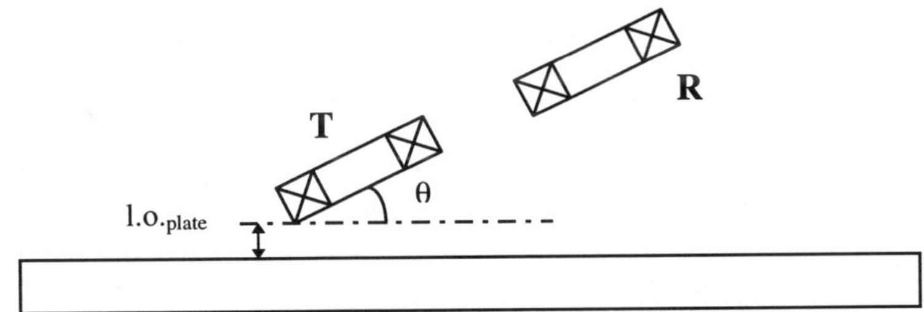

Figure 2 : Lift-off and tilt parameters in the no-defect situation

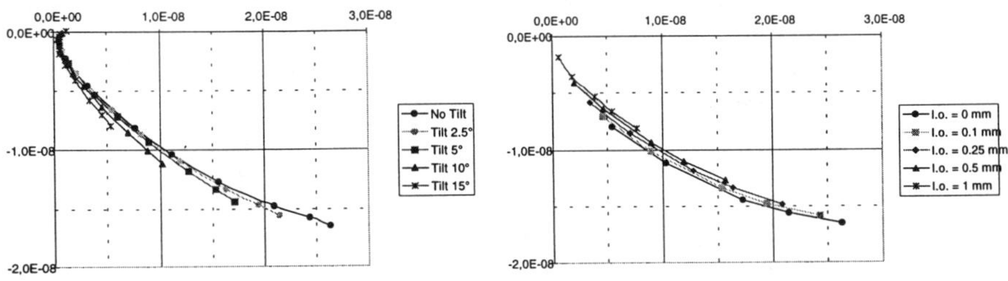

Figure 3 : Lift-off effect for different tilts

Figure 4 : Tilt effect for different lift-offs

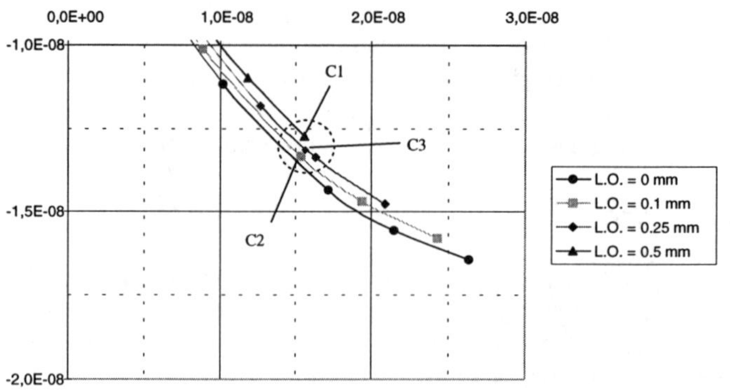

Figure 5 : Equivalent points in the impedance plane

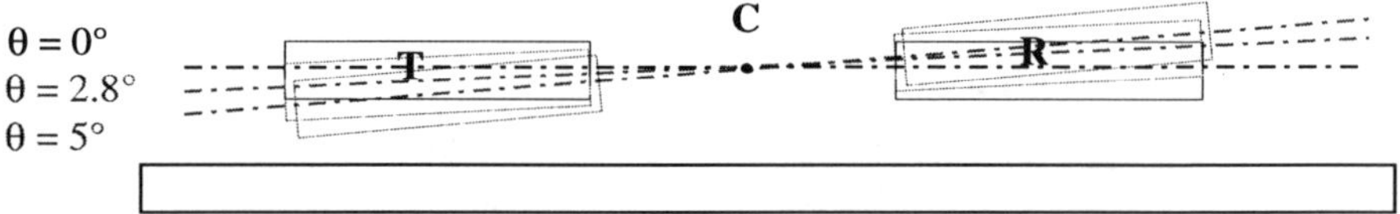

Figure 6 : Equivalent geometrical configurations

A closer analysis shows that several (lift-off / tilt) combinations lead to the same point in the impedance plane. For example, on figure 5, this is the case of the following points :

C1 : lift-off = 0.5 mm ; tilt = 0°

C2 : lift-off = 0.1 mm ; tilt = 5°

C3 : lift-off = 0.25 mm ; tilt = 2.8°

When the geometrical situations corresponding to points C1, C2 and C3 are represented (cf. figure 6), it can be observed that they have a crossing point C. This point is the middle of the segment linking the centers of the transmitter and of the receiver coils, and can be considered as the "center" of the probe. It is thus observed that, when the probe is tilted around its "center", almost no change in the eddy current response occurs. This does not necessarily mean that the probe is unsensitive to tilt, since, in real testing situations, it cannot be controlled that the probe is tilted around this particular point. Nonetheless, this is an important result, because it is shown that, for separate transmit-receive probes, the study of tilt effect is equivalent to the study of lift-off effect, with lift-off being measured at the "center" of the probe. This equivalence can be used to simplify to a large extent the experimental characterization of the probes, since an experimental study of lift-off effect is much easier (and cheaper) to carry on than the one of tilt effect.

2.2. Tubular geometry

In a second step, a lift-off variation for a tubular geometry has been simulated and compared to the one obtained for a plane geometry with a "saw-tooth" probe, in which the transmitter and the receiver are tilted in an opposite direction with an angle φ with the inspected plane (cf. figure 7).

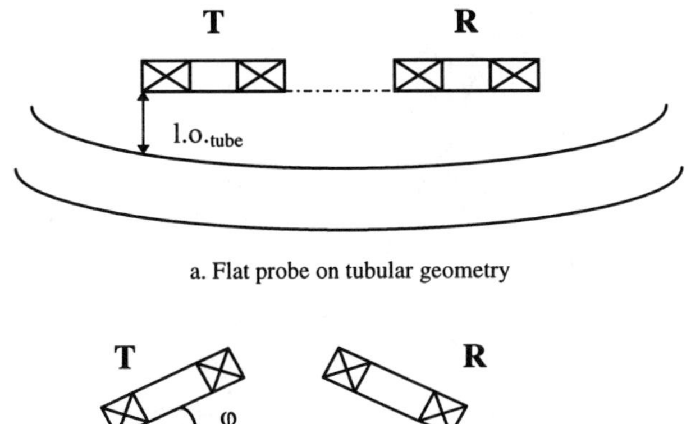

a. Flat probe on tubular geometry

b. "Saw-tooth" probe on plane geometry

Figure 7 : Modelled tubular and plane geometries

Figure 8 shows the trajectories obtained in the impedance plane when lift-off varies for the tubular geometry and for the plane geometries with values of φ between 0 and 15°. At first sight, these curves do not appear to match. Nonetheless, it can be observed that the points of the "plane probe on tube" curve are homothetic to the ones of the "13.5° saw-tooth probe on plate". This feature becomes more apparent by comparing these two curves after having applied a multiplicative factor of 0.544 to the points of the "13.5° saw-tooth probe on plate" (cf. figure 9). Thus, it appears that the two configurations drawn in figure 7 are equivalent, provided that a value of 13.5° is given to φ and that a multiplicative factor of 0.544, which can be assimilated to an instrument gain, is applied to the eddy current response of the " saw-tooth probe". The value of 13.5° is not purely coincidental, as can be seen in figure 10 : it corresponds to the angle between the probe plane and the line joining the extremity of the probe when it is in contact with the tube and the point under the "center" of the probe, as defined in § 2.1. So, as in § 2.1, the two equivalent configurations correspond to the same value of the lift-off of the "center" of the probe.

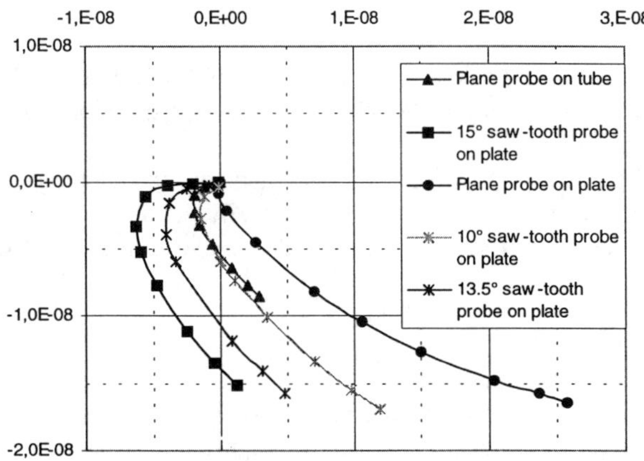

Figure 8 : Impedance plane trajectories for different configurations

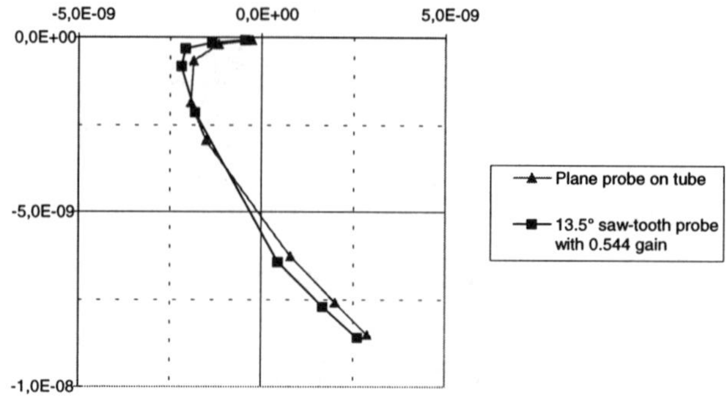

Figure 9 : Impedance plane trajectories after application of a gain to the "13.5° curved probe on plate" curve

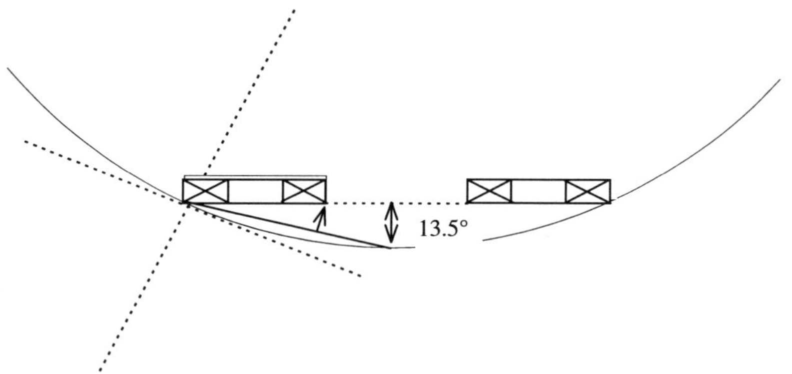

Figure 10 : Detailed view of the tubular geometry

This result is again highly interesting, since it shows that the more realistic, but harder to simulate, tubular geometry can be approximated by a plane geometry with a "saw-tooth" probe. Hence, this study led with a full-3D code shows that, for parametric studies where repetitive calculations are needed, a simpler model (e.g. an analytic one with a semi-infinite plane) can be used with no need to represent curved interfaces.

3. Effect on defect detection

This part of the study has been conducted using the FEM - BEM code developed by the Nuclear Engineering Research Laboratory (NERL) at the University of Tokyo, based on the A - Φ formulation [3].

The effect of lift-off and tilt has been studied on the detection of axial and circumferential outer notches in a tubular geometry, simulating a PWR steam generator tube, i.e. 22.22 mm outer diameter, 1.27 mm wall thickness, in Inconel 600. The notch dimensions were :

- length : 5 mm,
- width : 0.2 mm,
- depth : 20% of wall thickness.

The probe configuration was the actual commercial "Cecco C3" arrangement, i.e. with two differential receivers (cf. figure 11). This configuration is meant to be essentially sensitive to circumferential defects. Two frequencies were used for the calculation : 100 kHz and 400 kHz.

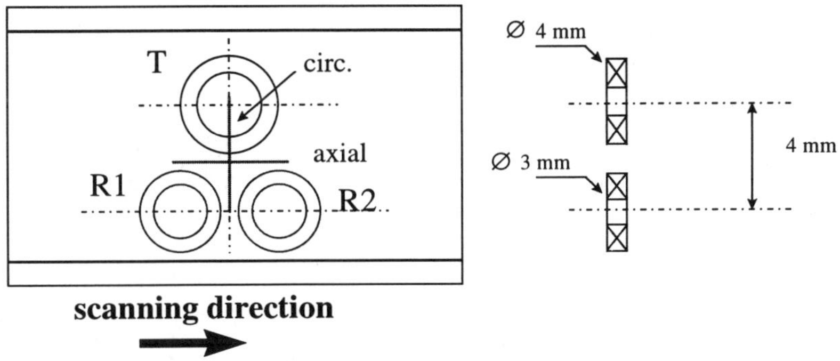

Figure 11 : Modelled geometry

The analyzed parameter was, like in a real eddy current inspection : $\Delta Z_{def} = Z - Z_0$, where Z_0 is the impedance in the no-defect case.

Figures 12 to 16 show the influence of different effects on the eddy current response to the defects in the impedance plane for both simulated frequencies.

Figure 12 shows the effect of the defect orientation on the non-lifted, non-tilted configurations. We have the confirmation of the better sensitivity of the probe to circumferential notches. Furthermore, it can be seen that the probe does not allow a very clear phase discrimination between the two orientations, whatever frequency is chosen.

Figures 13 and 14 show the effect of lift-off when it is increased from 1 mm (full lines) to 1.115 mm (dotted lines) : this effect is very small for both orientations, especially at 100 kHz. Once again, this confirms the expected behavior : low sensitivity to lift-off of separate transmit-receive probes with a preferred orientation ; increasing sensitivity when the frequency increases.

Figures 15 and 16 show the effect of a 2° tilt around a circumferential axis. If the response to the axial notch remains relatively unaffected, this is not the case of the circumferential one, where the signal amplitude is roughly divided by 2. The use of simulation shows here that lift-off and tilt act differently on the defect response. This has to be taken into account while studying the effect of the expansion transition zone, which can be seen as the combination of a lift-off and tilt variation.

To make the link with the study conducted in the no-defect case (section 2), it can then be concluded that, for Cecco probes, tilt gives a noise effect which is strongly comparable to the one of lift-off, but results in a much higher degradation of the defect signal. It is then demonstrated that, for the study of this last parameter, using a simplified model instead of the full-3D approach that was taken could lead to erroneous interpretations : thus, the use of a realistic representation of the actual geometry is necessary, whatever it concerns experiment or simulation.

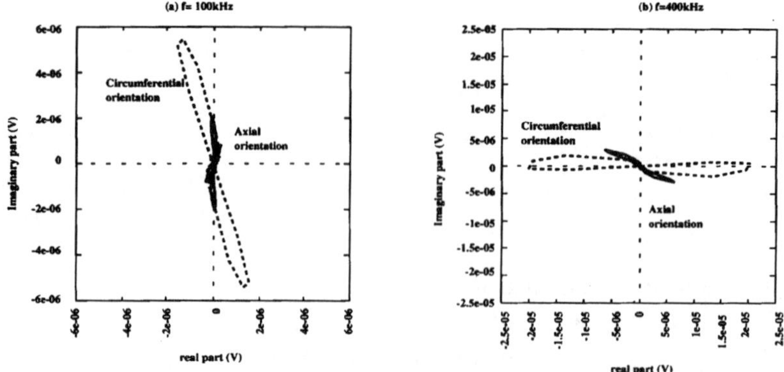

Figure 12 : Effect of defect orientation on eddy current response

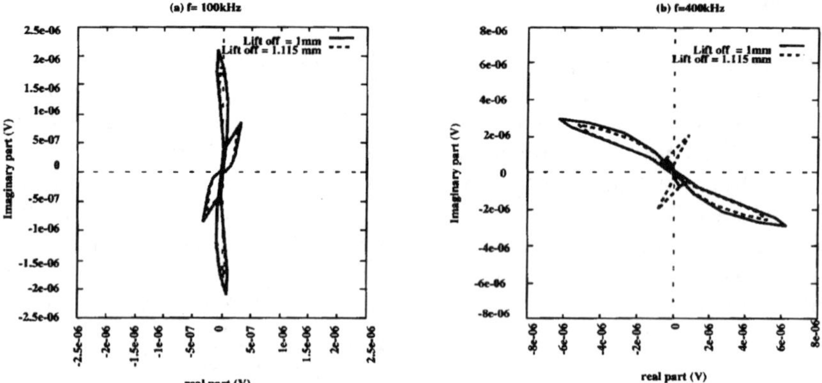

Figure 13 : Effect of lift-off on eddy current response - Axial notch

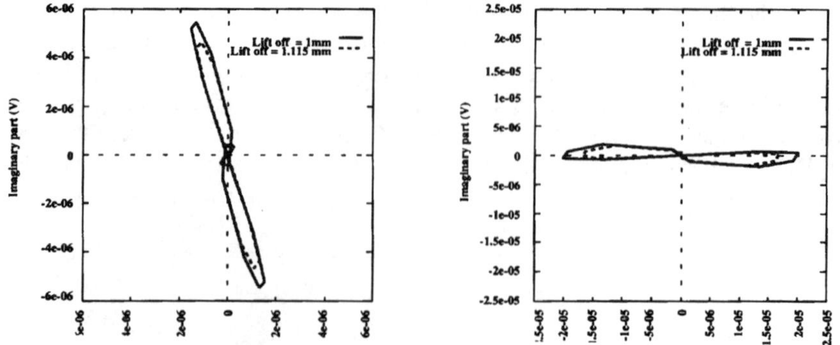

Figure 14 : Effect of lift-off on eddy current response - Circumferential notch

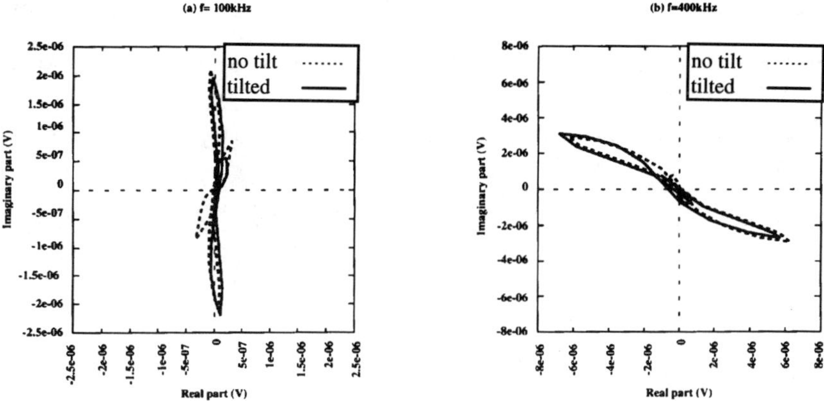

Figure 15 : Effect of tilt on eddy current response - Axial notch

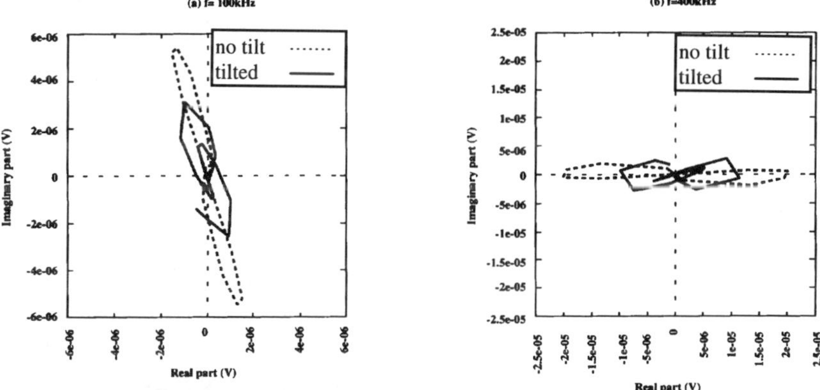

Figure 16 : Effect of tilt on eddy current response - Circumferential notch

4. Conclusions

The application of 3D-modelling to the case of steam generator tubing inspection with Cecco probes has shown the advantages that an eddy current user can expect from the use of simulation in the functional characterization of probes.

The study of noise evaluation shows that actual testing situations can be approximated by simplified configurations, resulting in the possibility to use both simpler experimental set-ups and models to lead parametric studies in which a high variability of the influencing parameters is sought. This leads to a significant gain in time, cost and operation complexity.

The study of defect response confirms the main expected features of the probe, but points out lesser-known characteristics, which can somehow impede its performances if they are not properly taken into account. In this case, it is thus preferable that a simplified approach is not substituted to the full 3D-simulation.

References

[1] V.S. Cecco, J.R. Carter, S.P. Sullivan, An eddy current technique for detecting and sizing surface cracks in carbon steels - *Materials Evaluation*, May 1993, pp. 572-577.

[2] S. Mastorchio, P.O. Gros, TRIFOU : 3D modelling for eddy current testing in steam generator tubes - *Electromagnetic Nondestructive Evaluation (II)*, R. Albanese *et al.* - IOS Press, Amsterdam, 1998, pp. 137-145.

[3] F. Matsuoka, Calculation of a three dimensional eddy current by the FEM-BEM coupling method, Proceedings IUTAM on electromagneto-mechanical interactions in deformable solids and structures, Elsevier Science, 1987, pp. 169-174.

Electromagnetic Nondestructive Evaluation (III)
D. Lesselier and A. Razek (Eds.)
IOS Press, 1999

Numerical Analysis of Eddy Current Testing for Steam Generator Tubes with a Support Plate

Haoyu Huang, Toshiyuki Takagi and Hiroyuki Fukutomi
Institute of Fluid Science, Tohoku University, Katahira 2-1-1, Aoba-ku, Sendai 980-8577, JAPAN

Abstract. A difficulty encountered in eddy current testing (ECT) arises in identifying crack profiles from their signals which may include noises caused by the variations of lift-off of a probe, the presence of the structures outside the tubes etc. An example of these structures is a support plate, which is made of ferromagnetic material. A method for the forward and inverse analysis for steam generator tubes with a support plate is introduced in this paper. For the elimination of the noises, the signal processing by a multi-frequency technique is used. The reduced magnetic vector potential method with edge-based finite elements can be used to evaluate the signals including a component due to the ferromagnetic material. A fast simulator based on this method using a pre-computed database has already been developed for the numerical simulation of ECT signals by the authors. Using this fast simulator, computational time in the inverse analysis can be reduced. This paper presents the noisy crack signals and the elimination effect of the multi-frequency technique as well as the reconstruction results of the crack depths.

1. Introduction

Eddy current testing (ECT) is used for the in-service inspection of steam generator (SG) tubes in nuclear or conventional power plants. For the development of this testing, three-dimensional numerical simulation has been recently used, substituted for experiments. The signal evaluations due to cracks are usually recognized as forward problems, and then the reconstructions of crack characteristics from measured signals are called inverse problems.

In the testing of steam generator tubes, a difficulty is the processing of the noisy ECT signals. On the whole the noises may be caused by the variation of the lift-off of the probe, the presence of the structures outside the tubes etc. One of these structures is a support plate, whose signals sometimes have higher magnitude in comparison with the signals due to the presence of a crack or cracks only. The signal processing by the multi-frequency technique is generally used since it is difficult to check the existence of cracks from the raw signals including these noises[1]. To deal with the support plate, the larger scale problem region is necessary to be analyzed. Besides the surface on the plate, its subsurface requires fine finite element mesh in consideration of shallow penetration of the eddy current. It turns out that the coefficient matrix from finite elements becomes larger. This makes the forward analysis difficult and time-consuming. To the authors' knowledge, this problem has been solved in a few papers[2,3], and there has been no report on the reconstruction of the crack considering the noise due to a support plate.

The objectives of this paper are (a) to evaluate the ECT signals in the presence of the support plates by numerical simulation; (b) to process the signals with the noise; (c) to reconstruct the crack depths from the processed signals. For the reconstruction and classification of cracks from their signals, neural networks[4] and reconstruction methods based on finite elements combined with optimization theory[5] have been introduced. The neural networks show some advantages in estimating the shapes and locations of cracks. However, they do not tell us the crack depth exactly, which is an important parameter in ECT. On the other hand, the reconstruction methods based on optimization theories require solving the forward problem for many times, and they have difficulty in obtaining a unique solution. Although the high accuracy of some simulation techniques has been demonstrated[6,7], an issue still remains in computational time. Hence, a very fast solver is necessary to solve the inverse problems in practical computational time. The pre-computed unflawed database approach[8] based on the reduced magnetic vector potential method and edge based finite elements is expedient to work out this issue. This method is used to solve the forward problems as well as the inverse problems in this paper. The computational time was reduced when comparing with that of a conventional method.

2. Forward Problem

The forward problem is solved using the reduced magnetic vector potential method and edge-based finite element[6,7]. The mesh of this model is shown in Fig.1. The support plate here is 20mm long and 4mm thick. Because of shallow penetration, great care must be

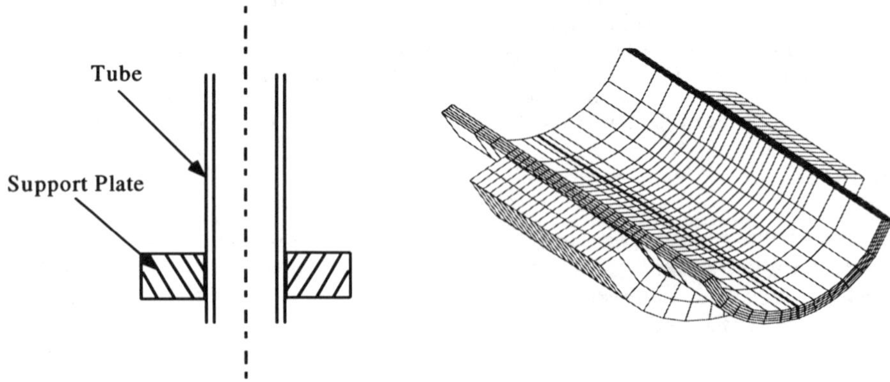

Fig.1 Mesh of the SG tubes with a support plate

Table 1 Summary of the conditions on a numerical test

Coil		
Height:	0.8mm	Current: 1/140 A
Inner diameter:	1.2mm	Turn: 140 turns
Outer diameter:	3.2mm	Lift-off: 0.5mm
Support plate		
Length:	20mm	$\mu_r = 100$
Thickness:	4mm	$\sigma = 8\times10^6$ S/m
Distance from tube:	0.19mm	
Tube		
Length:	40mm	$\mu_r = 1$
Thickness:	1.27mm	$\sigma = 1\times10^6$ S/m
Inner radius:	9.85mm	
Outer radius:	11.12mm	
EDM Crack		
Width:	0.2mm	$\mu_r = 1$
Length:	≤10mm	$\sigma = 0$ S/m
Depth:	20%, 40%	

taken to model the surface of the support plate in the direction of the wall. The crack of the tube is supposed to be axial direction and located at the center of the support plate region. The cracks here include both inner defects(IDs) and outer defects(ODs) with depths of 20% and 40% through the wall. We assumed that there are 21 measured scan points along the axial direction, which are located from –10mm to 10mm at regular intervals. Some other data about the setting of the model is summarized in table 1.

Results of ID40% and OD40%, compared with the crack-free case, are shown in Fig.2. Left side of the figure is the magnitude of the signals as the coil moves above the crack. It can be noticed that the crack signals almost overlap with the noise of the support plate, especially in the OD case. Even for the ID case, three peaks can be seen, which may indicate that there are three cracks. Computation shows that when the probe reaches the point 16mm, the signal becomes quite small. A right side figure shows the signal trajectories and indicates the phase difference among the signals. It can be seen that a subtle difference exists between the OD case and crack-free case from the phase. From these noisy signals, it may be difficult to identify cracks exactly.

3. Multi-frequency technique

The signals of the support plate at 100kHz and 200kHz are shown in Fig.3. It can be

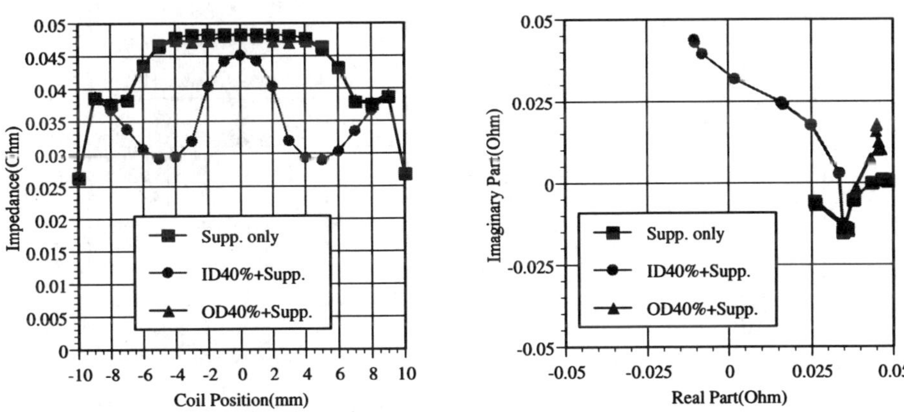

Fig.2 Signal of crack and support plate (100kHz)

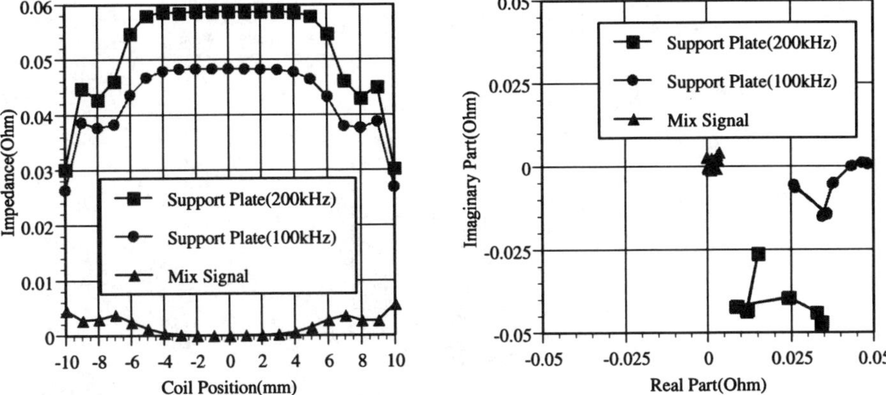

Fig.3. Subtraction of the support plate signals at two frequencies

seen that these two curves are very similar in shape to each other. Based on this similarity, a signal processing technique has been developed using these two sets of signals to obtain a mix signals which can reduce the support plate noise and, at the same time, exhibit the crack information only. This is so-called a multi-frequency technique[9]. This calculation can be described simply as follows:

$$S_{\mathrm{Mix}} = S_{200\mathrm{kHz}} - S_{100\mathrm{kHz}} \cdot \alpha \cdot e^{j\theta}, \qquad (1)$$

where S is the signal of the support plate, α and θ are constants. Constants used here are set to α=1.2158 and θ=−55.47°. The mix signals from 100kHz and 200kHz signals are also shown in Fig.3. The result shows that the support plate noise is decreased in some degree (about 1/100). This algorithm with the same constants is used to process the signals due to both cracks and a support plate. In Fig.4, examples of ID40% and OD40% cases are shown.

Signals due to many kinds of cracks are computed by this forward analysis method, and are processed by the multi-frequency technique. The results indicate that the OD20% may be the limit of this method, where the signals are nearly as small as the noise left, as shown in Fig.5. There are two peaks of noise at the position –10mm and 10mm because of the edge effect of the support plate. These mix signals computed from two frequencies are used to reconstruct the crack shapes.

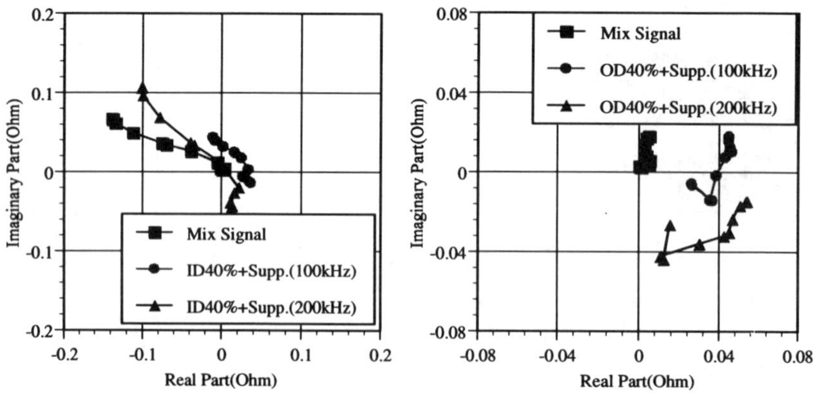

Fig.4 Mix signals of ID40% and OD40% (length of crack: 10mm)

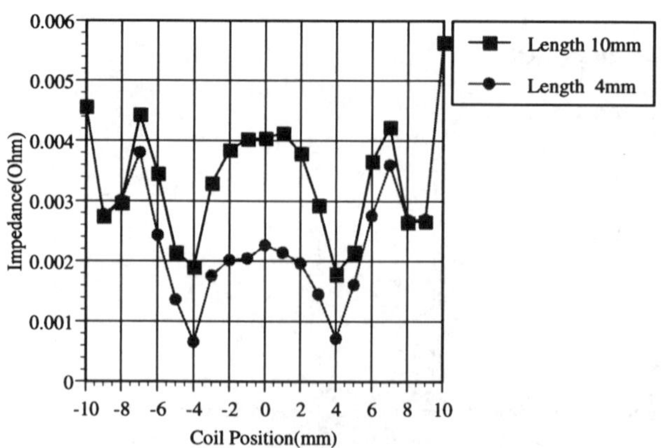

Fig.5 Mix signals of OD20%

4. Inverse Problem

The pre-computed unflawed database approach[8] based on the reduced magnetic vector potential method and edge based finite elements is used in this paper. In this database approach, a potential crack region is defined as a suspect region. Based on the reciprocity theory, the ECT signals can be predicted using only the unknowns in the suspect region. Using the pre-computed database, equations on the unknowns in the suspect region are established. The number of unknowns becomes much smaller compared with that of the conventional methods. The signals due to cracks are computed within about 1/100 of computational time of the conventional method. This method makes it possible to directly predict the difference of signals in cracked and crack-free cases even with a support plate. On the other hand, the ECT signals acquired from experiments include the support plate noise. To apply the pre-computed unflawed database method to the inverse analysis, the processed signals must be used. In this study, the signals are not obtained from experimental results but from the numerical results using the reduced magnetic vector potential method.

In the database method, there is a suspect region where the crack is supposed to be in. A cube suspect region is used whose length is 10mm, depth (height) is the thickness of the tube (1.27mm) and width is 0.2mm. This region consists of 50 finite elements. The least squares method is applied and the following evaluation function is used:

$$J(\boldsymbol{x}) = \sum_{i=1}^{N_{pos}} \left|Y^i(\boldsymbol{x}) - Y^i_{obs}\right|^2 \Big/ \sum_{i=1}^{N_{pos}} \left|Y^i_{obs}\right|^2, \tag{2}$$

where J is an evaluation function, $\boldsymbol{x}$ is the vector characterizing the shape of the cracks, $Y(\boldsymbol{x})$ is the predicted signal related to the vector $\boldsymbol{x}$, and Y_{obs} is the measurement signal.

Here the width of the crack is assumed to be 0.2mm, and the parameters are associated with the depth in this suspect region. The inverse results of ID40%, ID20%, OD40% and OD20% are shown in Fig.6. The upper side of the figure is the inner side of the tube, so the ID appears on the upper side of the figures. The original crack length is 4mm, and depth is shown in each figure. Fig.8 shows the convergence performance of these four cases. The results are satisfactory because the identified depths are the same as those of real cracks. As shown in Fig. 5, there is still a vivid noise in the case of OD20%, compared with the crack signal. That is the reason why the evaluation function does not converge to a very small value. Although the evaluation function does not become small enough in the OD20% case, it converges to the real crack shape as shown in Fig.

Considering the fact that the numerical data is used to reconstruct the crack, there is no guarantee to be successful when using experimental data. In the place of experimental data, numerical data with white noise is used. Random white noises (10 percent of the maximum magnitude of the signal) are added to the numerical data for 100kHz and 200kHz, and same signal processing is employed to obtain the mix signals. Fig.7 and 9 show the results and the convergence performance of each case. The depth of the OD20% case can be estimated, but the result of estimated length is not satisfactory. As for the further work the crack shape will be predicted when the crack exists near the edge of the support plate.

5. Conclusion

The forward and inverse problems on ECT, where a support plate exists around a SG tube, are solved in this paper:

1. The forward signals can be computed, using the reduced magnetic vector potential method. The raw signals include the support plate noises, which is very large in comparison with the signals of outside cracks.

2. Multi-frequency technique was employed to reduce the noise of the support plate and proved to be successful.
3. Pre-computed unflawed database method was used to solve the inverse problem. Inverse analysis from mix signals is possible and gives good results for inner defects.

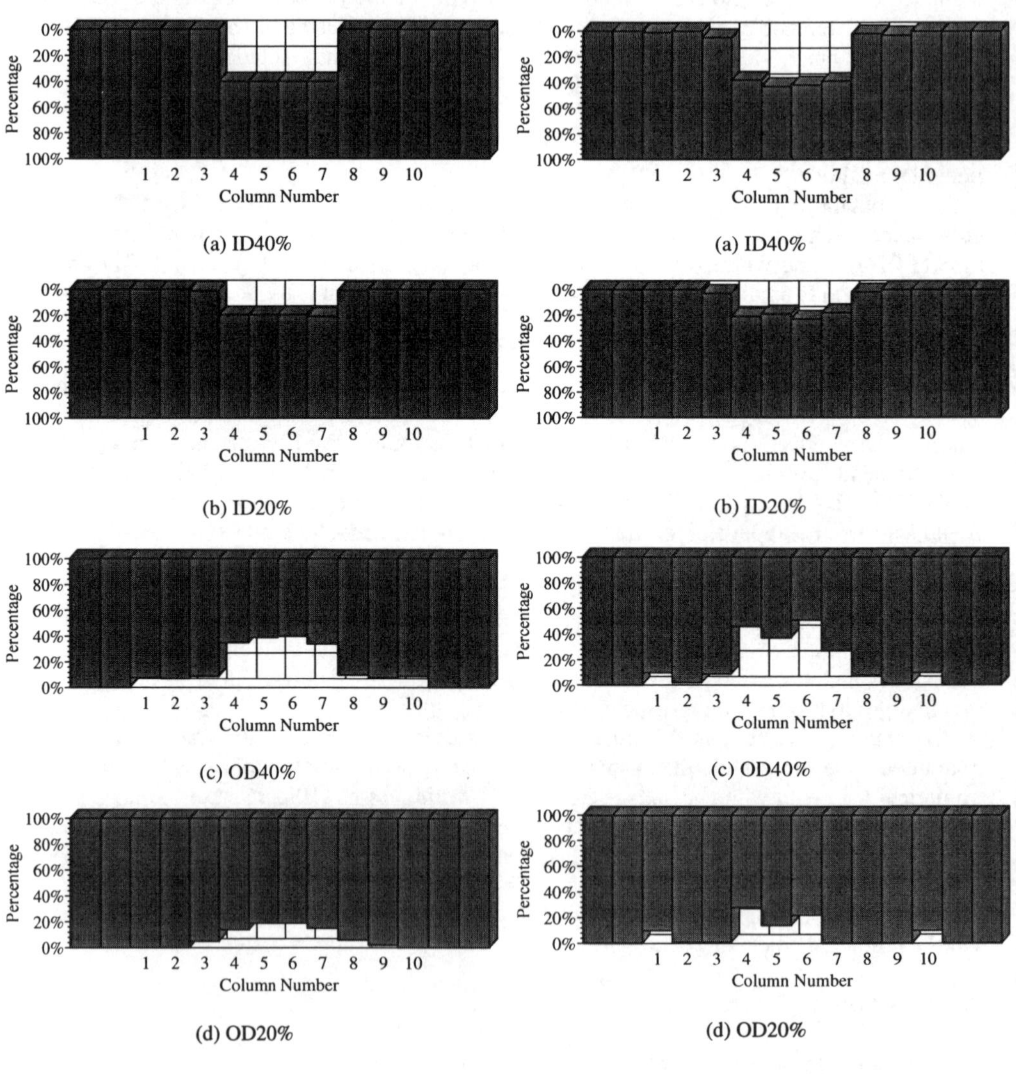

(a) ID40% (a) ID40%

(b) ID20% (b) ID20%

(c) OD40% (c) OD40%

(d) OD20% (d) OD20%

Fig.6 Reconstruction results

Fig.7 Reconstruction results (10% noise)

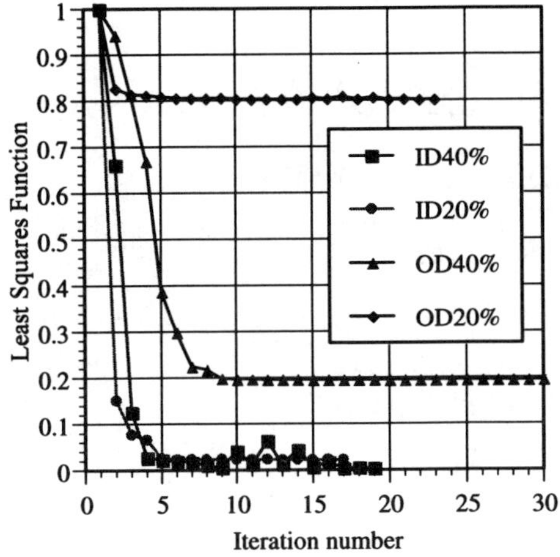

Fig.8 Convergence performance

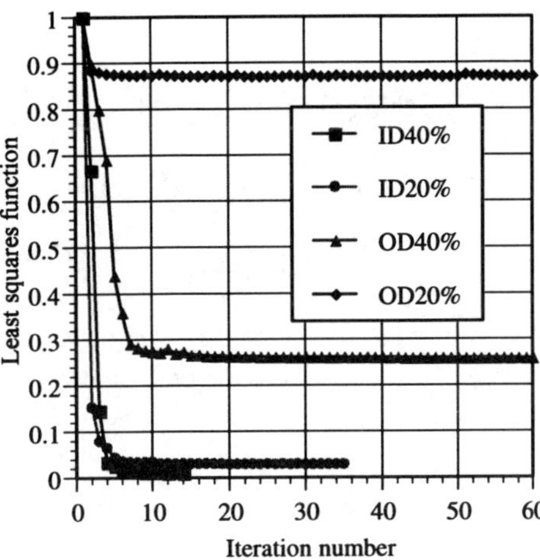

Fig.9 Convergence performance (10% noise)

Acknowledgments

This study was supported in part by the Research Committee on Advanced Eddy Current Testing Technology of the Japan Society of Applied Electromagnetics and Mechanics through a grant from 5 PWR utilities and Nuclear Engineering Ltd.

References

[1] Z. Badics, Y. Matsumoto, K. Aoki and F. Nakayasu, "Effective probe response calculation using impedance boundary condition in eddy current NDE problems with massive conducting regions present", *IEEE Transactions on Magnetics*, **32**, 1996, pp.737-740.

[2] H. Fukutomi, T. Takagi, J. Tani and G. Chen, "Consideration of ECT signal of SG tube with copper deposit", *Electromagnetic Nondestructive Evaluation*, IOS press, 1997, pp.79-86.

[3] R. Palanisamy and Whuang. Lord, "Finite element simulation of support plate and tube defect eddy current signals in steam generator NDT", *Materials Evaluation*, **39**, June, 1981, pp.651-655.

[4] L. Udpa and S. S. Udpa, "Application of neural Networks for classification of eddy current NDT data", *Review of Prog. QNDE*, **9**, Plenum press, 1990, pp.673-680.

[5] H. Fukutomi, T. Takagi, J. Tani and F. Kojima, "Crack shape characterization in eddy current testing," *Electromagnetic Nondestructive Evaluation (II)*, IOS press, 1998, pp.305-312.

[6] H. Fukutomi, T. Takagi, J. Tani, M. Hashimoto, J. Shimone and Y. Harada, "Numerical evaluation of ECT impedance signal due to minute cracks," *IEEE Transactions on Magnetics*, **33**, 1997, pp.2123-2126.

[7] A. Kameari, "Solution of asymmetric conductor with a hole by FEM using edge-element," *COMPEL*, **9**, 1990, pp.230-232.

[8] H. Huang, H. Fukutomi, T. Takagi and J. Tani, "Forward and inverse analyses of ECT signals based on reduced vector potential method using database," *Electromagnetic Nondestructive Evaluation (II)*, IOS press, 1998, pp.313-321.

[9] D. E. Bray and R. K. Stanley, "Nondestructive evaluation: A tool in design, manufacturing and service," CRC Press, 1997, pp.418-419.

Electromagnetic Nondestructive Evaluation (III)
D. Lesselier and A. Razek (Eds.)
IOS Press, 1999

Development of A Numerical Analysis Technique for Alternating Current Magnetic Flux Leakage Testing

Toshiyuki SUZUMA, Hirotsugu FUJIWARA, and Takahide SAKAMOTO
System Division, Sumitomo Metal Industries Ltd.
1-8 Fuso-cho Amagasaki 660-0891 Japan

Abstract. A two dimentional numerical simulation code intended for non-linear electromagnetic field analysis has been developed using the finite element method for an Alternating Current Magnetic Flux Leakage Testing system (AC-MFLT). Suitable analysis conditions are determined in order to promote accuracy and reduce calculation time. To demonstrate the validity of this code, experimental results are compared with numerical ones and they show good agreement. Furthermore, in order to investigate characteristics of inspection noise, magnetic flux leakage from the area which has different magnetic properties on the specimen's surface is calculated. Based on this analytical results and experimental ones, it is revealed that increasing magnetization is effective to suppress the inspection noise.

1. Introduction

MFLT has been utilized for the detection of flaws in ferromagnetic materials[1]-[5]. Recently AC electromagnets have been applied to MFLT to improve the detectability of surface flaws by increasing the magnetic flux density in the material

Numerical simulation techniques are effective to investigate the characteristics of magnetic flux leakage and to design electromagnets, magnetic sensors and so on. However, there are not so many studies on it because it requires non-linear and transient analysis to calculate the magnetic flux in the magnetically saturated region, and this calculation takes a long time.

In this study, a two-dimensional non-linear electromagnetic field analysis technique has been developed for MFLT using the finite element method. Numerical results are compared with experimental ones to confirm its validity. Furthermore, inspection noise caused by partial unevenness of magnetic properties is investigated.

2. Analysis technique

A finite element method is applied to solve electromagnetic fields governed with following equation obtained from Maxwell's law.

$$\nabla \times \frac{1}{\mu} \nabla \times A = J_0 - \sigma \frac{\partial A}{\partial t} \qquad (1)$$

where A , μ , J_0 and σ are the magnetic vector potential, permeability, external current density and conductivity respectively. Backward difference method is used for time-discretization, and the time interval Δt is one sixteenth of a current cycle. The Newton-Raphson method is applied non-linear calculation.

3. Influence of analysis conditions

In FEM method, it is important to set up (1) **the element size** and (2) **the time interval of calculation** Δt small enough for accurate calculation, but that means an increase of calculation time. Therefore, it is necessary to decide on a minimum number of division for (1) and (2) needed to keep good accuracy.

The 2-D analytical model shown in Fig. 1 is used to decide analytical conditions under which solutions converge. The magnetization curve of 0.45% Carbon steel (maximum relative permeability is about 100, conductivity is 6.0×10^6 S/m) is used as magnetic characteristic of the steel sheet model, and the magnetizing frequency is 2 kHz.

3.1 Influence of finite element size

Finite element thickness on the specimen's surface affects accuracy because the gradient of magnetic potential is larger near the surface because of skin effect. The horizontal component of magnetic flux density at a measurement point A in Fig. 1, is calculated with surface elements of various sizes.

Results are shown in Fig. 2. The value of the magnetic flux at point A is affected by the thickness of the elements, and the value converges when the element thickness is 0.083 mm. This specimen's skin depth is about 0.5 mm, and taking these facts into account, surface elements should be divided until the thickness become a fifth of the skin depth or less.

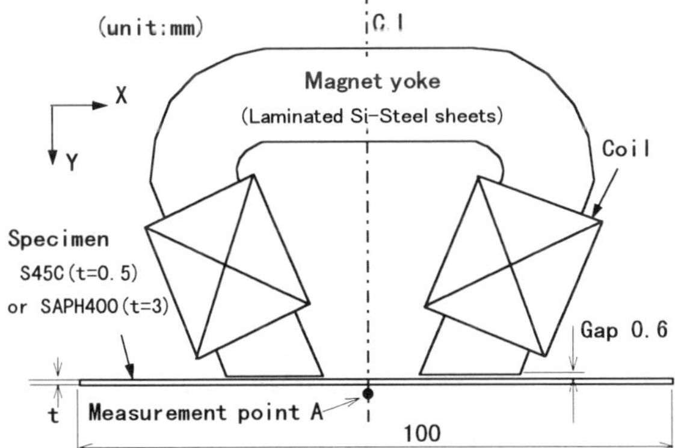

Fig. 1 Analytical model

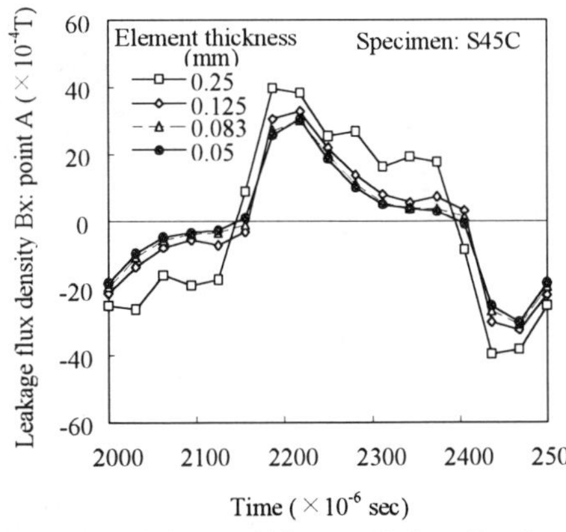

Fig. 2 Effect of element thickness on leakage flux density at measurement point A

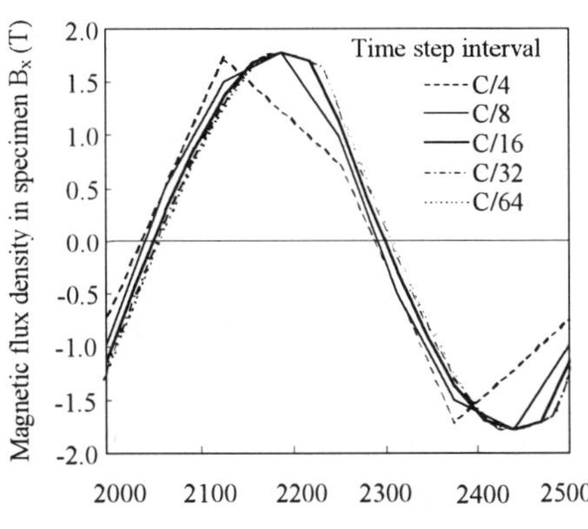

Fig. 3 Effect of time step interval on magnetic flux density in a specimen

3.2 Influence of time step interval

The time interval at each step of the calculation should be small enough for a good solution. But too many steps take a long time to be solved. Therefore, magnetic flux density in a specimen's surface are calculated with various intervals (from C/4 to C/64 where C is the duration of the magnetizing cycle).

Results are shown in Fig. 3. The solution converges when the interval is C/16, and it is clear that the time interval should be divided under C/16 for a stable result.

4. Verification of the model's results

In order to verify our analysis technique, calculations and experiments are carried out with a model consisting of a thin sheet specimen and an electromagnet, as shown in Fig. 4 and Table 1. Three materials, as shown in Table 2, are utilized to investigate the influence of the magnetic properties.

The magnetic flux density in the specimen and the leakage flux density from a flaw are important values to help designing electromagnet shape, related to magnetizing force, and designing sensors. Comparisons of calculations and experiments about the average flux density in the specimen are shown in Fig. 6. The average flux density decreases with increase of the frequency, because of skin effect. Although the saturation flux density of SUS430 is less than the others, the average flux density of SUS430 is higher. The reason is that the conductivity of SUS430 is lower and the flux can penetrate deeper into the specimen.

Fig. 7 shows the comparison of the normal component of leakage flux from the flaw. The leakage flux, on the other hand, increases with increase of frequency. The leakage flux of SAPH400 is higher than the others because of its larger permeability and conductivity. Fig. 6 and Fig. 7 show that the numerical results agree with the experimental ones in three specimens which have different magnetic property, which confirms the validity of the analysis technique.

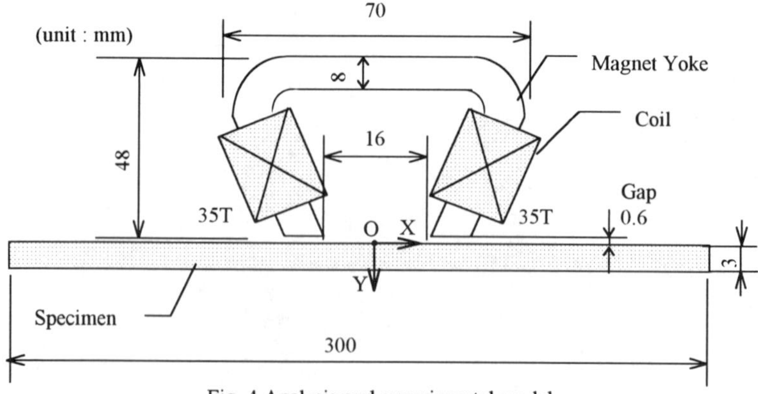

Fig. 4 Analysis and experimental model

Table 1 Electromagnet

Magnet yoke	Laminated structure, B-H (equal to Si-Steel)
Coil	35Turns × 2
Current	Peak Value = 2.5(A) Frequency =100, 500, 1k, 2k, 5k (Hz)

Table 2 Characteristics of specimens

Specimen	SAPH400	SK5	SUS430
Conductivity (S/m)	6.34×10^6	5.81×10^6	1.53×10^6
B-H property	Shown in Fig. 5		
Geometry (mm)	Specimen: Width 300, Thickness 3 Flaw : Width 0.2, Depth 0.5, Length 20		

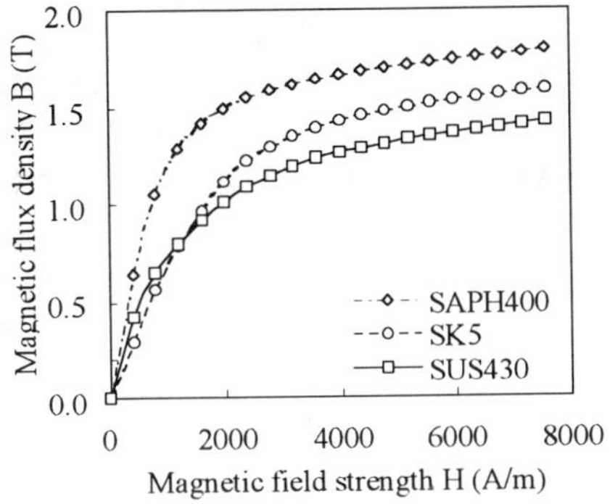

Fig. 5 Magnetization curve

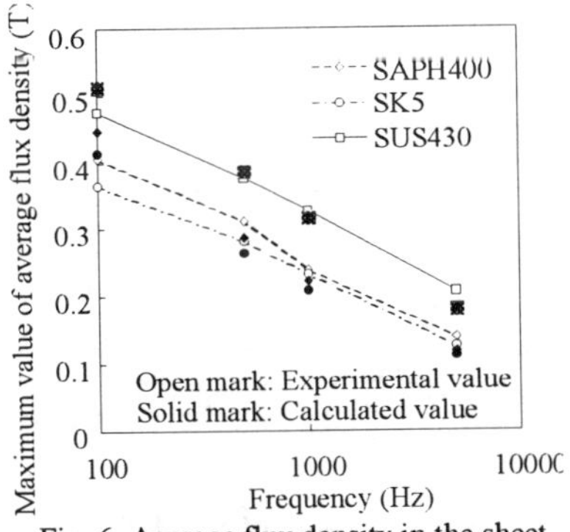

Fig. 6 Average flux density in the sheet

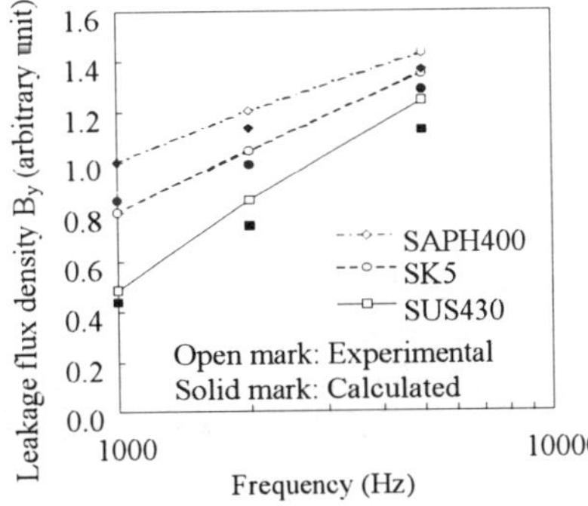

Fig. 7 Normal component of leakage flux

5. Analysis of inspection noise

Supressing inspection noise is important to improve flaw detectability. One of the sources of inspection noise is the magnetical unevenness of the surface such as local difference of heat treatment. A simplified model shown in Fig. 8 is analysed to investigate the leakage flux from a quenched part and a flaw, both in an annealed area. A calculated result is shown in Fig. 9. The magnetizing frequency is 8 kHz. The leakage flux from a flaw increases with increase of magnetic flux density B_x, however, the one from a quenched part decreases when B_x exceeds 1.7T, because the difference of the magnetic flux density between a quenched part and an annealed area becomes insignificant.

Based on the numerical result, signal of an artificial flaw and signal of a base noise are measured with various magnetizing current values, and compared with the numerical result.

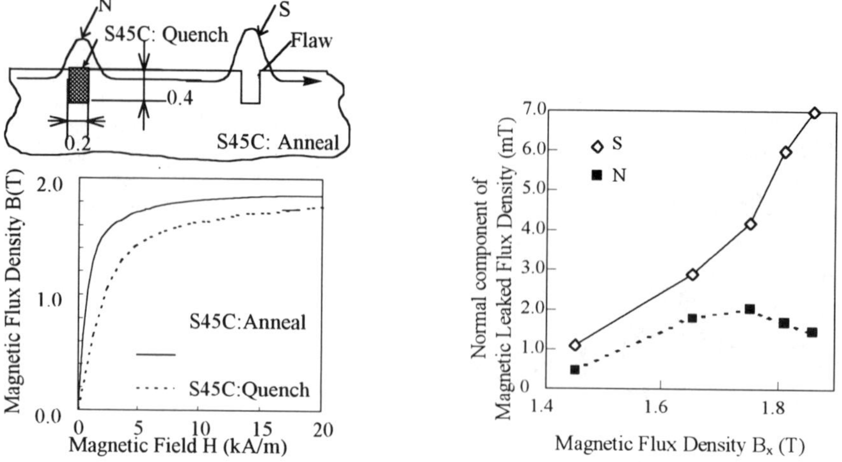

Fig. 8 Model of unevenness of magnetic properties

Fig. 9 Leakage flux from a flaw and an unevenness (numerical result)

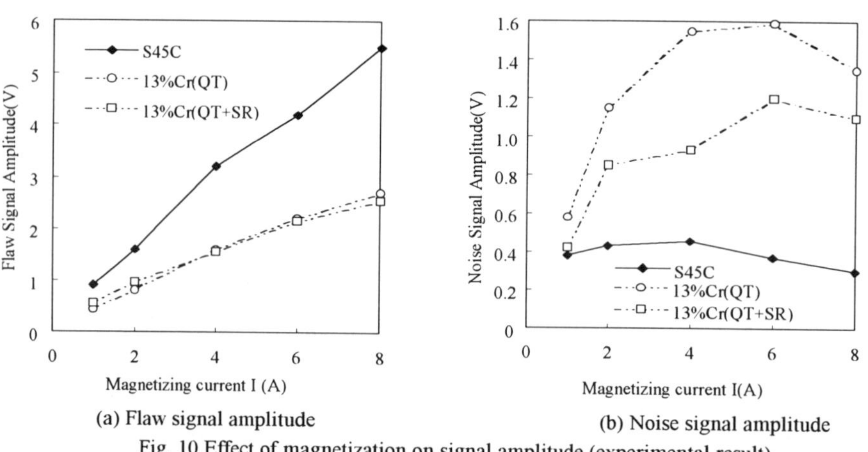

(a) Flaw signal amplitude

(b) Noise signal amplitude

Fig. 10 Effect of magnetization on signal amplitude (experimental result)

Specimens are 0.45 % Carbon steel (S45C) and 13% Cr high-alloy steel of which base noise are higher than Carbon steel, furthermore, two types of heat treatment are processed in 13% Cr steel, quench temper process and stress reducing process.

Fig. 10 shows the effects of magnetization on flaw and base noise signal. Flaw signal amplitude increases as magnetizing current increases, on the other hand, base noise signal amplitude of each material has a peak between 4 and 6 (A) in magnetizing current.

Considering the circumstances mentioned above, experimental results about noise signal are similar to the numerical results about leakage flux from a quenched part. Therefore the magnetic unevenness is likely to be one of the causes of inspection noise.

6. Conclusion

Numerical results have been compared with experimental ones to validate the numerical analysis technique. They showed a good agreement, provided that the finite element thickness is a fifth of the skin depth or less and that the time step is one sixteenth of the period or less.

The inspection noise caused by partial unevenness in magnetic property of the specimen has been calculated. And it is revealed that increasing magnetization is effective to suppress the flux leakage due to local unevenness.

References

[1] Nathan Ida and William Lord, 3-D Finite Element Predictions of Magnetostatic Leakage Fields, *IEEE Transactions on Magnetics*, Vol. 19, No. 5 (1983) pp. 2260-2265

[2] I. Uetake, H. Itoh, Measurement of Magnetic Leakage Flux due to Oblique Surface Cracks and Estimation Methods of Its Oblique Angle, *Journal of JSNDI (in Japanese)*, Vol. 37, No. 5 (1988) pp. 381-387

[3] S. Mukae, M. Katoh, and K. Nishio, Investigation on Quantitization of Flaw and Effect of Factors Affecting Leakage Flux Density in Magnetic Leakage Flux Testing Method, *Journal of JSNDI (in Japanese)*, Vol. 37, No. 11, (1988) pp. 885-894

[4] P C Charlton and K E Donne, Computer modelling of magnetic flux leakage methods, *British Journal of NDT*, Vol. 36, No. 3, (1994) pp. 128-133

[5] Eduardo Altschuler and Alberto Pignotti, Nonlinear model of flaw detection in steel pipes by magnetic flux leakage, *NDT&E International*, Vol. 28, No. 1 (1995) pp. 35-40

Defect Evaluation with Emphasis on Deterministic Methods

Electromagnetic Nondestructive Evaluation (III)
D. Lesselier and A. Razek (Eds.)
IOS Press, 1999

An Approach to Reconstruction of Natural Cracks using Signals of Eddy Current Testing

Zhenmao CHEN, Kenzo MIYA[†] and Masaaki KUROKAWA[‡]
Structure Safety Engineering Group, OEC/JNC
4002 Narida-chio, Oarai-machi, Ibaraki, 311-1393, Japan
[†] *Nuclear Engineering Research Laboratory, University of Tokyo*
2-22 Shirakata-shirane, Tokai-mura, Naka-gun, Ibaraki, 319-1106 Japan
[‡] *Takasago R&D Center, Mitsubishi Heavy Industries, Ltd.*
2-1-1 Shinahama Arai-cho, Takasago, Hyogo, 676, Japan

Abstract. As an effort to reconstruct natural cracks, an inverse scheme is proposed in this paper on the basis of a model-based optimization method and a novel rapid forward solver. Firstly, formulation for the shape reconstruction is carried out with the natural cracks modeled as multiple planar slits of non-vanishing conductivity. Then, a special case of the idealized cracks, i.e. an EDM artificial crack, is reconstructed from ECT signals of high S/N ratio for validating the proposed inverse scheme. Finally, the new method is applied to simulated ECT signals including artificial noise and to testing signals acquired under a more practical environment. Several typical results for the EDM crack are presented for signals with noise up to 20% that, in turn, verified the robustness and efficiency of the proposed approach. The validation of the scheme for multiple cracks and a crack of non-vanishing conductivity, on the other hand, is scheduled to be done in a near future.

1. Introduction

Reconstruction of crack shape is one of the important issues demanded in the upgrade of safety ensurance for various critical structural components (e.g. SG tubing of a PWR power plant). In addition, a reliable prediction of crack profiles is also a prerequisite for considering extension of the ISI (In-Service Inspection) period that, as well known, will lead to a huge economic benefit. From these points of view, the conventional means of nondestructive evaluation still need to be improved in order that such a typical inverse evaluation can be efficiently realized. For this purpose, a reconstruction scheme which uses databases was recently developed for cracks embedded in a conductor of complicated geometry by incorporating a novel fast forward scheme into a conventional optimization algorithm of the conjugate gradient method[1],[2],[3]. The validity of this approach has been demonstrated by applying it to the reconstruction of EDM artificial cracks from measured or simulated ECT signals. In such studies, however, only the case of a single EDM crack and the observing data of high S/N ratio were treated.

In this paper, two work are undertaken for extending the proposed reconstruction strategy and investigating its feasibility in more practical problems. At first, the scheme for an EDM crack is applied to signals including artificial noise and those acquired under a more practical environment, namely, measured data employing a SG mock-up. Several typical results for both an inner and outer crack are presented, which reveal that even for signals with noise up to 20% the approach can gives a reasonable reconstruction but with a little consuming of computer resources. As the second subject of this paper, the suitability of the proposed scheme in the reconstruction of natural cracks is discussed by idealizing the natural cracks as a model of multiple slits with non-vanishing-conductivity. The formulae for predicating the crack shape and the corresponding distribution of conductivity are also described in that part.

2. Scheme for Fast Forward Analysis

2.1 Basic Formulation

Subtracting formulae of the $A-\phi$ method for unknown potentials $\mathbf{A}, \phi$ of a conductor with flaw present from those about potentials $\mathbf{A}^u, \phi^u$ of an unflawed conductor, the deduced governing equations for the field perturbations can be written as,

$$\frac{1}{\mu_0}\nabla^2\mathbf{A}^f - \sigma_0(\dot{\mathbf{A}}^f + \nabla\phi^f) = -[\sigma_0 - \sigma(\mathbf{r})](\dot{\mathbf{A}} + \nabla\phi), \tag{1}$$

$$\nabla\cdot\sigma_0(\dot{\mathbf{A}}^f + \nabla\phi^f) = \nabla\cdot[\sigma_0 - \sigma(\mathbf{r})](\dot{\mathbf{A}} + \nabla\phi), \qquad (in\ conductor) \tag{2}$$

$$\frac{1}{\mu_0}\nabla^2\mathbf{A}^f = 0, \qquad (in\ air) \tag{3}$$

where, $\mathbf{A}^f = \mathbf{A} - \mathbf{A}^u$, $\phi^f = \phi - \phi^u$ are the potential perturbations due to the presence of a crack. As variable $[\sigma_0 - \sigma(\mathbf{r})]$ vanishes in the conducting area, Eqs.(1) $\sim$ (3) can be replaced by the following system of linear equations after an FEM-BEM discretization,

$$\begin{bmatrix} \overline{K}_{11} & \overline{K}_{12} \\ \overline{K}_{21} & \overline{K}_{22} \end{bmatrix}\begin{Bmatrix} q_1^f \\ q_2^f \end{Bmatrix} = \begin{bmatrix} \widetilde{K}_{11} & 0 \\ 0 & 0 \end{bmatrix}\begin{Bmatrix} q_1^f + q_1^u \\ q_2^f + q_2^u \end{Bmatrix}. \tag{4}$$

In Eq.(4), $\{q^f\} = \{\mathbf{A}^f, \phi^f\}^T, \{q^u\} = \{\mathbf{A}^u, \phi^u\}^T$ are respectively the unknown potential vectors and unflawed field, $\overline{K}$ is the coefficient matrix for the unflawed conductor, and $\widetilde{K}$ is the coefficient matrix corresponding to the term including variable $[\sigma_0 - \sigma(\mathbf{r})]$. In the equation, the potential vector $\{q\}$ was divided into two parts: $\{q_1\}$ is the potential vector at the nodes of crack element and $\{q_2\}$ the remained unknowns. As $[\sigma_0 - \sigma(\mathbf{r})]$ vanishes in region of base material, matrix $\widetilde{K}$ was written as the form in Eq.(4) with the nonzero sub-matrix $\widetilde{K}_{11}$ representing the part related to the crack region.

Multiplying $[H](\equiv[\overline{K}]^{-1})$ to Eq.(4), we find,

$$\begin{Bmatrix} q_1^f \\ q_2^f \end{Bmatrix} = \begin{bmatrix} H_{11} & H_{12} \\ H_{21} & H_{22} \end{bmatrix}\begin{bmatrix} \widetilde{K}_{11} & 0 \\ 0 & 0 \end{bmatrix}\begin{Bmatrix} q_1^f + q_1^u \\ q_2^f + q_2^u \end{Bmatrix}. \tag{5}$$

The equations related to the unknowns $\{q_1^f\}$ can be simply separated from Eq.(5). If we denote $[H_{11}][\widetilde{K}_{11}]$ as $[G]$, the relation connecting the field perturbation to the unflawed field is finally found as,

$$[I - G]\{q_1^f\} = [G]\{q_1^u\}. \tag{6}$$

Since the unknowns of Eq.(6) are limited at the crack region, solving the disturbed field $\{q^f\}$ from this equation is much faster than using the conventional FEM code because of the significantly reduced number of unknowns. The information about the conductor and exciting coils is contained in the coefficient matrices $[H_{11}]$ and unflawed field $\{q^u\}$, which can be calculated *a priori* considering their independence with the crack.

2.2 Establishment of Databases

From ECT signals, some features of the crack such as the position, length and inner/outer property usually can be estimated by using a classification method. Thus, it is not difficult to choose an area (Ω_0) in which a crack possibly occurs, and to subdivide it into a grid of small cells. The unflawed field at Ω_0 can be calculated with an FEM code for excitations of a probe or a unit potential at a node. Once these results were put into the databases, the matrices $[H_{11}]$ and vector $\{q^u\}$ in Eq.(6) can be directly extracted in practical calculation for any crack contained in Ω_0. In addition, if the conductor to be inspected can be considered as a tube with infinite length, the size of the databases will be very small as the shift property of electromagnetic field is valid. In such a case, we only need to calculate and store the fields excited by a unit source arranged at some positions on a radial line which is located at the central cross section of the tube.

2.3 Introducing of a New FEM Element

As the databases have to be established under a given set of cells *a priori*, only the cracks possibly being made up of the cells can be treated with use of these databases if homogeneous conductivity is a necessary condition in the FEM analysis. However, in case of a low frequency eddy current problem, as the permeability coefficient of a SG tube is the same as that of the free space, the electric charge at the bottom edge of the crack can be considered negligible especially when the excitation coil is arranged just over the crack. Hence the electromagnetic fields are approximately continuous and the corresponding potentials are one order differentiable at the bottom edge of crack. In addition with the fact that the grid cells are usually small, it is reasonable to apply the shape function of a normal FEM element to interpolate the field in the whole cubic cell even it contains different media. Introducing such a new element will enable the rapid evaluation of a crack of arbitrary shape with a given mesh. In fact, the difference of the material has been taken into account in the coefficient matrices as the edge curve of the crack was applied for the definition of the integration region.

3. Algorithm of Inverse Problem

Usually, the microscopic structure of a natural crack is not important in the practical applications even for serving-life evaluation of the structural component. Therefore, we can consider the aim of the crack reconstruction as to determine its edge curve and other major parameters, such as the opening, number and the distances between the cracks. Under this premise, it is reasonable to establish a numerical model for the natural cracks at first, and then to reconstruct shapes of the idealized crack models. By observing the detailed structure of the stress corrosion cracks, it is not difficult to find that the two side surfaces of a crack is not separated completely and interaction between different cracks also may occurs. On the basis of these phenomena, we propose a numerical model for the natural cracks as multiple thin EDM notches with non-zero conductivity and try to establish a scheme to reconstruct their shapes and conductivity distributions from the measured signals.

3.1 Model-based scheme using an optimization approach

In this section, we introduce the major formulae of an first order inverse algorithm by considering the observed data as the impedance signals of a pancake probe stored as a function of the probe positions and exciting frequencies. The selection of the crack parameters and the formulae for the prediction of gradients will be given latter.

Similar with the case of an EDM crack, we choose the object function as follows,

$$\varepsilon(\mathbf{c}) = \sum_{m=1}^{M} |Z_m(\mathbf{c}) - Z_m^{obs}|^2, \tag{7}$$

where, $\mathbf{c}$ is a vector of crack shape parameters. $z_m(\mathbf{c})$ and z_m^{obs} are respectively the predicted and observed impedance signals at the m-th sampling point. M, the total number of data, is equal to the product of the scanning steps and the number of exciting frequencies.

We solve the shape vector $\{c\}$ by minimizing the mean-square error with the following iteration algorithm,

$$\{c\}_n = \{c\}_{n-1} + a_n\{\delta c\}_n, \tag{8}$$

where $\{\delta c\}_n$ is the updating direction at the n-th iteration, which is chosen as the direction parallel to the gradient vector $\{\partial\varepsilon/\partial c_i\}$ for the steepest descent algorithm. a_n is a step-size parameter which reduces the residual most significantly. In the conjugate gradient method, the vector $\{\delta c\}$ in Eq.(8) is adjusted to a direction considered the convergent history in order to accelerate the convergence speed.

The gradients $\partial\varepsilon/\partial c_i$ used in the iteration can be obtained from the derivatives $\partial Z/\partial c_i$ by using the equation,

$$\frac{\partial\varepsilon}{\partial c_i} = 2Re\left\{\sum_{m=1}^{M}\{Z_m(\mathbf{c}) - Z_m^{obs}\}^* \frac{\partial Z_m(c)}{\partial c_i}\right\}. \tag{9}$$

For derivatives $\partial Z/\partial c_i$, the method using adjoint field is valid[4], i.e.,

$$\frac{\partial Z_m(\mathbf{c})}{\partial c_i} = -\sigma_0 \int_{flaw} \mathbf{E}_{mt}(\mathbf{r}) \cdot \widetilde{\mathbf{E}}_{mt}(\mathbf{r}) \frac{\partial v(\mathbf{c},\mathbf{r})}{\partial c_i} d\mathbf{r}, \tag{10}$$

where, $\mathbf{E}_{mt}$ is the tangential component of the electric field when the probe is at m-th location, $\widetilde{\mathbf{E}}_{mt}$ is its corresponding adjoint field, and $v(c,\mathbf{r}) = (\sigma_0 - \sigma(\mathbf{r}))/\sigma_0$.

For SG tubing with crack perpendicular to its surface, the field is self-adjoint that can simplify Eq.(10) further. The detailed formulae of the derivatives are related to the selection of the crack parameters and the property of the crack. In the following parts, the crack parameters for three types of idealized cracks and their corresponding derivatives will be presented separately.

3.2 Reconstruction of a single EDM crack[2],[3]

For an artificial EDM crack, the variation of the crack parameters depends on the change of the crack edge uniquely as the conductivity at the crack region is imposed zero. In this case, the derivatives with respect to the crack shape parameter are in a following expression,

$$\frac{\partial Z_m(\mathbf{c})}{\partial c_i} = -\sigma_0 \int_S \mathbf{E}_{mt}(\mathbf{r}) \cdot \widetilde{\mathbf{E}}_{mt}(\mathbf{r}) \frac{\partial s(\mathbf{c},\mathbf{r})}{\partial c_i} \mathbf{n} \cdot \mathbf{k}_i dS, \tag{11}$$

where, $s(\mathbf{c}, \mathbf{r})$ is the function and $\mathbf{n}$ the normal unit vector of the crack edge. $\mathbf{k}_i$ is the unit vector parallel to the variation direction of δc_i. The integral region S is a band shaped surface along the crack edge with a width equal to the crack opening h_0.

We focus our problem further to the surface breaking cracks and assume the inner/outer and axial/circumferential properties were known from the phase property of ECT signal prior to the inversion. In such a case, it is reasonable to discretize the crack edge curve with an open piecewise line defined by n_c discrete points. Some calculations show that the derivatives of the impedance Z_m with respect to coordinates (x_{p_i}, z_{p_i}) of the discrete points can be expressed in term of the electric field by,

$$\frac{\partial Z_m}{\partial x_{p_i}} = -\sigma_0 \int_{\frac{-h_0}{2}}^{\frac{h_0}{2}} \left\{ \int_{\Gamma_i} E_{mt}^2 \frac{(t_{i+1} - t)}{(t_{i+1} - t_i)} \frac{(z_{p_{i+1}} - z_{p_i})}{(t_{i+1} - t_i)} dt + \int_{\Gamma_{i-1}} E_{mt}^2 \frac{(t - t_{i-1})}{(t_i - t_{i-1})} \frac{(z_{p_i} - z_{p_{i-1}})}{(t_i - t_{i-1})} dt \right\} dy, \tag{12}$$

$$i = 1, 2, \ldots\ldots n_c$$

$$\frac{\partial Z_m}{\partial z_{p_i}} = -\sigma_0 \int_{\frac{-h_0}{2}}^{\frac{h_0}{2}} \left\{ \int_{\Gamma_i} E_{mt}^2 \frac{(t_{i+1} - t)}{(t_{i+1} - t_i)} \frac{(x_{p_i} - x_{p_{i+1}})}{(t_{i+1} - t_i)} dt + \int_{\Gamma_{i-1}} E_{mt}^2 \frac{(t - t_{i-1})}{(t_i - t_{i-1})} \frac{(x_{p_{i-1}} - x_{p_i})}{(t_i - t_{i-1})} dt \right\} dy, \tag{13}$$

$$i = 2, 3, \ldots\ldots, n_c - 1$$

If a crack parameter c_i is chosen as step length of the modification along the normal direction at discrete point P_i, the derivative with respect to this parameter can be obtained by imposing the point p_i updating along the normal direction. Thus, the derivative with respect to c_i is equal to the projection at the normal direction, i.e.,

$$\frac{\partial Z_m}{\partial c_i} = \nabla Z_m \cdot \mathbf{n}_{pi} = \frac{\partial Z_m}{\partial x_{p_i}} n_{x_{p_i}} + \frac{\partial Z_m}{\partial z_{p_i}} n_{z_{p_i}}. \tag{14}$$

3.3 Reconstruction of an EDM crack with non-zero conductivity

In case of an EDM crack with a non-zero conductivity, the variation of crack parameters $\delta v(\mathbf{c}, \mathbf{r})$ contains not only the parameters of the crack edge, but also the parameters representing the conductivity distribution. In addition to the discretization of crack edge by a piecewise line as adopted for a single EDM crack, we choose parameters for conductivity as the discrete coefficients by which the distribution function was expanded within the selected region. The base functions for such an expansion can be chosen as a set of box function or other bases such as the wavelet functions. In this case, the crack of non-zero conductivity can be expressed uniquely by two sets of parameter, viz, edge parameters $\{c_i\}$ and conductivity parameters $\{\sigma_i\}$. Therefore, the variation of the crack function can be written as,

$$\delta v(\mathbf{c}, \sigma, \mathbf{r}) = \sum_i^{n_c} \frac{\partial v(\mathbf{c}, \sigma, \mathbf{r})}{\partial c_i} \delta c_i + \sum_i^{n_\sigma} \frac{\partial v(\mathbf{c}, \mathbf{r})}{\partial \sigma_i} \delta \sigma_i \tag{15}$$

where, n_c, n_σ are respectively the number of parameter c_i and σ_i.

Some calculations show that the derivatives with respect to the crack parameters are expressed as,

$$\frac{\partial Z_m(\mathbf{c})}{\partial c_i} = - \int_S \sigma(\mathbf{r}) \mathbf{E}_m(\mathbf{r}) \cdot \widetilde{\mathbf{E}}_m(\mathbf{r}) \frac{\partial s(\mathbf{c}, \mathbf{r})}{\partial c_i} \mathbf{n} \cdot \mathbf{k_i} dS, \tag{16}$$

$$\frac{\partial Z_m(\mathbf{c})}{\partial \sigma_i} = -\sigma_i \int_{V_i} \mathbf{E}_m(\mathbf{r}) \cdot \widetilde{\mathbf{E}}_m(\mathbf{r}) \phi_i(\mathbf{r}) dv, \tag{17}$$

where, V_i is a joint region of the $i-th$ cell with the zone of crack, ϕ_i is the $i-th$ basis function employed for discretization of distribution of conductivity. It needs to point out that the tangential components used in Eq.(11) were replaced by the total electric field vector here as the normal components at the crack surface is not zero further in case of a non-zero conductivity.

For cracks perpendicular to the tube surface, the derivatives $\partial Z_m(\mathbf{c})/\partial c_i$ can be calculated using a formula similar to that for a single EDM crack. However, as the conductivity variates in different position of the crack zone, it needs to be taken into account in the integration.

3.4 Reconstruction of the multiple EDM cracks

In this part, we consider the reconstruction of the multiple cracks by dealing with an example of two axial cracks. In this case, we also assume that the number, directions, and the basic properties of the cracks were known *a priori*. In other words, we will reconstruct the distance between the cracks in addition to their shapes. We choose the crack parameters as $\{c_{1i}\}$, $\{c_{2i}\}$ for describing the crack shapes and d_1, d_2 for representing the distance between the cracks and a reference point. In such a case, the variation $\delta v(\mathbf{c}, \mathbf{r})$ becomes,

$$\delta v(\mathbf{c}, \mathbf{r}) = \sum_{i}^{n_{c1}} \frac{\partial v(\mathbf{c}, \mathbf{r})}{\partial c_{1i}} \delta c_{1i} + \sum_{i}^{n_{c2}} \frac{\partial v(\mathbf{c}, \mathbf{r})}{\partial c_{2i}} \delta c_{2i} + \frac{\partial v(\mathbf{c}, \mathbf{r})}{\partial d_1} \delta d_1 + \frac{\partial v(\mathbf{c}, \mathbf{r})}{\partial d_2} \delta d_2 \tag{18}$$

Substituting Eq.(18) into Eq.(10), we can find that the derivatives with respect to the parameters c_{1i} and c_{2i} are the same with those for the signal EDM crack if we consider the integral surface of Eq.(12) and Eq.(13) as the surface of the corresponding crack. For the parameters d_1 and d_2, some calculations show,

$$\frac{\partial Z_m}{\partial d_j} = -\sigma_0 \left\{ \int_{S_{j1}} |\mathbf{E}_{mt}|^2 dS - \int_{S_{j2}} |\mathbf{E}_{mt}|^2 dS \right\} \tag{19}$$

where $j = 1, 2$ represents the different crack, S_{j1} is the outer side surface of the $j-th$ crack referring to the given point, and S_{j2} is the inner side surface.

4. Numerical Results

We choose a self induction pancake coil of 100 turns as the probe for ECT data which is winded with a outer radius 1.6 mm, an inner radius 0.5 mm and a thickness 1 mm. The probe is arranged in parallel with the inner tube surface and with a lift-off equal 1. mm. The excitation frequency is chosen as 400kHz which is a typical one in the practical inspection of SG tubing. The scanning is axially performed along the crack from -10 mm to 10 mm and with the step length chosen as 0.5 mm. These 41 data of impedance signal are used for the reconstruction of all the crack parameters. As a need of the fast forward solver, two data-bases are established with the possible crack region selected as a $10mm \times 1.27mm \times 0.2mm$ cubic zone and subdivided into $20 \times 8 \times 1$ cells.

4.1 ECT signals with artificial noise

To consider the robustness of the present inverse scheme, the reconstruction is carried out for observing data containing artificial noise. By adjusting the maximum value of the white noise component added into the signals, we found that the the proposed method can give a good estimation of crack shapes even for noise as large as over 20%.

Fig.1 depicts the reconstructed results for a crack in elliptic shape. The noise level in this case was adopted as 20% of the maximum absolute value of the used signal. Although the robustness is not as good as the case for a square crack, the reconstruction converged to the true crack shape even for such a large noise level. In addition, the computational burden is almost the same as the case of high S/N signals (about 20 minutes, Silicon Graphics, Indigo 2).

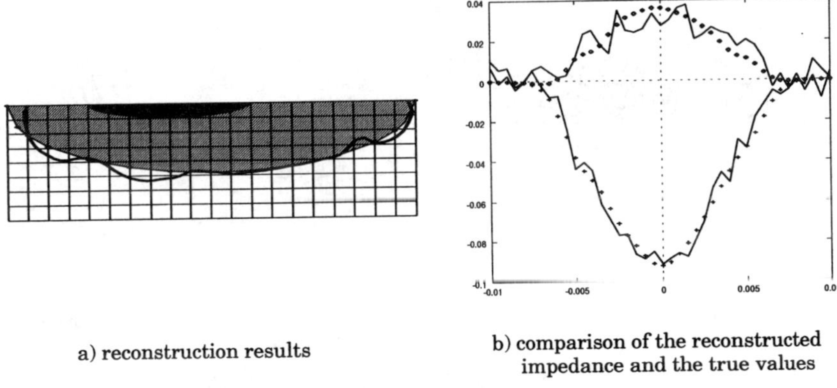

a) reconstruction results

b) comparison of the reconstructed impedance and the true values

Figure 1: Reconstructed results of an elliptic crack (elliptic crack, 400kHz, tube, noise level = 20%)

4.2 Signals of practical inspection

For validating the present inverse approach further, a tubing work-piece with 10 EDM cracks is fabricated and detected by setting it to a SG mock-up in order to acquire data in a practical environment. The depths of the ten cracks are respectively, 100%, ID80%, ID60%, ID40%, ID20%, 100%, OD80%, OD60%, OD40%, and OD20%, while the length and the opening of all the cracks are chosen as 5 mm and 0.2 mm respectively. A rotating probe of a pancake coil was employed for the inspection of 8 frequencies ranged from 100kHz to 1000kHz. The size and lift-off of the probe are used as the same value as those for the bench-mark models[5]. After data processing, the signals when the probe just pass over the cracks are extracted and stored for reconstruction.

Fig.2 shows the reconstruction results for the OD40% crack. The S/N ratio of an OD40% crack is very small and even smaller than an ID20% crack. However, the predicted profiles of the crack agree well with the true values. As the S/N of an OD20% crack is very small, to reconstruct such a shallow crack using a pancake type probe seems very difficult. Therefore, some probes of high performance are needed in such a case.

5. Conclusions

In this paper, formulae for the reconstruction of natural cracks are proposed by idealizing the cracks as multiple planar slits of non-vanishing conductivity. The scheme is applied to the signals including artificial noise, as well as those acquired under a more practical environment. Several typical results of an EDM crack are presented for signals with noise up to 20%, which in turn, verified the robustness and efficiency of the proposed approach. The validation of the scheme for multiple cracks and a crack of non-vanishing conductivity, on the other hand, is left as a task for the near future.

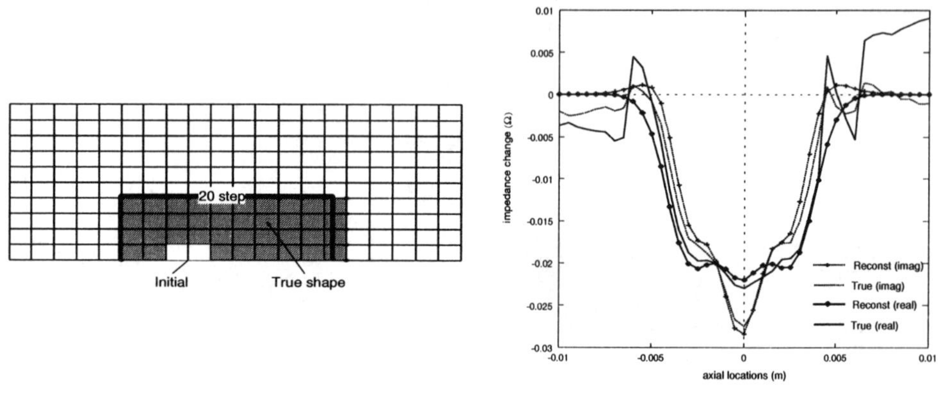

a) Reconstruction results

b) comparison of impedance

Figure 2: Reconstruction results using practical ECT data (OD40%, 300kHz, tube)

References

[1] Z.Badics, and et.al., Rapid flaw reconstruction scheme for 3-d inverse problems in eddy current NDE, in Stud. Appl. Electromagn. Mech., Vol.12, 303-309, eds. T.Takagi et al., IOS press, 1996.

[2] Z.Chen, K.Miya, and M.Kurokawa, Rapid prediction of eddy current testing signals using $\mathbf{A} - \phi$ method and database, NDT&E international, Vol.32, No.1, 29-36, 1998.

[3] Z.Chen and K.Miya, ECT inversion using a knowledge based forward solver, J. Nondestr. Eval., Vol.17, No.3, 167-175, 1998.

[4] S.Norton and J.Bowler, Theory of ECT inversion, J. Appl. Phys. Vol.73, No.2, 501-512, 1993.

[5] T.Takagi, and et al., Benchmark models of eddy current testing for steam generator tube: experiment and numerical analysis, Int. J. Appl. Electromagn. in Materials, Vol.4, 149-162, 1994.

Electromagnetic Nondestructive Evaluation (III)
D. Lesselier and A. Razek (Eds.)
IOS Press, 1999

Reconstruction of Cracks With Multiple ECT Coils Using a Database Approach

Weiying CHENG and Kenzo MIYA

Nuclear Engineering Research Laboratory,
Graduate school of Engineering, The University of Tokyo
2-22 Shirakata-shirane, Tokai-mura, Naka-gun, Ibaraki, 319-1106 Japan

Zhenmao CHEN
Structure Safety Engineering Section, Oarai Engineering Center,
Power Reactor & Nuclear Fuel Development Corporation, 311-13 Japan

Abstract — This paper presents some recent studies connected with flaw reconstruction. The location of a planar crack in an Inconel 600 plate specimen is determined and the shape of it is reconstructed using the signals from a multi-pancake coil probe. The reconstruction strategy, with help of a priori data based fast forward solver and a first order optimization inverse algorithm, is based on the minimization of the nonlinear least square error function. Computational characteristics of the inverse analysis method are investigated with respect to the influence of the arrangement of scanning paths. The efficiency of the computation is also discussed.

1 Introduction

ECT (Eddy Current Testing) technique plays an important role in the in-service inspection of heat exchanger tubes in steam generators of nuclear power plants. Many methods have been proposed for the technical enhancement of this classical means of NDT (Non-Destructive Testing). In recent years, with strong requirements for structural safety, ECT has progressed to a quantitative detection stage. Shapes and locations of defects are strongly required for assessments of defect propagation and determination of critical stage. Encouraging results were achieved for the reconstruction of a single crack[1] in the case when the detecting probe scans just over the crack.

However, in view of the real inspection situation, signals are not measured just over the crack, and the crack is not necessarily parallel to the scanning paths, too. Problems have to be dealt in a more practical way.

Considering the actual situation, a crack found in a SG (Steam Generator) tube usually appears as an axial or a circumferential one perpendicular to the tube surface, while the inspecting probe scans along the axis or rotates circumferentially. Therefore, at the first step, it is reasonable to assume that the cracks are either parallel or perpendicular to a scanning path. Since the width of a crack varies very little, the crack can be considered as a planar one with fixed width. An orientation of a crack is assumed known beforehand. The conductivity of a crack is assumed to be zero.

A multi-pancake coil probe is used. The probe is composed of 12×2 self-induction pancake coils with the arrangement of Fig. 1. Considering the diameter of real SG tube, two neighboring coils are about 2.7mm apart. Parameters of single pancake coil are listed in Fig. 2. Table 1 gives the parameters of the specimen. A three dimensional crack reconstruction code was developed for a plate geometry.

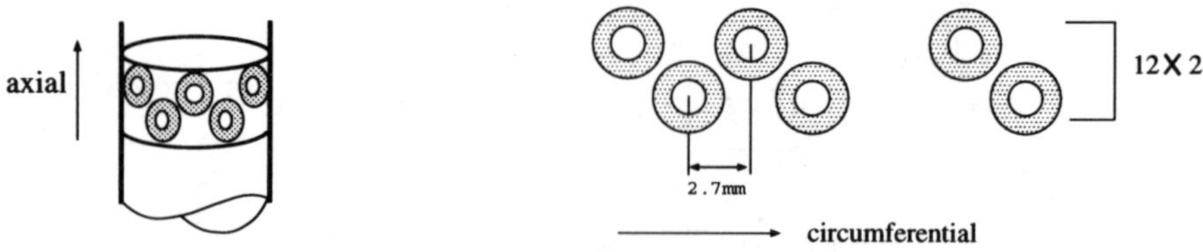

Fig. 1 The arrangement of a multi-pancake coil probe

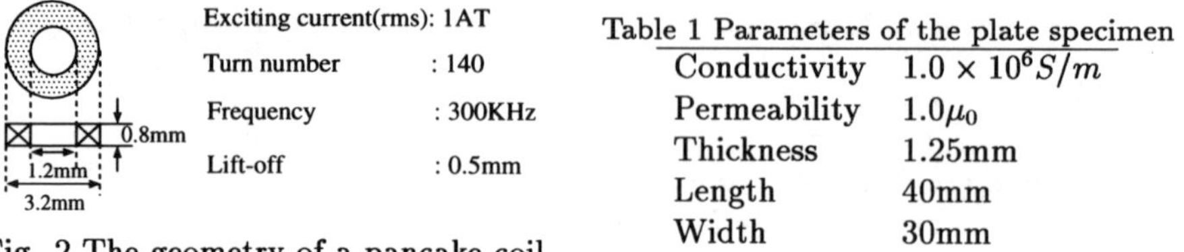

Fig. 2 The geometry of a pancake coil

Table 1 Parameters of the plate specimen

Conductivity	$1.0 \times 10^6 S/m$
Permeability	$1.0\mu_0$
Thickness	1.25mm
Length	40mm
Width	30mm

A surface breaking crack with arbitrary shape was reconstructed. The position of it was also determined using the multi-pancake coil probe.

2 Basic Concepts of Crack Reconstruction

A deterministic algorithm is chosen for crack shape reconstruction. The flowchart of the basic inversion algorithm is described in Fig. 3. At each iteration, the estimated signals have to be calculated and compared with the measured set. The forward problem, the computation of signals corresponding to a certain location and profile of crack, is very time consuming. A fast forward solver is required.

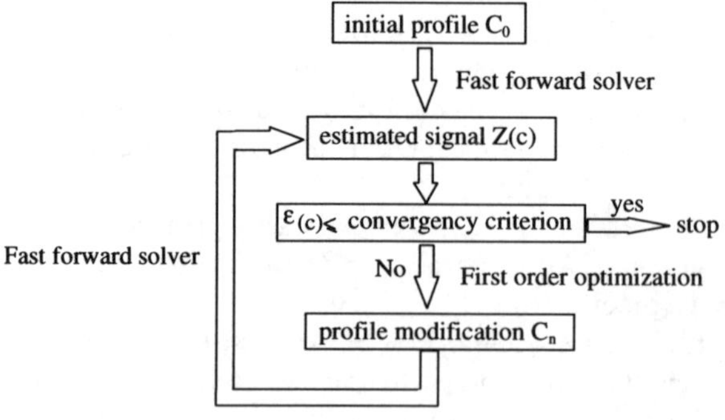

Fig. 3 The flowchart of the basic inversion algorithm

The method used here is based on the method proposed by Z. Chen et.al.,[1]. It is model-based, however, by employing of a priori prepared database, which is flaw independent, the forward problem can be solved very fast and the optimization is significantly speeded up.

2.1 Fast Forward Solver

The following governing equations can be derived based on the conventional FEM-BEM algorithm[1].

$$\frac{1}{\mu_0}\nabla^2 A^f = j\omega\sigma_0(A^f + \nabla\Phi^f) + j\omega[\,\sigma(r) - \sigma_0](A + \nabla\Phi), \tag{1}$$

$$j\omega\nabla\cdot\sigma_0(A^f+\nabla\Phi^f)=j\omega\nabla\cdot[\sigma_0-\sigma(r)](A+\nabla\Phi),\qquad \text{(in conductor)} \tag{2}$$

$$\frac{1}{\mu_0}\nabla^2A^f=0.\qquad \text{(in air)} \tag{3}$$

Symbols σ_0 is the conductivity of the host conductor, and $\sigma(r)$ is defined as,

$$\sigma(r)=\begin{cases}\sigma_0 & \text{in conductor}\\ 0 & \text{in crack}\end{cases} \tag{4}$$

The potentials with crack present and absent are denoted by symbols A, Φ and A^u, Φ^u, respectively. A^f and Φ^f are defined as the change of potentials due to the flaw.

$$A^f=A-A^u,\qquad \Phi^f=\Phi-\Phi^u. \tag{5}$$

Since $\sigma(r)-\sigma_0$ vanishes outside the flaw region, the following equation can be obtained using the FEM-BEM discretization[2].

$$\begin{bmatrix}\tilde{K_{11}} & \tilde{K_{12}}\\ \tilde{K_{21}} & \tilde{K_{22}}\end{bmatrix}\begin{bmatrix}q_1^f\\ q_2^f\end{bmatrix}=\begin{bmatrix}K_{11} & 0\\ 0 & 0\end{bmatrix}\begin{bmatrix}q_1^f+q_1^u\\ q_2^f+q_2^u\end{bmatrix}. \tag{6}$$

Symbols $q_1^f=\{A_1^f,\Phi_1^f\}^T$, $q_1^u=\{A_1^u,\Phi_1^u\}^T$. Subscripts 1 and 2 denote the region within or out of the flaw. Finally, the required potential in the flaw region can be deduced from,

$$[I-[H_{11}][K_{11}]]\{q_1^f\}=[H_{11}][K_{11}]\{q_1^u\}, \tag{7}$$

where q_1^u is the potential in the absence of flaw in region 1, and $[H_{11}]$, which is only flaw region related, is a sub-matrix of $[H]$, with

$$[H]=\begin{bmatrix}\tilde{K_{11}} & \tilde{K_{12}}\\ \tilde{K_{21}} & \tilde{K_{22}}\end{bmatrix}^{-1}=\begin{bmatrix}H_{11} & H_{12}\\ H_{21} & H_{22}\end{bmatrix}. \tag{8}$$

Both $q_1{}^u$ and $[H]$ are flaw independent and can be calculated a priori by the conventional FEM-BEM method.

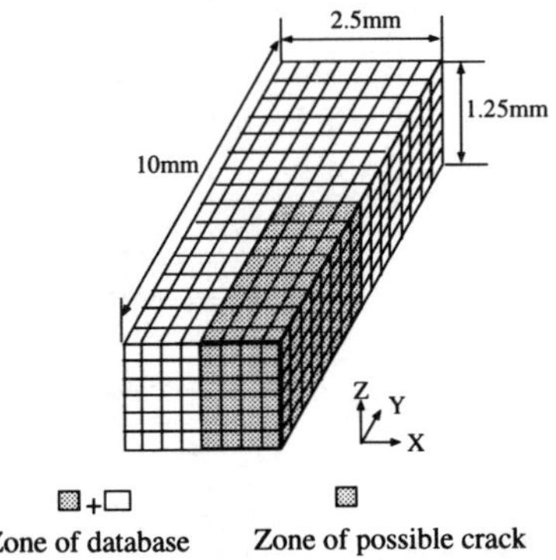

Fig. 4 1/4 region of database and possible crack

In actual testing, the flaw region can be inferred from signals beforehand. In this work, the flaw is assumed to be in a region measuring $5mm\times10mm\times1.25mm$. Databases q_1^u and $[H]$ were established in a region of $10mm\times20mm\times1.25mm$ in advance.

In case of self-induction probe, the signal is expressed by the impedance change of the testing coil. According to the reciprocity theorem[3],

$$\Delta Z=\frac{1}{I^2}\int_{flaw}E^u\cdot(E^f+E^u)\sigma_0dv, \tag{9}$$

where I is the current in the testing probe. With the aid of equation

$$E = -\frac{\partial A}{\partial t} - \nabla\Phi = -iwA - \nabla\Phi, \tag{10}$$

E^u, the electric field density in the absence of flaw, can be obtained from database q_1^u directly, and E^f, the electric field density due to flaw, can be calculated from Equ. (7). The integration of Equ. (9) is limited in the flaw region. This also contributes to the reduction of computational requirements.

2.2 Inverse Solver

2.2.1 Modeling

As there are no deviated parallel scanning or perpendicular scanning experimental data yet, in this work, the reconstruction is carried out from simulated ECT signals. The probe scans along the Y direction. A typical Y axis oriented crack is sketched in Fig. 6(a). Fig. 6(b) shows the cross section of the crack in the Y-Z plane. The crack is considered 0.2mm in width.

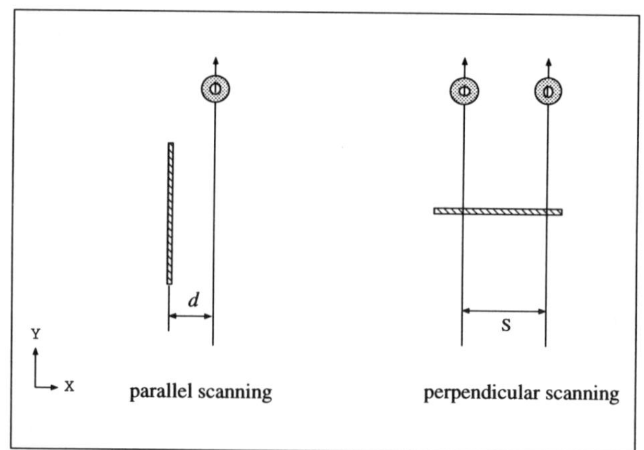

Fig. 5 The arrangement of scanning paths and crack

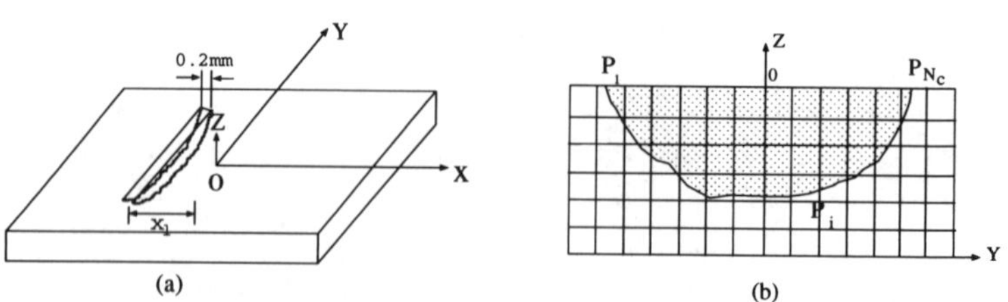

Fig. 6 Sketch of a crack

The problem of reconstructing the shape of a surface breaking crack is equivalent to a problem of determining its edge curve. The crack boundary is discretized into N_c points. Parameter x_l indicates the location of a crack as is described in Fig. 6(a). Hence, a crack can be designated by a parameter vector

$$\{c\} = \{t_i, x_l\}^T \qquad i = 1, N_c, \tag{11}$$

where t_i is defined as the coordinate of point P_i in the Y-Z plane.

2.2.2 Inverse Algorithm

Here we use the least-square method as the inversion analysis of crack reconstruction.

The objective function is defined by,

$$\varepsilon(c) = \sum_{m=1}^{M} w_m |Z_m(c) - Z_m^{obs}|^2, \tag{12}$$

where Z_m^{obs} and $Z_m(c)$ are observed and estimated impedance respectively. Both are functions of probe position r_0, flaw function v, and frequency f. w_m is a weight of each testing point. The summation of equation (12) is taken for the whole set of inspection.

An iterative algorithm is selected for parameter optimization (Fig. 3). At each iteration, the crack parameters are modified and the corresponding impedances $Z(c)$ are estimated. The parameters are considered optimized when the error defined by Equ. (12) decreases to a small tolerance value. The parameter vector is updated as follows,

$$\{c\}_n = \{c\}_{n-1} + a_n\{\delta c\}_n, \tag{13}$$

where $\{\delta c\}_n$ indicates the optimization direction at the n-th iteration and a_n designates the n-th updating step size. According to the steepest descent algorithm, the optimization is chosen to be in parallel with the derivative vector $\partial\varepsilon/\partial c_i$,

$$\frac{\partial \varepsilon}{\partial c_i} = 2\mathcal{R}e\{\sum_m w_m\{Z_m(c) - Z_m^{obs}\}^* \frac{\partial Z_m(c)}{\partial c_i}\}. \tag{14}$$

The convergence can be accelerated by the conjugate-gradient algorithm. In this case, the updating direction vector is expressed in the form of

$$\{\delta c\}_n = -\frac{\partial \varepsilon}{\partial c_i} + \beta_n\{\delta c\}_{n-1}, \tag{15}$$

with the initial condition $\{\delta c\}_1 = -\frac{\partial \varepsilon}{\partial c_i}$, and $\beta_1 = 0$.

$$\beta_n = \sum_i |\frac{\partial \varepsilon}{\partial c_i}|_n^2 / \sum_i |\frac{\partial \varepsilon}{\partial c_i}|_{n-1}^2 \tag{16}$$

Equations (14), (15) and (16) indicate that in order to find the updating direction $\{\delta c\}_n$, the derivative of signal Z with respect to crack parameter c_i has to be found firstly. By the approach of adjoint-field [3], a key equation is derived,

$$dZ[v, \delta v] = -\sigma_0 \int \tilde{E}(r) \cdot E(r)\delta v(r) dr, \tag{17}$$

where, $E(r)$ is the electric field density in the absence of flaw and $\tilde{E}(r)$ the adjoint field of $E(r)$. The integration is over the flaw volume v.

Since $\delta v(r)$ is nonzero only in the region between the surface of the original flaw and the surface after the variation, for an infinitesimal variation, the integration of $E(r)$ and $\tilde{E}(r)$ with respect to flaw function v is changed to the integration of the tangential component of $E(r)$ and $\tilde{E}(r)$ with respect to flaw surface S (Fig. 7).

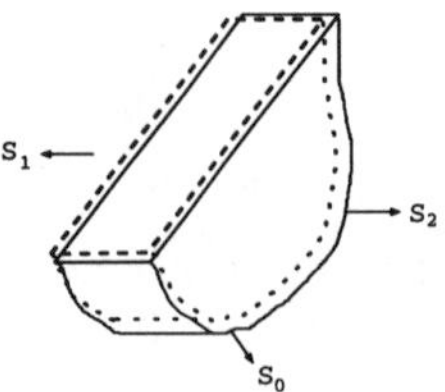

Fig. 7 An infinitesimal variation of the flaw

Fig. 7 shows that a surface breaking crack consists of three surfaces which are denoted as S_0, S_1 and S_2, respectively. Since the crack is assumed Y axis oriented and the width of it is fixed, S_1 and S_2 are facing each other.

A variation of impedance due to a small deformation of the surface is expressed as,

$$\begin{aligned}
\delta Z &= -\sigma_0 \int \tilde{E}(r) \cdot E(r) \delta v(r) dr \\
&= -\sigma_0 \int_{S_1+S_2+S_0} \tilde{E}_t(r) \cdot E_t(r) \frac{\partial s(c,r)}{\partial c_i} n \cdot k_i ds \delta c_i \\
&= -\sigma_0 \int_{S_0} \tilde{E}_t(r) \cdot E_t(r) \frac{\partial s(c,r)}{\partial t_i} n \cdot k_i ds \delta t_i \\
&\quad -\sigma_0 \int_{S_1} (\tilde{E_{1t}}(r) \cdot E_{1t}(r) - \tilde{E_{2t}}(r) \cdot E_{2t}(r)) \delta[s(r)] dx.
\end{aligned} \tag{18}$$

$\partial Z / \partial c_i$ can be obtained by differentiating Equ. (18) with respect to parameter c_i.

The step-size parameter a_n is derived using the conjugate-gradient algorithm,

$$a_n = P/Q, \tag{19}$$

with

$$P = Re\{\sum_m w_m (Z_m^{n-1} - Z_m^{obs})^* \frac{\partial Z_m^{n-1}}{\partial a_n}\}, \tag{20}$$

$$Q = \sum_m w_m |\frac{\partial Z_m^{n-1}}{\partial a_n}|^2. \tag{21}$$

The required estimated impedance $Z_m(c)$ at each step is calculated by the fast forward solver described in section 2.1.

3 Simulation Results

3.1 Detectability Analysis

Fig. 8(a) shows the change of detectability for a Y axis oriented inner half elliptical crack in the X direction of multi-pancake coil probe. It is clear that the detectability is highest when the probe is just over the crack, while when the probe is about 2.7mm far from the crack, the signal decreases to a very small value. It is almost impossible to detect cracks from such small signals, let alone to reconstruct them. As described in Fig. 8(b), for a crack located between coils, only the signals issued by the nearest coil B and coil C are significant.

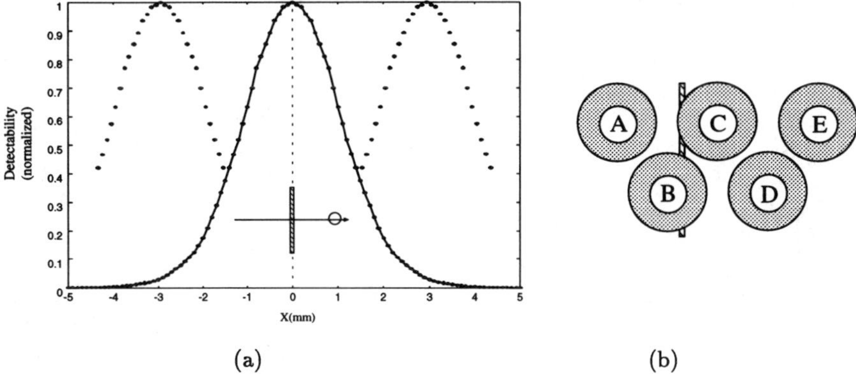

(a) (b)

Fig. 8 The detectability around position of probe

3.2 Reconstruction from parallel scanning signals

First, we consider the case when the probe scans along paths which are parallel to the crack. The probe scans from $Y = -10mm$ to $Y = 10mm$ with a pitch of 0.5mm. This corresponds to the reconstruction of an axially oriented crack from axial scanning probe or a circumferential crack from rotating probe.

3.2.1 Reconstruction from one set of signals

Only one set of signals is used. two cases are considered,

(a) the deviation of the scanning path from the crack, d, is 1.0mm,

(b) the deviation is 1.4mm.

Figures 9, 10 and 11 show the reconstruction results corresponding to the two cases mentioned above. Fig. 9 shows profiles of the initial, the original and the reconstructed crack. Figures 10 and 11 describe the change of parameter x_l and the residual which is defined by Equ. (12) during iteration.

Fig. 8(a) shows that when the probe is 1.0mm far from the crack, the signal is about 63% of the one when the probe is just over it, while when the crack is 1.4mm deviated, the signal decreases to only about 45% of the maximum value. Figures 10 and 11 demonstrate that the higher the detectability, the faster the convergence. It is easier to reconstruct crack from data with higher detectability.

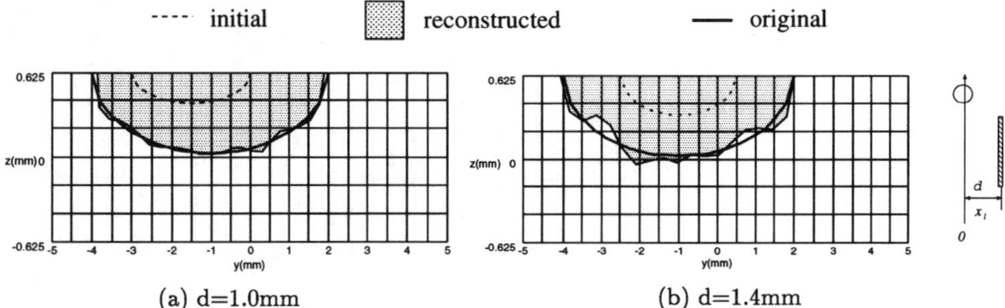

(a) d=1.0mm (b) d=1.4mm

Fig. 9 The profile of the initial, original, and the reconstructed crack

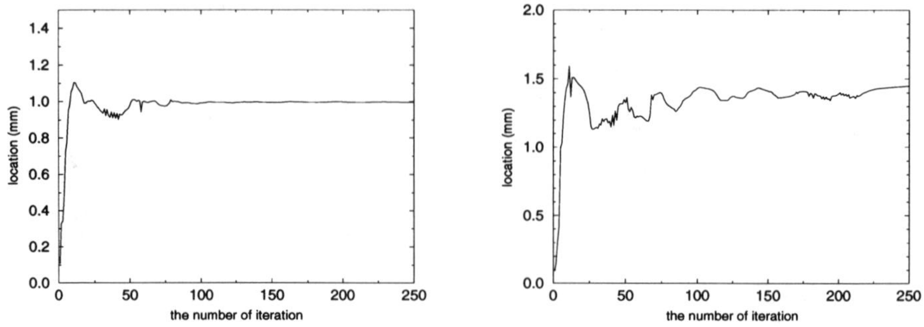

(a) d=1.0mm (b) d=1.4mm

Fig. 10 The change of location

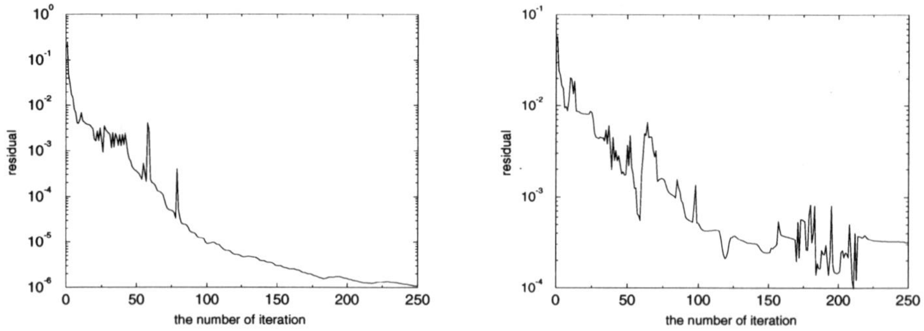

(a) d=1.0mm (b) d=1.4mm

Fig. 11 The change of residual during iteration

3.2.2 Reconstruction from two sets of signals

Fig. 8(b) indicates that for a single crack, signals collected by both coil B and coil C are significant. The same half-elliptical crack as described in the previous section is reconstructed from two sets of signals. Deviation $d_1 = 1.4mm$ and $d_2 = 1.3mm$. (a) is the profile of the initial, the original and the reconstructed crack, (b) the convergence of parameter x_l, the initial of it is set to 0.1mm, and (c) the change of residual during iteration.

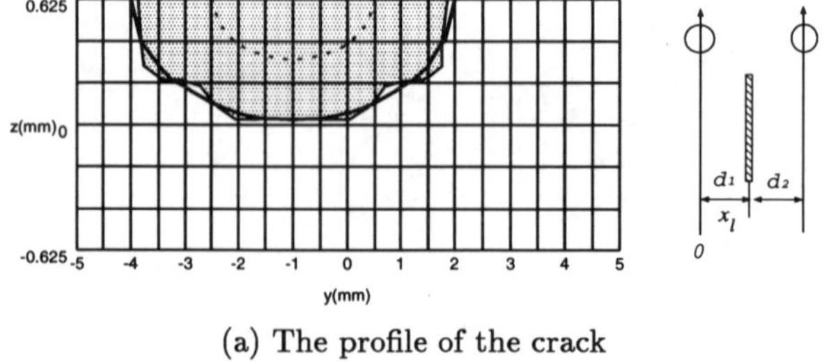

(a) The profile of the crack

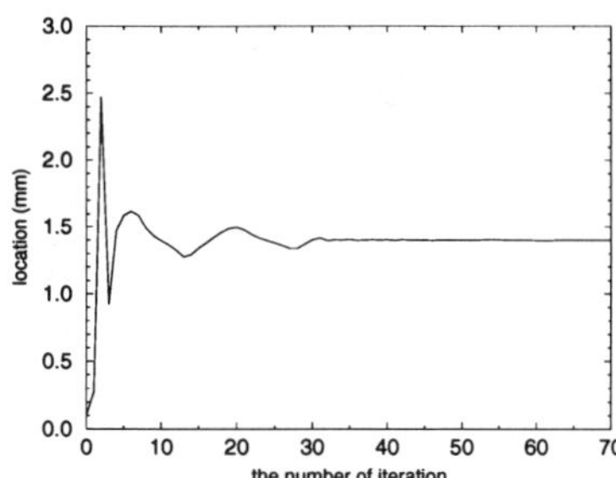

(b) The change of location

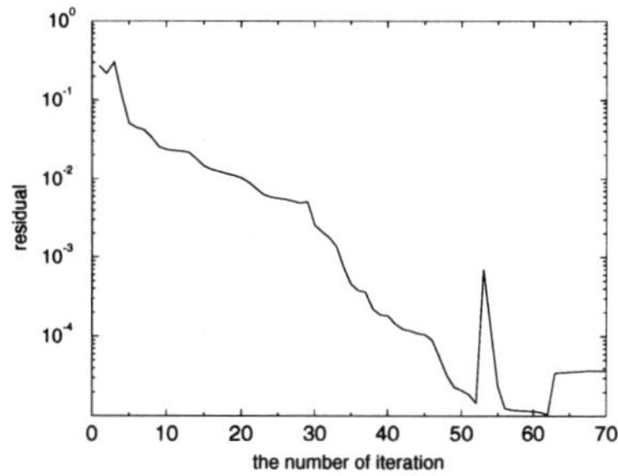

(c) The change of residual

Fig. 12 Reconstruction of a half-elliptical crack from two sets of signals

The comparison of Fig. 9(b)–Fig. 11(b) and Fig. 12 demonstrates that with two sets of signals, the reconstruction is significantly speeded up. The computational stability is improved. The computation is much more efficient. In dealing with practical problems, as much information as possible should be used efficiently with the multi-pancake coil probe.

3.3 Reconstruction from signals of perpendicular scanning

In case that the crack is circumferentially oriented and the probe scans along the axial direction, or the crack is axially oriented and the probe rotates circumferentially, the crack is perpendicular with the scanning paths.

3.3.1 Recontruction of a short crack (L=2mm)

A short crack is taken into consideration. It is 2mm long, shorter than the distance between two neighboring coils. The crack will cut across at most one scanning path (Fig. 14(a)).

(1) Reconstruction from one set of signals

Fig. 13 shows the reconstruction from one set of perpendicular scanning signals. The crack is assumed to be located at $x_l = 0$, and only the shape is reconstructed. Two cases are considered,

(a) the scanning path cuts across the center of the crack,

(b) the scanning path does not cut across the crack.

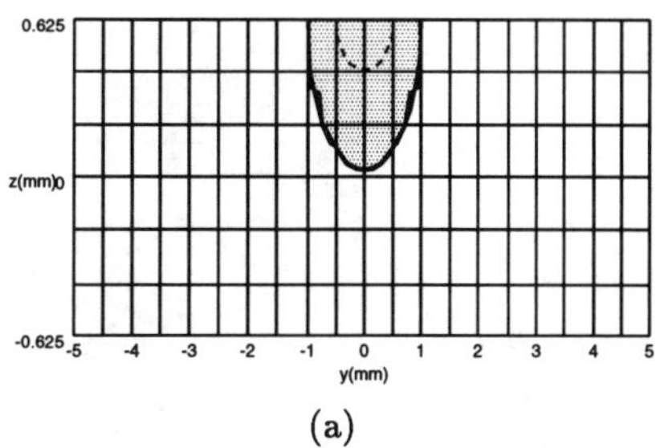

(a)

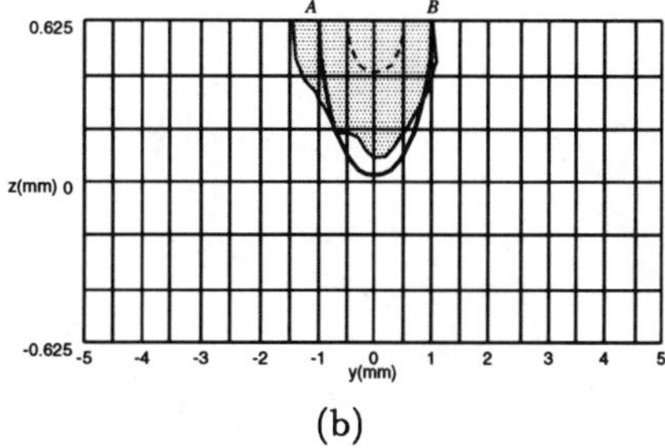

(b)

Fig. 13 Reconstruction of a short crack from one set of perpendicular signals

Fig. 14 shows the signals corresponding to the two cases mentioned above. The scanning path is perpendicular with the crack. The probe scans from $X = -3mm$ to $X = 3mm$ with a pitch of 0.5mm. The signals described in Fig. 14(b) are taken from path SX_1 only.

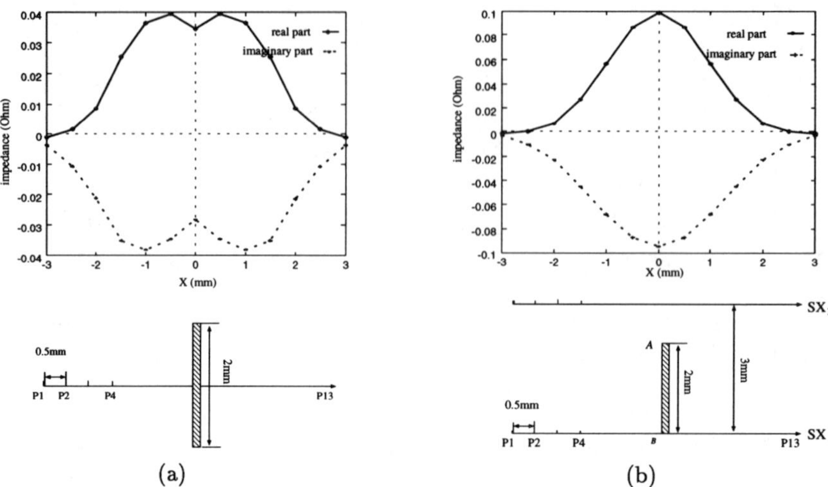

Fig. 14 The signals issued by a perpendicular scanning probe

It is found from Fig. 8(a) that the active range of the employed pancake coil at 300kHz and for the 0.5mm lift-off extends to about 3mm from the position of its center, however the signal at 3mm is only about 3% of the center one. The analysis of Fig. 14 shows that in case of (b), when the probe goes further away from the crack, signal declines quickly. Some informations of side A are lost, the reconstruction is not satisfactory.

(2) Reconstruction from two sets of signals

To improve the reconstruction of the case described in Fig. 13(b), signals sampled from scanning path SX_2 are also used. Both the location and the shape are to be reconstructed. The crack is located at $x_l = 0$, while the initial is set to $x_l = 1.0mm$. The two scanning paths are 3mm separated.

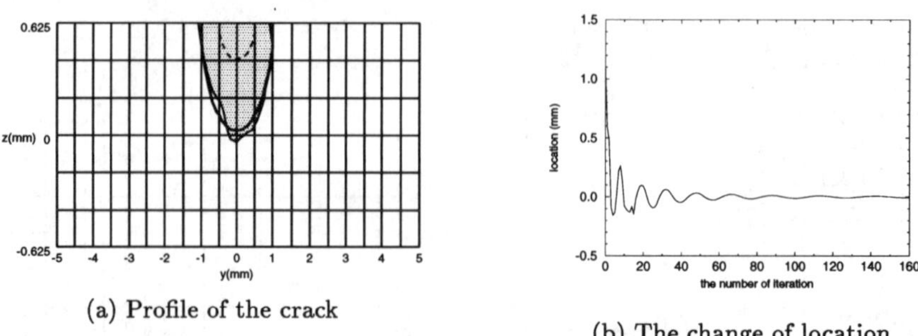

(a) Profile of the crack

(b) The change of location

Fig. 15 Reconstruction of a short crack from two sets of perpendicular signals

Fig. 15 shows that the reconstruction is improved using two sets of signals.

3.3.2 Reconstruction of a long crack (L=6mm)

A crack cuts across one or more than one scanning paths when it is longer than the separation between two coils.

Fig. 16 shows the reconstruction results for a 6mm long crack. Three sets of signals are used. Each set includes three sampling points only, and the pitch is 1mm. Fig. 16(a) shows the arrangement of the 9 sampling points and the profile of the crack. The change of location and residual during the iteration are described in Figures 16(b) and (c) respectively.

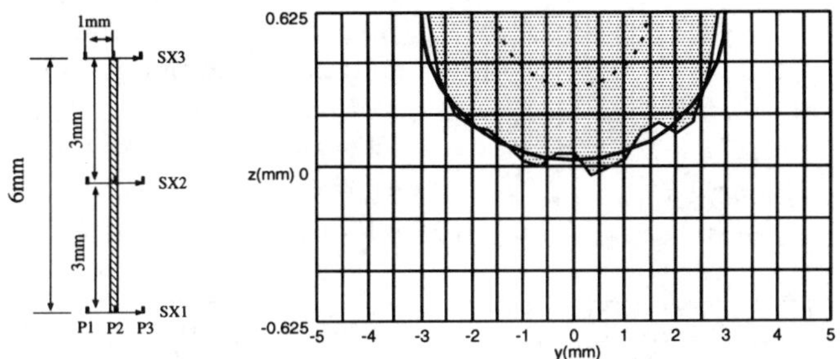

(a)The sampling of signals and the crack shape

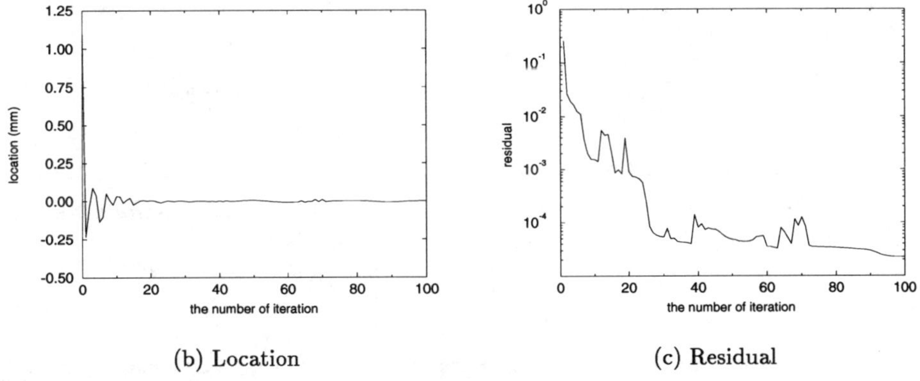

(b) Location (c) Residual

Fig. 16 Reconstruction of a long crack from perpendicular signals

Although the signals are sampled very coarsely, the reconstruction is fairly well. The 9 sampling points are with high detectability. Comparing with the case described in section 3.3.1, it is clear that the significance of data plays an important role in crack reconstruction. Sampling points should be well chosen in order to improve the computational efficiency.

3.4 Reconstruction from noise-polluted signals

In practical problems, signals are, without exception, polluted by noise. To verify how robust this code is, reconstruction is carried out from signals with artificial noise.

The noise is defined in the following way,

$$Z_m^{sn} = Z_m^{sim} + \alpha |Z_{max}^{sim}| \zeta_m, \tag{22}$$

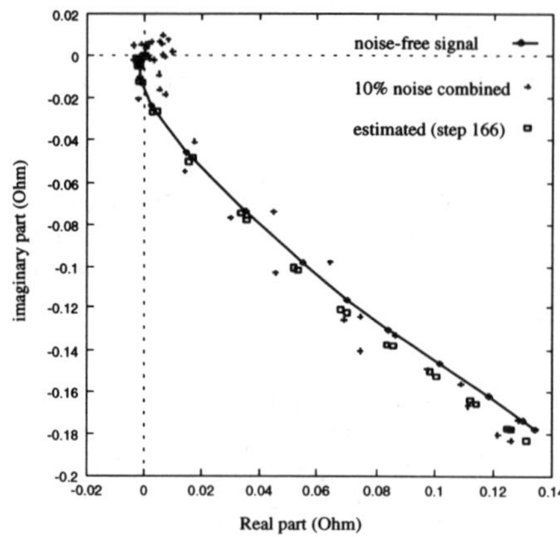

(a) Comparison of the noise-free, the noise-polluted and the estimated signals

where Z_m^{sn} is noise-polluted signal, Z_m^{sim} is simulated noise-free signal and $|Z_{max}^{sim}|$ is the maximum simulated signal for a complete scan set. α is a positive coefficient which indicates how many percentage of noise is imposed in the composed signal. ζ_m is a random value ranged from -1.0 to 1.0. The distribution of the noise is white.

Fig. 17 shows the reconstruction results from signals with 10% noise. The computation condition is as same as that of section 3.2.1(a). Only one set of parallel scanning data is used. The deviation of the scanning path from the crack is 1.0mm.

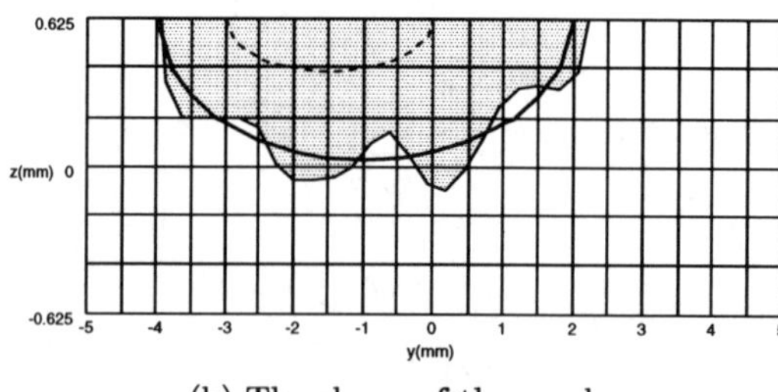

(b) The shape of the crack

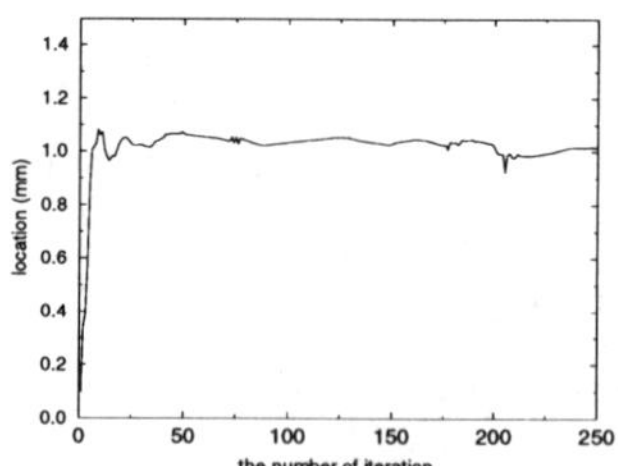

(c) The change of location

Fig. 17 Reconstruction from noise-polluted signals (noise level=10%)

Acceptable result is obtained from the noise-polluted data. The interested length and depth of crack can be approximately estimated.

4 Conclusion

Numerical simulations are carried out for some inversion problems. Satisfactory reconstruction results are achieved by using the inversion scheme presented in this paper. The location of crack is determined and the crack shape is also reconstructed with the multi-pancake coil probe using a database approach. Signals with high detectability are favorable for the efficient and accurate reconstruction of cracks. As much information as possible from the multi-pancake coil probe should be used sufficiently.

References

[1] Z. Chen, K.Miya and M. Kurokawa, Reconstruction of Crack Shapes Using a Newly-Developed ECT Probe, *E'NDE '97 Italy.*

[2] F.Matsuoka, A.Kameari, Calculation of Three Dimensional Eddy Current by FEM-BEM Coupling Method, *IEEE Trans. Mag. Vol.24, No.1,* January, 1988.

[3] S.J.Norton and John R.Bowler, Theory of eddy current inversion, *J.Appl.Phys.73(2),* 15 January, 1993.

Electromagnetic Nondestructive Evaluation (III)
D. Lesselier and A. Razek (Eds.)
IOS Press, 1999

Inversion of Eddy Current NDE Signals

K. Ng, S. S. Udpa, Y. S. Sun, L. Udpa, and P. Sacks
Materials Characterization Research Group
Department of Electrical and Computer Engineering
Iowa State University, Ames, IA, USA

Abstract. This paper presents an eddy current NDE inverse model that uses the finite element method for simulating the underlying physical process. The strategy attempts to estimate the material distribution associated with the test specimen that results in an eddy current signal which is close in the mean square sense to the experimental signal. This is accomplished by starting with an initial mesh configuration and updating the coordinates of the nodal points in the defect region until a cost function is minimized. Results showing the validity of the approach with simulated data are presented.

1. Introduction

Eddy current methods are used extensively for the inspection of a wide variety of industrial goods and materials. Conventional eddy current techniques typically involve the measurement of changes in the impedance of a coil, excited by an alternating current source, as it scans the surface of the test specimen. The inverse problem is related to the task of interpreting the impedance trajectory to obtain the size, shape and location of the defect. One of the more widely studied applications involves the use of eddy current methods for the inspection of steam generator tubing in nuclear power plants. The steam generators serve as heat exchange units transferring thermal energy from the nuclear reactor to the turbine side of the system. The harsh environmental conditions within the unit causes tubes that carry the radioactive primary coolant to suffer from degradation in such forms as pitting, cracking and wastage. The steam generators are periodically inspected using an eddy current probe to detect such defects in their incipient stages. This allows initiation of appropriate remedial action before more catastrophic events occur.

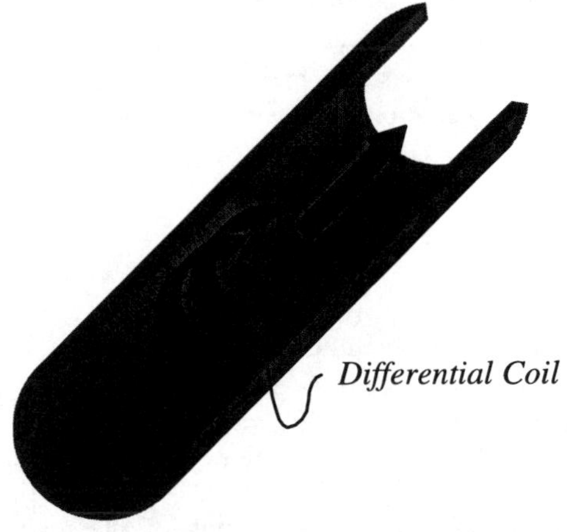

Figure 1 Differential bobbin coil probe

Although several different eddy current probes are used for inspecting steam generator tubing, one of the more commonly used transducers is the differential bobbin coil probe as shown in Figure 1. The variation in the differential impedance of the coil as it moves past a defect is measured and interpreted. Most of the methods that are currently in use rely on either calibration approaches or a combination of signal processing and pattern recognition methods [1]. Recent years have witnessed the emergence of methods that employ a model of the underlying physical process to arrive at the inverse solution. Sabbagh and Sabbagh [2] use a volume integral model that is continuously updated until the measured and predicted signals match. Hoole, et al. [3] differentiate the finite element stiffness matrices to obtain gradient information. The gradient information is utilized in the search process that is designed to arrive at the inverse solution. Monebhurrun, et al. [4], who use a vector integral formulation to model the forward problem, employ an iterative method to predict binary flaw profiles in metal tubes. Huang, et al. [5] use the reduced magnetic vector potential method to create a database that contains a portion of the inverse of the coefficient matrix and potentials of the unflawed conductor model. Since the unknowns that are updated are confined to a small region, the inverse model is computationally efficient to implement.

This paper presents an approach for inverting eddy current tube inspection signals that is an extension of the technique proposed earlier [6] for characterizing defects using the magnetostatic method. The model uses finite element techniques to model the eddy current NDE geometry. The strategy involves identification of a material distribution that results in an eddy current signal that is close in the mean square sense to the measured signal. This is accomplished by minimizing the cost function

$$F = \sum_{i=1}^{N} \left\| (Z_i - Z_{mi}) \right\|^2 \tag{1}$$

where Z_i is the impedance signal generated by the forward model at position i and Z_{mi} is the corresponding measured signal. N represents the total number of measurement points.

2. Governing Equation and Optimization Procedure

Although defects are truly three-dimensional in nature, computational considerations force us to limit our attention to an axisymmetric geometry. Since the vector potential $\overline{A}$ is continuous in two dimensions, the electric scalar potential V is omitted and, consequently, the governing equation for the steady state eddy current NDE process is given by:

$$\nabla \times \left(\frac{1}{\mu} \nabla \times \overline{A} \right) = j\omega\sigma\overline{A} - \overline{J} \ , \tag{2}$$

where $\overline{A}$ is the magnetic vector potential, $\overline{J}$ is the impressed current density, μ is the magnetic permeability, σ is the electrical conductivity and ω is the angular frequency. We impose the Neumann boundary condition along the axis of symmetry and the homogeneous Dirichlet conditions on the remaining three boundaries.

The finite element method calls for discretization of the region under study with a suitable mesh. Linear interpolation functions are then used to relate the magnetic vector potential at any point with the potential at the nodes. The solution is obtained indirectly by minimizing an energy functional that is related to the governing equation [7]. Minimization of the energy function yields the matrix equation

$$[K][A]=[S] \tag{3}$$

where K, called the stiffness matrix, is a matrix with complex valued elements and A and S are complex vectors representing the vector magnetic potentials at the nodes and source terms, respectively. The procedure proposed in this paper begins with an initial mesh configuration and involves adjusting the nodal coordinates of the defect in a manner that minimizes the cost function given in equation (1). Figure 2 shows how the mesh configuration around the defect and the corresponding impedance plane trajectory evolves as the iterative procedure converges to the final solution.

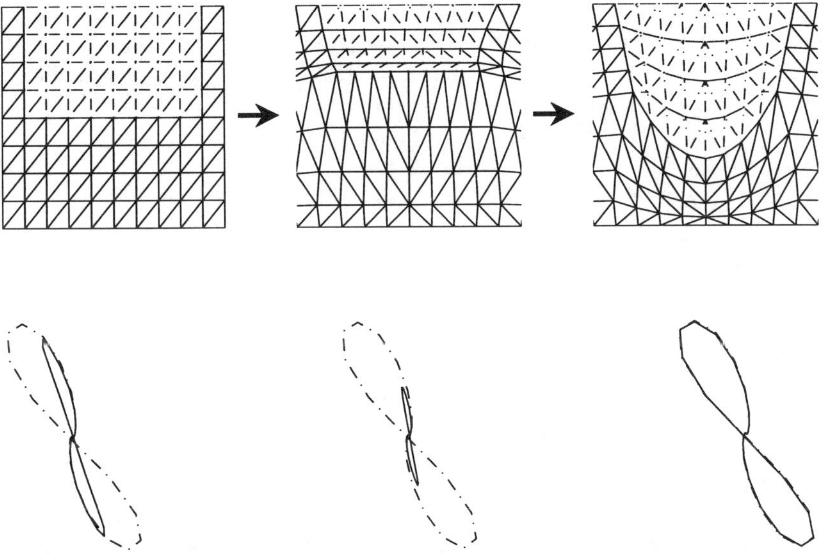

Figure 2 Evolution of the mesh as the iterative procedure progresses

The procedure for minimizing the cost function uses its derivative with respect to the coordinates of the nodal points. If we assume an axisymmetric geometry and represent the coordinates generically by the variable d_i (representing either r_i or z_i) then,

$$\frac{\partial F(d_i)}{\partial d_i}=\sum_{k=1}^{P}\left[2(\alpha_k-\alpha_k')\frac{\partial \alpha_k(d_i)}{\partial d_i}+2(\beta_k-\beta_k')\frac{\partial \beta_k(d_i)}{\partial d_i}\right], \tag{4}$$

where $Z_m = \alpha + j\beta$ is the measured impedance and $Z = \alpha' + j\beta'$ is the impedance calculated using the FEM. P denotes the number of measurement samples.

The derivatives of α and β with respect to d_i are obtained as follows. The differential impedance of the probe is calculated from values of the magnetic vector potential at the nodes of the elements comprising the eddy current coils.

$$z_{probe} = \frac{j\omega 2\pi \bar{J}_s}{I_s^2}\left[\sum_{k=1}^{Q_u}(r_{ck}\Delta_k)A_{ck} - \sum_{k=1}^{Q_v}(r_{ck}\Delta_k)A_{ck}\right], \tag{5}$$

where I_s is the current in the coil.

Differentiating z_{probe} with respect to d_i yields,

$$\frac{\partial z_{probe}}{\partial d_i} = \frac{\partial \alpha(d_i)}{\partial d_i} + j\frac{\partial \beta(d_i)}{\partial d_i} = \frac{j\omega 2\pi \bar{J}_s}{I_s^2}\left[\sum_{k=1}^{Q_u}\left(r_{ck}\Delta_k \frac{\partial A_{ck}}{\partial d_i}\right) - \sum_{k=1}^{Q_v}\left(r_{ck}\Delta_k \frac{\partial A_{ck}}{\partial d_i}\right)\right], \tag{6}$$

where A_{ck} and r_{ck} are the magnetic vector potential at the centroid and the centroid respectively of the element k, Q_u and Q_v denote the elements in the two bobbin coils comprising the differential probe and Δ_k is the area of the element k. Equation (6) is derived based on the fact that $\partial r_{ck}/\partial d_i = \partial \Delta_k/\partial d_i = 0$ in the case of elements comprising the coils. The term $\partial A_{ck}/\partial d_i$ can be calculated by taking the derivative of both sides of equation (3). Realizing that $[\partial S/\partial d_i] = 0$, we obtain,

$$\frac{\partial A}{\partial d_i} = -[K]^{-1}\left[\frac{\partial K}{\partial d_i}\right][A]. \tag{7}$$

If $[K] = [Q] + j[R]$, (8)

then the derivatives of the source and the stiffness contributions from each element with respect to r and z coordinates can be derived:

$$\frac{\partial Q_e}{\partial r_i} = \frac{r_c}{4\Delta^2 \mu}\frac{\partial \Delta}{\partial r_i}\begin{bmatrix} b'^2_l + c_l^2 & b'_l b'_m + c_l c_m & b'_l b'_n + c_l c_n \\ b'_m b'_l + c_m c_l & b'^2_m + c_m^2 & b'_m b'_n + c_m c_n \\ b'_n b'_l + c_n c_l & b'_n b'_m + c_n c_m & b'^2_n + c_n^2 \end{bmatrix} \tag{9}$$

$$\frac{r_c}{4\Delta\mu}\begin{bmatrix} g_{11} & g_{12} & g_{13} \\ g_{21} & g_{22} & g_{23} \\ g_{31} & g_{32} & g_{33} \end{bmatrix} + \frac{1}{4\Delta\mu}\frac{\partial r_c}{\partial r_i}\begin{bmatrix} b'^2_l + c_l^2 & b'_l b'_m + c_l c_m & b'_l b'_n + c_l c_n \\ b'_m b'_l + c_m c_l & b'^2_m + c_m^2 & b'_m b'_n + c_m c_n \\ b'_n b'_l + c_n c_l & b'_n b'_m + c_n c_m & b'^2_n + c_n^2 \end{bmatrix},$$

where

$$g_{11} = \frac{4}{3}\left(\frac{\Delta}{r_c^2}\frac{\partial r_c}{\partial r_i} + \frac{1}{r_c}\frac{\partial \Delta}{\partial r_i}\right)b_l + \frac{4}{9}\left(\frac{\Delta^2}{r_c^3}\frac{\partial r_c^2}{\partial r_i} + \frac{1}{r_c^2}\frac{\partial \Delta^2}{\partial r_i}\right) + 2c_l(N_{ni} - N_{mi}),$$

$$g_{22} = \frac{4}{3}\left(\frac{\Delta}{r_c^2}\frac{\partial r_c}{\partial r_i} + \frac{1}{r_c}\frac{\partial \Delta}{\partial r_i}\right)b_m + \frac{4}{9}\left(\frac{\Delta^2}{r_c^3}\frac{\partial r_c^2}{\partial r_i} + \frac{1}{r_c^2}\frac{\partial \Delta^2}{\partial r_i}\right) + 2c_m(N_{li} - N_{ni}),$$

$$g_{33} = \frac{4}{3}\left(\frac{\Delta}{r_c^2}\frac{\partial r_c}{\partial r_i} + \frac{1}{r_c}\frac{\partial \Delta}{\partial r_i}\right)b_n + \frac{4}{9}\left(\frac{\Delta^2}{r_c^3}\frac{\partial r_c^2}{\partial r_i} + \frac{1}{r_c^2}\frac{\partial \Delta^2}{\partial r_i}\right) + 2c_n(N_{mi} - N_{li}),$$

$$g_{12} = \frac{2}{3}(b_l + b_m)\left(\frac{\Delta}{r_c^2}\frac{\partial r_c}{\partial r_i} + \frac{1}{r_c}\frac{\partial \Delta}{\partial r_i}\right) + c_l(N_{li} - N_{ni}) + c_m(N_{ni} - N_{mi}),$$

$$g_{13} = \frac{2}{3}(b_l + b_n)\left(\frac{\Delta}{r_c^2}\frac{\partial r_c}{\partial r_i} + \frac{1}{r_c}\frac{\partial \Delta}{\partial r_i}\right) + c_l(N_{mi} - N_{li}) + c_n(N_{ni} - N_{mi}),$$

$$g_{21} = \frac{2}{3}(b_m + b_l)\left(\frac{\Delta}{r_c^2}\frac{\partial r_c}{\partial r_i} + \frac{1}{r_c}\frac{\partial \Delta}{\partial r_i}\right) + c_m(N_{ni} - N_{mi}) + c_l(N_{li} - N_{ni}),$$

$$g_{23} = \frac{2}{3}(b_m + b_n)\left(\frac{\Delta}{r_c^2}\frac{\partial r_c}{\partial r_i} + \frac{1}{r_c}\frac{\partial \Delta}{\partial r_i}\right) + c_m(N_{mi} - N_{li}) + c_n(N_{li} - N_{ni}),$$

$$g_{31} = \frac{2}{3}(b_n + b_l)\left(\frac{\Delta}{r_c^2}\frac{\partial r_c}{\partial r_i} + \frac{1}{r_c}\frac{\partial \Delta}{\partial r_i}\right) + c_n(N_{ni} - N_{mi}) + c_l(N_{mi} - N_{li}), \quad \text{and}$$

$$g_{32} = \frac{2}{3}(b_n + b_m)\left(\frac{\Delta}{r_c^2}\frac{\partial r_c}{\partial r_i} + \frac{1}{r_c}\frac{\partial \Delta}{\partial r_i}\right) + c_n(N_{li} - N_{ni}) + c_m(N_{mi} - N_{li}).$$

Similarly,

$$\frac{\partial Q_e}{\partial z_i} = \frac{r_c}{4\Delta^2 \mu}\frac{\partial \Delta}{\partial z_i}\begin{bmatrix} b_l'^2 + c_l^2 & b'_l b'_m + c_l c_m & b'_l b'_n + c_l c_n \\ b'_m b'_l + c_m c_l & b_m'^2 + c_m^2 & b'_m b'_n + c_m c_n \\ b'_n b'_l + c_n c_l & b'_n b'_m + c_n c_m & b_n'^2 + c_n^2 \end{bmatrix} + \frac{r_c}{4\Delta\mu}\begin{bmatrix} h_{11} & h_{12} & h_{13} \\ h_{21} & h_{22} & h_{23} \\ h_{31} & h_{32} & h_{33} \end{bmatrix}, \quad (10)$$

where

$$h_{11} = 2b_l(N_{mi} - N_{ni}) + \frac{4}{3}\left[\frac{\Delta}{r_c}(N_{mi} - N_{ni}) + \frac{b_i}{r_c}\frac{\partial \Delta}{\partial z_i}\right] + \frac{4}{9r_c^2}\frac{\partial \Delta^2}{\partial z_i},$$

$$h_{22} = 2b_m(N_{ni} - N_{li}) + \frac{4}{3}\left[\frac{\Delta}{r_c}(N_{ni} - N_{li}) + \frac{b_i}{r_c}\frac{\partial \Delta}{\partial z_i}\right] + \frac{4}{9r_c^2}\frac{\partial \Delta^2}{\partial z_i},$$

$$h_{33} = 2b_n(N_{li} - N_{mi}) + \frac{4}{3}\left[\frac{\Delta}{r_c}(N_{li} - N_{mi}) + \frac{b_i}{r_c}\frac{\partial \Delta}{\partial z_i}\right] + \frac{4}{9r_c^2}\frac{\partial \Delta^2}{\partial z_i},$$

$$h_{12} = b_l(N_{ni} - N_{li}) + b_m(N_{mi} - N_{ni}) + \frac{2\Delta}{3r_c}(N_{mi} - N_{li}) + \frac{2}{3r_c}\frac{\partial \Delta}{\partial z_i}(b_l + b_m) + \frac{4}{9r_c^2}\frac{\partial \Delta^2}{\partial z_i},$$

$$h_{13} = b_l(N_{li} - N_{mi}) + b_n(N_{mi} - N_{ni}) + \frac{2\Delta}{3r_c}(N_{li} - N_{ni}) + \frac{2}{3r_c}\frac{\partial \Delta}{\partial z_i}(b_l + b_n) + \frac{4}{9r_c^2}\frac{\partial \Delta^2}{\partial z_i},$$

$$h_{21} = b_m(N_{mi} - N_{ni}) + b_l(N_{ni} - N_{li}) + \frac{2\Delta}{3r_c}(N_{mi} - N_{li}) + \frac{2}{3r_c}\frac{\partial \Delta}{\partial z_i}(b_m + b_l) + \frac{4}{9r_c^2}\frac{\partial \Delta^2}{\partial z_i},$$

$$h_{23} = b_m(N_{li} - N_{mi}) + b_n(N_{ni} - N_{li}) + \frac{2\Delta}{3r_c}(N_{ni} - N_{mi}) + \frac{2}{3r_c}\frac{\partial\Delta}{\partial z_i}(b_m + b_n) + \frac{4}{9r_c^2}\frac{\partial\Delta^2}{\partial z_i},$$

$$h_{31} = b_n(N_{mi} - N_{ni}) + b_l(N_{li} - N_{mi}) + \frac{2\Delta}{3r_c}(N_{li} - N_{ni}) + \frac{2}{3r_c}\frac{\partial\Delta}{\partial z_i}(b_n + b_l) + \frac{4}{9r_c^2}\frac{\partial\Delta^2}{\partial z_i}, \quad \text{and}$$

$$h_{32} = b_n(N_{ni} - N_{li}) + b_m(N_{li} - N_{mi}) + \frac{2\Delta}{3r_c}(N_{ni} - N_{mi}) + \frac{2}{3r_c}\frac{\partial\Delta}{\partial z_i}(b_n + b_m) + \frac{4}{9r_c^2}\frac{\partial\Delta^2}{\partial z_i}$$

Also,

$$\frac{\partial R_e}{\partial r_i} = \frac{\omega\sigma r_c}{12}\frac{\partial\Delta}{\partial r_i}\begin{bmatrix}2 & 1 & 1\\ 1 & 2 & 1\\ 1 & 1 & 2\end{bmatrix} + \frac{\omega\sigma\Delta}{12}\frac{\partial r_c}{\partial r_i}\begin{bmatrix}2 & 1 & 1\\ 1 & 2 & 1\\ 1 & 1 & 2\end{bmatrix}, \quad \text{and} \tag{11}$$

$$\frac{\partial R_e}{\partial z_i} = \frac{\omega\sigma r_c}{12}\frac{\partial\Delta}{\partial z_i}\begin{bmatrix}2 & 1 & 1\\ 1 & 2 & 1\\ 1 & 1 & 2\end{bmatrix}, \tag{12}$$

where μ is the permeability, Δ is the area of the triangle in the element, ω is the angular frequency, σ is the conductivity of the coils, and,

$$b'_k = b_k + \frac{2\Delta}{3r_c}, \qquad k = l,m,n \qquad \text{and} \qquad \begin{matrix} N_{ij} = 1 & if \quad i = j \\ N_{ij} = 0 & if \quad i \neq j \end{matrix}$$

Finally, $\frac{\partial\Delta}{\partial r_i} = \frac{1}{2}\left[b_l(N_{li} - N_{ni}) - b_m(N_{ni} - N_{mi})\right]$ and

$$\frac{\partial\Delta}{\partial z_i} = \frac{1}{2}\left[c_m(N_{mi} - N_{ni}) - c_l(N_{ni} - N_{li})\right],$$

where

$$\begin{aligned} b_l &= Z_m - Z_n; \quad c_l = R_n - R_m; \\ b_m &= Z_n - Z_l; \quad c_m = R_l - R_n; \\ b_n &= Z_l - Z_m; \quad c_n = R_m - R_l; \end{aligned}$$

and

$$\begin{aligned} R_l &= \sum_{i=1}^{8} N_{li} r_i, \quad Z_l = \sum_{i=1}^{8} N_{li} r_i, \\ R_m &= \sum_{i=1}^{8} N_{mi} r_i, \quad Z_m = \sum_{i=1}^{8} N_{mi} r_i, \\ R_n &= \sum_{i=1}^{8} N_{ni} r_i, \quad Z_n = \sum_{i=1}^{8} N_{ni} r_i. \end{aligned}$$

The elemental matrices are combined to obtain the global stiffness and derivative matrices so that the information could be used by an optimization algorithm such as the SQP method [8] to arrive at the final solution.

3. Test Results

A 22.225 mm diameter Inconel tube was modeled using the finite element method. Each coil in the differential probe is 1.27 mm thick. Signals from a variety of O. D. defects were used to evaluate the inverse model. Figure 3 shows some typical defect profiles and predictions generated by the inverse model. The lack of symmetry in the signals is an artifact of the mesh employed for modeling the geometry. The robustness of the method was evaluated by superimposing white additive Gaussian noise. Results obtained with 0%, 10%, 15% and 20% noise energy levels are indicated in Figure 4. The inverse model initial mesh configuration and density was modified relative to the forward model. Typical results obtained are shown in Figure 5.

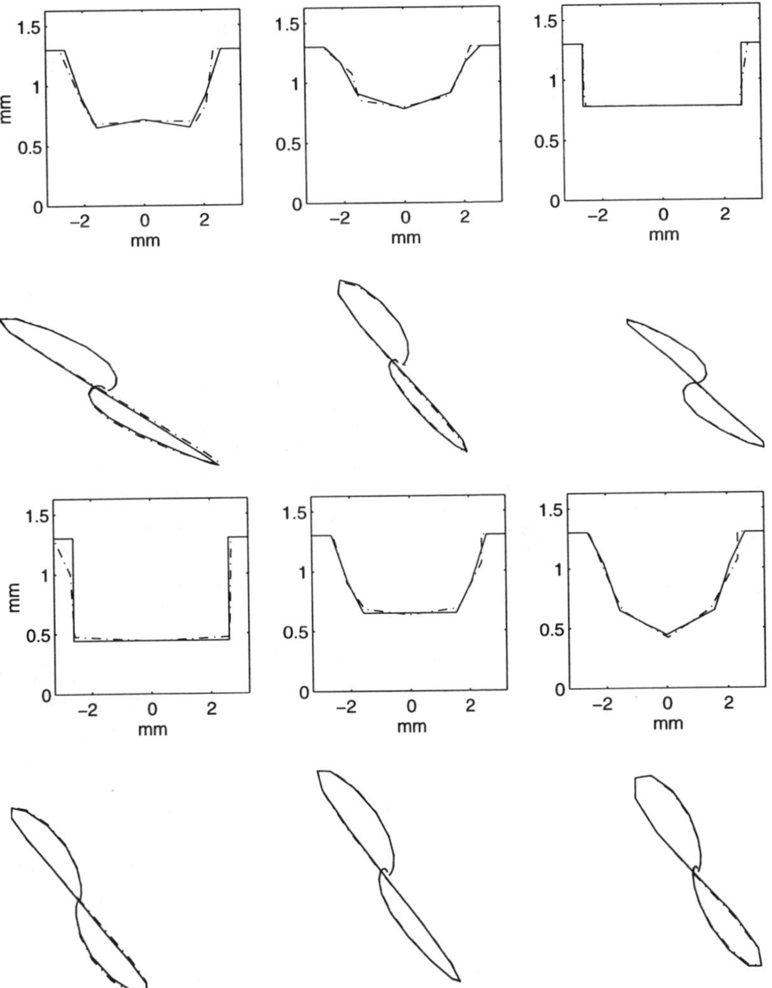

Figure 3 Typical results [- true profile, -.- reconstructed profile]

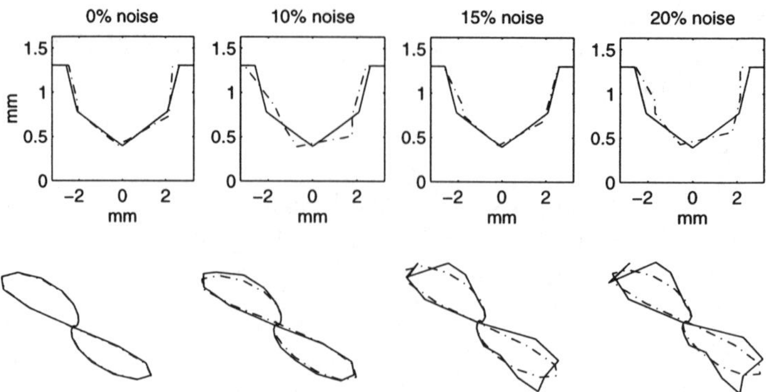

Figure 4 Effect of noise on the reconstructed result [- true profile, -.- reconstructed profile]

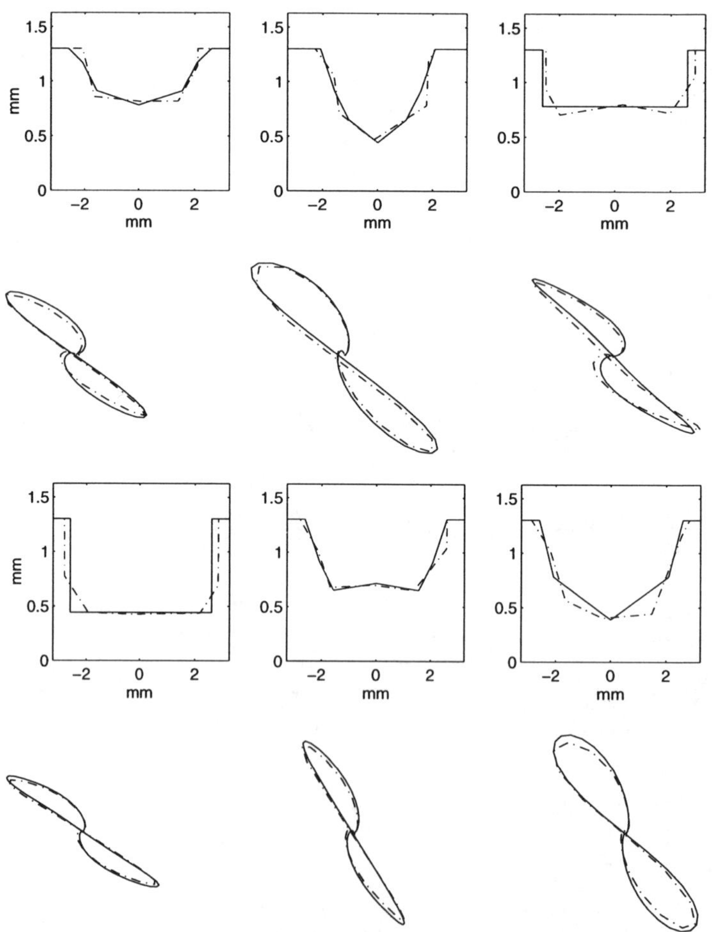

Figure 5 Effect of different meshes on the reconstructed result [- true profile, -.- reconstructed profile]

4. Conclusions

A 2D eddy current inverse model that uses the finite element method to simulate the underlying physical process is presented. The inverse model was used for characterizing eddy current signals obtained from differential probes used in nuclear steam generator tubing inspection. Preliminary results obtained to date show that the approach has significant potential. A major advantage of the approach lies in the ability to characterize defects with complex shapes. Extension of the current work to three-dimensional problems is currently underway.

References

[1] L. Udpa and S. S. Udpa, "Application of Signal Processing and Pattern Recognition Techniques to inverse Problems in NDE," International Journal of Applied Electromagnetics and Mechanics, 8 (1997), pp. 99-117.

[2] H. A. Sabbagh and L. D. Sabbagh, "An Eddy Current Model for Three-Dimensional Inversion," IEEE Transactions on Magnetics MAG-22 (1986), pp. 282-291.

[3] S. R. H. Hoole, S. Subramanium, R. Saldanha, J. L. Coulomb and J.C. Sabonnadiere, "Inverse Problem Methodology and Finite Elements in the Identification of Cracks, Sources, Materials and Their Geometry in Accessible Locations," IEEE Transactions on Magnetics, MAG-27 (1991), pp. 3433-3443.

[4] V. Monebhurrun, D. Lesselier and B. Duchene, "Eddy Current Nondestructive Evaluation of a 3D Bounded Defect in a Metal Tube Using Volume Integral Methods and Nonlinearized Inversion Schemes," Electromagnetic Nondestructive Evaluation II, R. Albanese, et. al., Eds., IOS Press, 1998, pp. 261-270.

[5] H. Huang, T. Takagi, H. Fukutomi and J. Tani, "Forward and Inverse Analysis of ECT Signals Based on Reduced Vector Potential Method Using a Database,", Electromagnetic Nondestructive Evaluation II, R. Albanese, et. al., Eds., IOS Press, 1998, pp. 313-321.

[6] M. Yan, M. Afzal, S. S. Udpa, S. Mandayam, Y. Sun, L. Udpa and P. Sacks, "Iterative Algorithms for Electromagnetic NDE Signal Inversion," Electromagnetic Nondestructive Evaluation II, R. Albanese, et. al., Editors, IOS Press, 1998, pp. 287-296.

[7] R. Palanisamy and W. Lord, "Prediction of Eddy Current Probe Signal Trajectories," IEEE Transactions on Magnetics, MAG-38, (1980), pp. 39-43.

[8] K. Shittkowski, "On the Convergence of a Sequential Quadratic Programming Method with a Augmented Lagrangian Line Search Functions," Mathematische Operationsforschung und Statistil Series Optimization, Vol. 14, (1983).

Electromagnetic Nondestructive Evaluation (III)
D. Lesselier and A. Razek (Eds.)
IOS Press, 1999

Eddy Current Imaging : a New Approach Using Attenuation

Aude CORVAISIER-RICHE and Bernard BEAUZAMY
Société de Calcul Mathématique
111, rue du Fbg Saint Honoré, 75008 Paris, France

Riadh ZORGATI
EDF/Etudes et Recherches, EP/SDM
6, quai Watier - BP 49, 78401 Chatou Cedex, France

Abstract. In Eddy Current (E.C.) inversion, attenuation phenomenon perceptibly impairs quality of reconstruction in deep zone. Thanks to an extensive study of the Green's function, we propose an efficient and fast recursive E.C. inversion method, based on both a layer-by-layer flaw reconstruction and a reconstruction on a single column of the reconstruction mesh.

1. Introduction

Eddy current testing is used for checking steam generator tubes in nuclear power plants. For safety, it is of prime importance to locate and analyze all the defects of the tubes : the defects on the inner surface as well as the deeper ones within.

Until now, there exist two cases of reconstruction. The first one consists in verifying only the internal surface of the tube (about 0.3 mm depth), in which case the attenuation is not critical. Several methods yield a reconstruction of these internal flaws [3]. The second case consists in verifying all the depth of the tube (about 1 mm). Unfortunately, the attenuation prevents a good reconstruction of the flaws and at the present time no algorithm reconstructs the defects correctly [1]. This paper investigates the properties of the attenuation term in the eddy current and provides an explicit using of this attenuation in order to obtain better reconstructions for deep defects.

For this purpose, we model in the second section a time harmonic current induced by a nonmagnetic, axisymmetric coil source by a plane wave. The eddy current is induced inside a tube damaged by 2-D defects (see figure 1). There could be several bounded flaws in the considered region. Using the Maxwell's equations and the Green's theorem, we obtain some continuous equations. Using the Born approximation and after discretization, we have to solve a matrix equation. In the third section, we present some results about the attenuation. The Green's function can be rewritten into the product of a part almost constant with respect to the depth and a part exponentially decreasing with the depth. We use these important results in a new algorithm. It reconstructs the flaws with a mesh of only one column of cells. Finally, we present interesting numerical reconstruction results with this algorithm.

2. Forward Problem

We have to reconstruct the flaws of conductivity in the tube and for the sake of simplicity we only consider now a 2-D geometry [4]. In this problem (see figure 1), we denote by D_2 the metallic half plane (permittivity ε_0, permeability μ_0 and conductivity

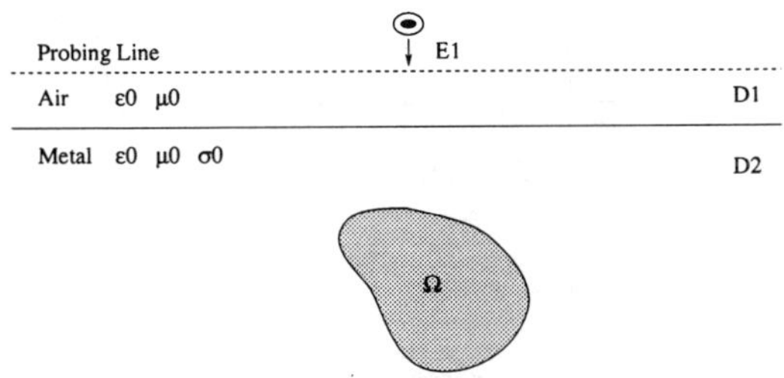

Figure 1: 2D Geometry

σ_0) which stands for a cross section of the tube. We assume that the metal is linear, isotropic non-magnetic and its conductivity is high (about 10^7 S/m). We also denote by D_1 the other half plane which represents the medium of air (with the same permittivity ε_0 and permeability μ_0). Let us consider an incident plane wave with frequency f and angular frequency ω (a time dependence $exp(-j\omega t)$ being dropped from now on). With this plane wave, the propagation constants k_i in the medium D_i are denoted by $k_1 = \omega(\mu_0\varepsilon_0)^{1/2}$, $k_2 = (\omega^2\mu_0\sigma_0\varepsilon_0)^{1/2} \approx (1+j)/\delta$, where δ is the skin depth. Flaws of conductivity are bounded and are located in the set denoted by Ω. Conductivity of defect $\sigma(\mathbf{r})$ is assumed to be zero for simplicity (crack). We define a conductivity contrast function $\chi = 1 - \sigma/\sigma_0$, equal to 1 inside the defect and 0 outside.

Under conditions of low frequencies and small contrast of conductivity, we approximate the field inside the defect, denoted by E_2, by the field existing in its absence, denoted by E_{02} in all the medium d_2 : it is the Born approximation.
In order to reconstruct the conductivity σ in the area D_2 from the existing field, we define the Green function G_{ij} $(i,j=1,2)$: it is the solution of the wave equation :

$$\Delta \times \Delta \times G_{ij}(\mathbf{r},\mathbf{r}') - k_i^2 G_{ij}(\mathbf{r},\mathbf{r}') = I\delta(\mathbf{r}-\mathbf{r}'). \tag{1}$$

(where $\mathbf{r} \in D_i$, $\mathbf{r}' \in D_i$ and I is the identity matrix) and verifies appropriate conditions at the boundary of D_2 and at infinity. Let $E_i(\mathbf{r})$ be the total electric field at a point $\mathbf{r}$ and $E_{0i}(\mathbf{r})$ be the incident electric field at a point $\mathbf{r}$ in D_i, $i = 1,2$. From the Maxwell's equations and by applying the Green theorem, we have the equation linking the defect σ and the electric fields E_1 and E_{01} in D_1:

$$E_{01}(\mathbf{r}) - E_1(\mathbf{r}) = j\omega\mu_0 \int\int_{\Omega} G_{12}(\mathbf{r},\mathbf{r}')(\sigma(\mathbf{r}')-\sigma_0)E_{02}(\mathbf{r}')d\mathbf{r}'. \tag{2}$$

This continuous equation is then discretized by the method of moments [2]. We obtain N cells per row, M cells per column and M equidistant measurements on the line $\mathcal{L}$ above the top of D_2 (with the distance x_0). The conductivity σ and the electric field E_{02} are assumed to be constant inside each square cell. Thus the previous equation can be rewritten as a discretized equation:

$$E_{01}(\mathbf{r_m}) - E_1(\mathbf{r_m}) = -j\omega\sigma_0\mu_0 \sum_{n=1}^{MN} \chi(\mathbf{r'_n}) - E_{02}(\mathbf{r'_n}) \int\int_{\Delta_n} G_{12}(\mathbf{r_m},\mathbf{r}')d\mathbf{r}', \tag{3}$$

where r_m is the mth measurement, $m = 1,\ldots,M$ and r'_n, $n = 1,\ldots,NM$ is the center of each square cell Δ_n.

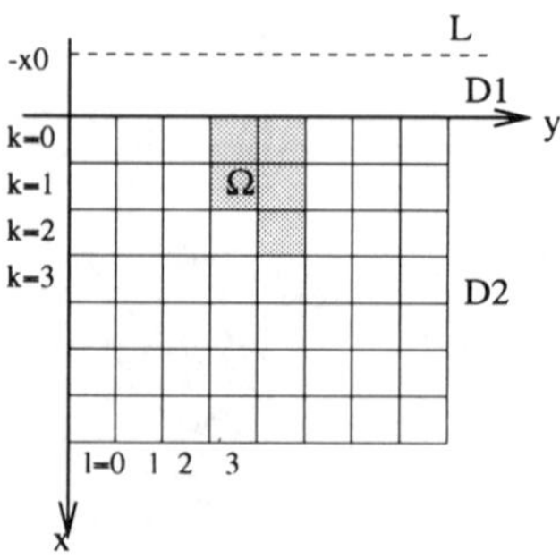

Figure 2: Discretization

3. Inversion taking into account attenuation

In order to take into account attenuation, we first simplify equations by introducing a matrix numbering instead of the classical numbering. This matrix numbering will be useful when analysing the Green's function. The attenuation is then highlighted by the mathematical study of the Green's function. We propose a lemma which separates the contribution of the exponentially damped terms and the others. Thanks to this lemma, we are able to retrieve the defect using an abacus. This abacus is somewhat analogous to impedance plane signal representation. Then our method does not require heavy computation.

3.1 Renumbering

Now we introduce a new numbering. The classical one goes over the cells linearly from the left to the right on each row, the first row first. We then use a matrix numbering with two indices. The first index determines the row of the cell and the second the column of this cell.

We denote by (x_n, y_m) the center of the cell $\Delta_{n,m}$, $n = 1, \ldots, N$ and $m = 1, \ldots, M$ and by $(-x_0, \tilde{y}_k)$ each measurement. Δ is the size of each cell. So we deduce from the previous equation the following equation:

$$E_{01}(-x_0, \tilde{y}_k) - E_1(-x_0, \tilde{y}_k) = -j\omega\sigma_0\mu_0$$

$$\sum_{n=1}^{N}\sum_{m=1}^{M} \chi(x_n, y_m) E_{02}(x_n, y_m) \int\int_{\Delta_{n,m}} G_{12}(-x_0, \tilde{y}_k; x, y)\,dx\,dy$$

1	2		M
M+1			2M
M(N-1)+1	M(N-1)+2		NM

a- Classical Numbering

1,1	1,2		1,M
2,1	2,2		2,M
N,1	N,2		N,M

b-New Numbering

Figure 3: A Matrix numbering

We can now deduce a matrix equation from this:

$$b = -j\omega\sigma_0\mu_0 \sum_{n=1}^{N} e_n^{02} H_n x_n, \tag{4}$$

for all $n = 1, \ldots, N$, $m = 1, \ldots, M$ and $k = 1, \ldots, M$. b is the vector, the components of which are $b_k = E_{01}(-x_0, \tilde{y}_k) - E_1(-x_0, \tilde{y}_k)$ $k = 1, \ldots, M$; for each $n = 1, \ldots, N$. e_n^{02} is the vector, the components of which are $E_{02}(x_n, y_m)$; x_n is the vector which defines the contrast function $\chi(x_n, y_m)$ and the components of the M by M matrix are:

$$h_{k,m}^n = \int\int_{\Delta_{n,m}} G_{12}(-x_0, \tilde{y}_k; x, y)dxdy.$$

3.2 Green's Function Analysis: Highlighting Attenuation

So far, the algorithms of inversion did not use the intrinsic properties of the Green's function. In particular they do not take the attenuation of this function into account. In this part, we highlight this attenuation. The components of the matrix H_n, $n = 1, \ldots, N$ are

$$h_{k,m}^n = -\int_{\Delta_{n,m}} G_{12}(-x_0, \tilde{y}_k; x, y)dxdy, \tag{5}$$

with $k = 1, \ldots, M$ and $m = 1, \ldots, M$. We deduce from [4]:

$$h_{k,m}^n = \frac{-1}{\pi} \int_{\mathbb{R}} \frac{1}{\beta_1 + \beta_2} e^{j\beta_1 x_0} e^{ja(\tilde{y}_k - y_m)} e^{j\beta_2(x_n - \Delta/2)} \frac{e^{j\beta_2\Delta} - 1}{\beta_2} \frac{\sin(a\Delta/2)}{a} da, \tag{6}$$

where $\beta_1 = \sqrt{k_1^2 - a^2}$ and $\beta_2 = \sqrt{k_2^2 - a^2}$. First, we calculate the values of the centers of cells:

$$\begin{aligned} \tilde{y}_k &= (k - 1/2)\Delta, \; k = 1, \ldots, M \\ x_n &= (n - 1/2)\Delta, \; n = 1, \ldots, N \\ y_m &= (m - 1/2)\Delta, \; m = 1, \ldots, M. \end{aligned}$$

Then, we obtain:

$$h_{k,n}^n = \frac{-1}{\pi} \int_{\mathbb{R}} \frac{1}{\beta_1 + \beta_2} e^{j\beta_1 x_0} e^{ja(k-m)\Delta} e^{j\beta_2(n-1)\Delta} \frac{e^{j\beta_2\Delta} - 1}{\beta_2} \frac{\sin(a\Delta/2)}{a} da. \tag{7}$$

$h_{k,m}^n$ is the $|k - m|^{th}$ Fourier's coefficient of the function under the integral. Thus, its modulus decreases when $|k - m|$ increases. There exists a more important attenuation phenomenon. Indeed we prove also that the coefficient $h_{k,m}^n$ decreases exponentially when n increases, because of the term $e^{j\beta_2(n-1)\Delta}$ (in which the imaginary part of β_2 is positive).

Lemma 1 *Let $\phi = \Delta\sqrt{\pi\mu_0\sigma_0 f}$, $\gamma = \frac{1-j}{\sqrt{2}}$ and $\beta = \sqrt{\tau^2 - j}$. Then $h_{k,m}^n$, $n = 1, \ldots, N$, $k, m = 1, \ldots, M$ can be separated into the following product:*

$$h_{k,m}^n = \frac{j\phi}{2\pi} e^{-(n-1)\phi\gamma} (e^{-\phi\gamma} - 1) I_{|k-m|}^n \tag{8}$$

where $I_{|k-m|}^n$ is almost constant with n, $n = 1, \ldots, N$ and

$$I_{|k-m|}^n = \int_{\mathbb{R}} (-|\tau| + \beta) e^{-|\tau|\phi x_0/\Delta} e^{j\tau\phi|k-m|-1} e^{-(n-1)\phi(\beta-\gamma)} \frac{e^{-\phi\beta} - 1}{\beta(e^{-\phi\gamma} - 1)} \frac{\sin(\tau\phi/2)}{\tau\phi/2} d\tau \tag{9}$$

We can use this lemma to separate the contribution of each mesh column. Instead of a matrix equation we get several scalar equations to invert. We can approximate each matrix H_n, $n = 1,\ldots,N$ by the product of a matrix by a scalar Kd_n where $d_n = \frac{j\phi}{2\pi}e^{-(n-1)\phi\gamma}(e^{-\phi\gamma}-1)$ and the matrix K approximates each matrix I_p^n, $n = 1,\ldots,N$ and $p = 0,\ldots,M-1$.Then we obtain from the equation (4) the following equation:

$$b = j\omega\mu_0\sigma_0 K \sum_{n=1}^{N} e_n^{02} d_n x_n. \tag{10}$$

3.3 The Algorithm

This paragraph describes the reconstruction algorithm for a mesh which contains only one column. The algorithm computes the surface cell first then iteratively reconstructs the cells at increasing depths under the material surface.

This method is well suited to this problem. Indeed the eddy current due to the surface cell is much greater than the one due to the bottom cell. Indeed, we prove that the Green's function and the electric field exponentially decrease with respect to the depth.

In a one column mesh, we have only one measurement and we use only one frequency. In this case $e_n^{02}d_n$ for $n = 1,\ldots,N$ is a complex scalar denoted by a_n. Figure 4 shows all the partial sums. The first branch represents all the partial sums with $e_1^{02}d_1 = 1$, the second branch all the partial sums with $e_1^{02}d_1 = 0$, $e_2^{02}d_2 = 1$ and so on. Before writing the algorithm, we have to define the convex hull and the abacus. First the abacus is the set of all the partial sums of a_n. Each partial sum corresponds to one measurement due to one discrete flaw. The branch number n denotes all the partial sums beginning with a_n. Next we define the vertices of the convex hull of the branch number n, as made of the following complex points:

$$a_n,\ a_n + a_{n+1},\ a_n + a_{n+1} + a_{n+2}, \ldots, a_n + a_{n+1} + a_{n+2} + \ldots + a_N \tag{11}$$

The algorithm is the following:

for n from 1 to N (number of rows) do

⇒ *Convex hull construction*
The convex hull of the branch number n+1 is computed. The result is a list of one vector. It describes the vertices 6

of this convex hull.

⇒ *Calculation of the corrected measurement*
We substract the measurement due to the defect lying in the $n-1$ first rows from the initial measurement.

⇒ *cell number n reconstruction*
If the corrected measurement is on the left of the convex hull then this cell is fine. Otherwise it is a defect.

end for

This algorithm does not require heavy computation.

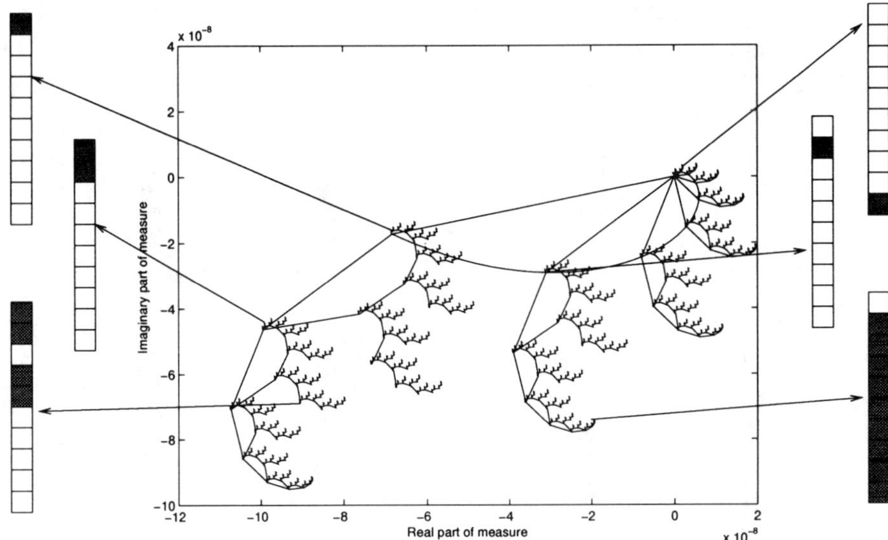

Figure 4: Abacus

3.4 Numerical Results

In order to test the capability of the algorithm regarding the reconstruction in deep zone (where the attenuation is not negligible), four cases are defined:

1. a 40% inner defect of 4×0.1 mm deep and 0.1 mm wide (see figure 5a),
2. a 40% outer defect of 4×0.1 mm deep and 0.1 mm wide (see figure 5b),
3. a throughwall defect (10×0.1 mm deep), (see figure 5c),
4. a critical case: a 20% outer defect (2×0.1 mm deep) and a 60% inner defect of 6×0.1 mm deep, (see figure 5d).

All defects are cracks, the conductivity equals zero. Reconstructions have been performed with the following parameters: $f = 100$ kHz (frequency), $\sigma_0 = 10^7$ S/m (conductivity of the sound metal) and thickness equal to 1.27 mm and on a reconstruction mesh of only one column and ten rows.
The input data of the inverse problem has been computed in the 2D configuration defined in section 2. Only the measurement above the column has been used. According to the layer-by-layer reconstruction feature, this basic measurement is used one time for reconstructing the first inner cell. Then a new data is obtained from the knowledge of the previous result. This new data is used for reconstructing the second cell and so on until the last cell.

Figure (5) shows the four original defects and the reconstruction obtained by our algorithm. For the four cases, the reconstruction is perfect even in the case (5d): the 20% outer defect, which generates a very small signal compared to the signal of a 60% inner defect, is perfectly retrieved. The attenuation has been well taken into account and does not impaired the quality of reconstruction.

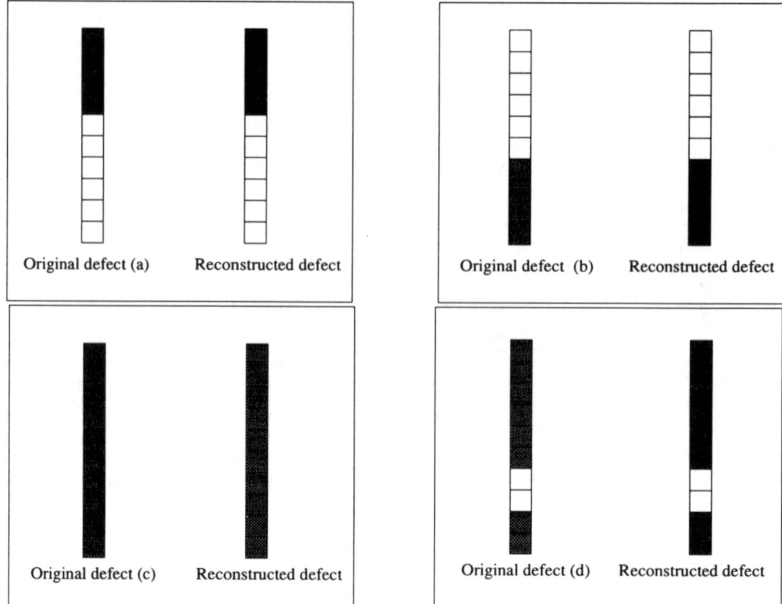

Figure 5: Tests of reconstruction

Due to the geometrical approach of the algorithm, the computation time is very low (algorithm implemented in Matlab language): no heavy computation is required because the algorithm involves logical tests and additions only. Thus, on a classical workstation the reconstruction needs less than one second.

4. Conclusion

Eddy current imaging is a possible way for improving non-destructive testing of steam generator tubes. The first hurdle to overcome attenuation of electromagnetic waves in metal for performing good reconstructions of defects in deep zone has not been taken into account in previous algorithms.

Thanks to a rigourous mathematical analysis of the Green's function, we showed that this function can be split into exclusively exponentially damped term and almost constant terms.

A recursive layer-by-layer reconstruction algorithm has been deduced. This algorithm, based on the construction of an abacus, somewhat analogous to impedance plane signal representation, performs perfect reconstructions of defects even in the difficult case of a small outer defect masked by an important inner defect because the attenuation has been explicitely used. It is very fast because it requires simple algorithmics (logical test and additions only).

The method is restricted to reconstruction mesh of one column. The generalisation to several columns is under progress. We expect some difficulties as bad condition number of the matrix K and the combinational: for one column (2^{10} possible cases), we can reduce this combinational to 10; for n columns, we have 2^{10n} possible cases wich are reducable to 2^{10} cases.Taking this combinational into account may prove difficult.

Annex: proof of *lemma*

In order to show the attenuation of the entries $h^n_{k,m}$ of matrices H_n with f (the frequency) and n (the depth) we are going to do two changes of variable. We see easily that both real and imaginary parts of β_2 must be positive in order to have the convergence of the integral. As the speed of light is much higher than the frequency f, β_1 is approximated by $j|a|$.

We work out the first change of variable: $F = 2\pi\mu_0\sigma_0\Delta^2 f$ and $t = \Delta a$.

$$h^n_{k,m} = \frac{-\Delta^2}{\pi}\int_{\mathbb{R}} \frac{1}{j|t| + \sqrt{jF - t^2}} e^{-|t|x_0/\Delta} e^{jt(k-m)} e^{j(n-1)\sqrt{jF-t^2}} \frac{e^{j\sqrt{jF-t^2}} - 1}{\sqrt{jF - t^2}} \frac{\sin t/2}{t} dt.$$

We see that:

$$\frac{1}{j|t| + \sqrt{jF - t^2}} = \frac{1}{F}\left(-|t| + \sqrt{t^2 - jF}\right)$$

We deduce from this equality:

$$h^n_{k,m} = \frac{j\Delta^2}{\pi\phi^2}\int_{\mathbb{R}} \left(-|t| + \sqrt{t^2 - jF}\right)\left(e^{-|t|x_0/\Delta}\right) e^{jt(k-m)} \times$$
$$e^{j(n-1)\sqrt{jF-t^2}} \frac{e^{-\sqrt{t^2-jF}} - 1}{\sqrt{t^2 - jF}} \frac{\sin t/2}{t} dt.$$

We perform a new change of variable $\phi = \sqrt{F}$ and $\tau = t\sqrt{F}$:

$$h^n_{k,m} = \frac{j\Delta^2}{\pi\phi^2}\int_{\mathbb{R}} \Psi(\tau)d\tau$$

with

$$\Psi(\tau) = \left(-|\tau| + \sqrt{\tau^2 - j}\right) e^{-|\tau|\phi x_0/\Delta} e^{j\tau\phi(k-m)} e^{-(n-1)\phi\sqrt{\tau^2-j}} \left(\frac{e^{\phi\sqrt{\tau^2-j}} - 1}{\sqrt{\tau^2 - j}}\right) \frac{sin(\tau\phi/2)}{\tau}.$$

In order to have a better understanding of the attenuation of $h^n_{k,m}$ with n, we factorize the Ψ function by its value at zero. Writing

$$h^n_{k,m} = \frac{j\Delta^2}{\pi\phi^2}\Psi(0)\int_{\mathbb{R}} \frac{\Psi(\tau)}{\Psi(0)} d\tau$$

we obtain the two equations that we wanted. Since the real part of γ is positive and ϕ is a real number, we see easily that there exists a term which modulus is exponentially decreasing with n (the depth). Now we have to prove that the matrix $I^n_{|k-m|}$ is almost constant with n. To show this we have made some numerical experiments.

For example, we chose a 10×10 mesh and a frequency of 100kHz. The figure 6 shows the results of this numerical experiment. It represents the variations of the real and imaginary part of $I^n_{|k-m|}$ with n. The curve number p represents the matrices I^n_p when n changes.

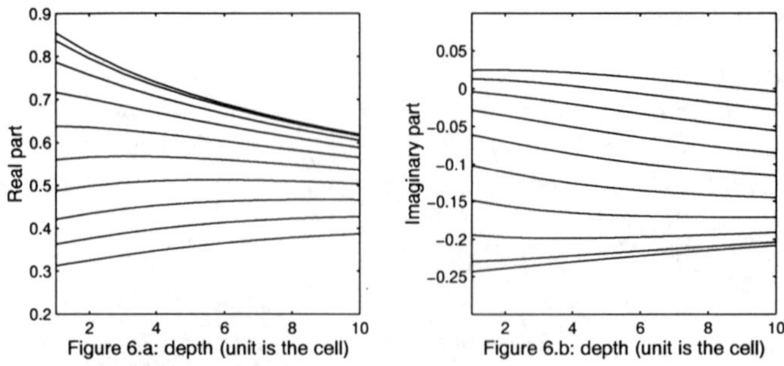

Figure 6: Variations of the real and imaginary part of $I^n_{|k-m|}$ with n

References

1 Monebhurrun V., *Formulation intégrale de volume et inversion non lineaire. Application au contrôle non destructif par courants de Foucault des tubes de générateurs de vapeur*, Thèse de doctorat, Paris VI, octobre 1997.

2 Richmond J.H. *Scattering by a dielectric cylinder of arbitrary cross-section schape*, IEEE Trans. On Antennas and Propagation, AP-13, pp.334-341, 1965

3 Sabbagh L. D. and Sabbagh H.A. *Eddy current modeling and flaw reconstruction*, J. Non Destructive Evaluation, Vol. 7, 1/2, pp. 95-110, 1988.

4 Zorgati R. *Imagerie par Courants de Foucault. Application au contrôle non destructif*, Paris VII, Paris, France, 1990.

5 Zorgati R., Lesselier D., Duchêne B., Pons F, *Eddy current testing of anomalies in conductive materials, Part I : Qualitative imaging via diffraction tomography techniques*, IEEE Trans. Magn., vol.27, n° 6, pp. 4416-4437, November 1991.

Electromagnetic Nondestructive Evaluation (III)
D. Lesselier and A. Razek (Eds.)
IOS Press, 1999

Shape and Conductivity Reconstruction of Cracks in Tube Specimens Using Eddy Current Impedance Data

József Pávó
Department of Electromagnetic Theory, Technical University of Budapest
H-1521 Budapest, Egry J. u. 18., Hungary

Zsolt Badics
Ansoft Corporation, Four Station Square, Suite 660, Pittsburgh, PA 15219, USA

Hidenobu Komatsu, Yoshihiro Matsumoto, Sota Kojima and Kazuhiko Aoki
Nuclear Fuel Industries, Ltd.,
950 Ohaza-noda, Kumatori-cho, Sennan-gun, Osaka-fu 590-0451, Japan

Abstract. An inversion strategy used for the reconstruction of the shape and conductivity of thin cracks in tube specimens is presented. The input data of the reconstruction is the impedance variation of a small pancake induction coil scanning above the crack. The inversion is done by finding the actual shape, width and conductivity of the crack by minimising the mean-square difference of the impedance change of the crack prediction and the measured data using a deterministic optimisation method. The impedance change prediction is obtained by the rapid calculation method called reaction variational method. The calculations are processed from a reaction data set created by 3D finite element solver. The same forward solution method and pre-calculated data set are used for the fast evaluation of the sensitivity information (i.e. the functional derivatives of the objective function with respect to the optimisation parameters). Using the sensitivity information efficient optimisation algorithm is applied based on the steepest descent scheme. The initial crack estimate is selected close to the optimum after a simple search in a data set containing the impedance variation predictions related to representative crack prototypes. At the end of the paper numerical examples are given demonstrating the performance of the presented method.

1. Introduction

The problem of reconstruction of the type and quality of defects detected by different eddy current testing (ECT) methods is becoming an important part of the research activity of the field. One possible way of targeting the problem is based on optimisation. In this case the measured ECT data are tried to be approximated by numerically simulated data obtained with the assumption of different crack properties. As the result of the optimisation, the crack properties yielding the best matching of the simulated and measured data are dedicated to the tested crack. Successful inversion results have been presented for the case of crack shape reconstruction of artificial cracks made with electric discharging machining (EDM) [1]-[5] (note that the theory presented in most of the cited references is applicable for the reconstruction of general conductivity distributions, not just for EDM notches). In the present work we try to make a step forward the consideration of more realistic crack models in the reconstruction procedure. In our approximation, the crack is assumed to be a volume that is bounded by two planes perpendicular to the surface of the conductor. The conductivity of the material in the crack volume is considered to vary between zero (void) and the conductivity of the host material. In the case of void cracks the thickness of the crack might also vary. Consequently our inverse problem is formulated to find the shape, thickness/conductivity of a volumetric crack if the impedance changes of an induction probe scanning above the crack are given.

Developing a method for the fast and accurate solution of the forward problem is very crucial for obtaining a successful inversion method. In our case, the so-called reaction variational [4] method is used for the calculation of the impedance change of the induction coil due to the presence of the crack in the tested tube specimen. This method is very suitable for inverse problems, because it gives a fast and accurate solution of the direct problem. The speed of the solution is increased by the application of a data set that is pre-calculated by a 3D finite element solver. The same data set might be used for the modelling of cracks with different shape and conductivity in the a given specimen. Moreover, the applied data set can be also used for calculating the gradient of the objective function with respect to the optimisation parameters [5], this gradient is frequently referred to as sensitivity. Using the solution of the direct problem and the sensitivity information, the steepest descent optimisation scheme is applied for the reconstruction of the crack. Although the sensitivity information provides sufficient information for the optimal implementation of the steepest descent algorithm [1], the unavoidable discretization of the direct and inverse problem causes minor difficulties in the numerical implementation of the theory. Contrary to these difficulties, the great advantage of the ability of the calculation of the objective function and its gradient by the solution of one forward problem [1], [2], [5] makes it possible to target the outlined inverse problem.

The speed of the reconstruction procedure can be further improved by starting the optimisation close to the optimum. In order to find a suitable starting point the impedance change due to crack prototypes are calculated and stored in a data set. The actual crack prototypes are selected to give a rough map of the space of the possible results of the inversion. After a simple search in the data set the best fitting crack prototype can be found that provides the starting point of the optimisation.

In the following, first the studied geometry is described. Then the overview of the calculation of the objective function and its gradient is presented. After the accurate definition of the inverse problem the optimisation procedure is discussed. Finally, numerical results are provided for the verification of the solution of the direct and inverse problem.

2. The Studied Geometry

The ECT problem studied in this paper is identical to one of the arrangements described in the E'NDE Benchmark Problem #3 (real steam generator tube with cracks) [6]. The only difference is that in our present case a whole tube is considered. In the experiment a pancake type induction coil scans above a crack located in a tube as it is shown in Fig. 1. The parameters of the arrangement are listed in Table I.

Table I. Parameters of the assumed experiment.

Specimen:		
	Geometry:	Tube, Outer radius: 11.1 mm, Wall thickness: 1.27 mm.
	Material:	Conductivity: 1.0×10^6 S/m, Relative permeability: 1.0.
Exciting coil:		
	Geometry:	Pancake type, Outer diameter: 3.2 mm, Inner diameter: 1.2 mm, Number of turns: 140, Lift-off: 1.0 mm.
	Excitation:	Input AC current (amplitude: 1/140 A) supplied by an impedance analyser, Frequency: 300 kHz.
Measurement:		Impedance change measured along the crack direction.

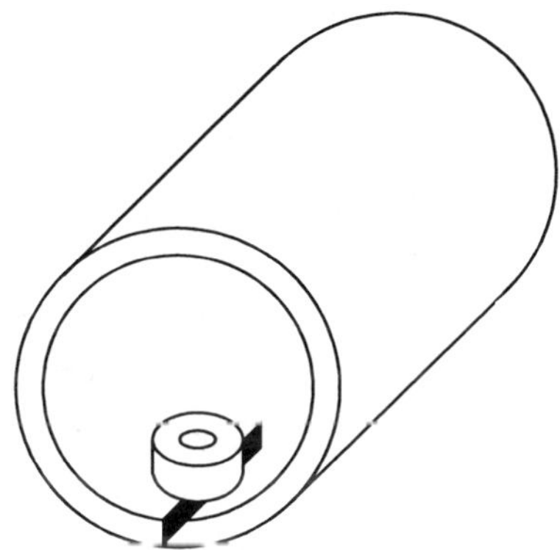

Fig. 1. Pancake type coil scans above a crack in a tube specimen.

3. Calculation of the Impedance Change and the Sensitivity

The impedance change due to flaws in the tube wall is recorded as the function of position. We use this measured data set to try to reconstruct the defects. In order to describe a flaw, we define a function, the *flaw function* η, as

$$\eta = \sigma - \sigma_u \tag{1}$$

where σ and σ_u are the conductivity of the flawed and unflawed arrangements, respectively. Since we consider arrangements with only linear material properties and sinusoidal excitations, the electromagnetic field is governed by a linear quasi-static subset of Maxwell's equations. The field with flaws present called *flawed field* is split into two parts

$$\mathbf{E} = \mathbf{E}^{(u)} + \mathbf{E}^{(f)} \tag{2}$$

where $\mathbf{E}^{(u)}$ and $\mathbf{E}^{(f)}$ are the *unflawed and flaw electric fields* and they are generated in the unflawed arrangement by sources $\mathbf{J}_S$ and $\mathbf{P}$, respectively. $\mathbf{P}$ is introduced to model the flaw region. We then replace the equation $\nabla \times \mathbf{H} = \sigma\mathbf{E} + \mathbf{J}_S$ with equations

$$\nabla \times \mathbf{H} = \sigma_u \mathbf{E} + \mathbf{P} + \mathbf{J}_S ; \qquad \mathbf{P} = \eta\mathbf{E} . \tag{3}$$

Let ζ be the reaction signal of the unflawed field on the flawed field at a given probe position $\mathbf{x}_p$ and measurement frequency ω_p [4], [5]. Thus the signal at *observation point* $(\mathbf{x}_p, \omega_p)$ is

$$\zeta_p = \int \mathbf{E}^{(u)}(\mathbf{x}_p, \omega_p, \mathbf{x}) \cdot \mathbf{P}(\mathbf{x}_p, \omega_p, \mathbf{x}) d\Omega = \left\langle \mathbf{E}^{(u)}, \mathbf{P} \right\rangle \tag{4}$$

where $\langle .,. \rangle$ denotes a complex reaction scalar product and $p = 1, \ldots, L_{obs}$. The integral is extended over the whole studied volume, Ω. For the sake of brevity, we will sometimes drop the subscript p of ζ_p.

The input of the flaw reconstruction procedure is a set of measured eddy current data $\{\zeta_{obs,p} | p = 1, \ldots, L_{obs}\}$. Using this set, we define a nonlinear error functional

$$\Phi(\eta) = \sum_{p=1}^{L_{obs}} w_p \left| \zeta_p(\eta) - \zeta_{obs,p} \right|^2 \tag{5}$$

that depends on the actual estimate of the flaw function η and indicates the quality of the flaw estimate compared to the true flaw. In order to find the true flaw shape, we minimize the error functional (5) by changing the flaw function η so that $\eta \in C^0(\Omega_A)$. The function space $C^0(\Omega_A)$ denotes the set of admissible flaw functions that are continuous and their support is contained in a fixed *anomalous region* Ω_A.

3.1. Solution of the forward problem

Rearranging (2) we obtain the operator equation

$$\mathcal{L}\mathbf{E} = \mathbf{E}^{(u)}, \qquad \mathcal{L} = \mathcal{I} - \mathcal{E}\eta \tag{6}$$

where $\mathcal{E}$ denote a linear operator that can be constructed by using Green's dyads or represented by finite elements [4]. $\mathcal{I}$ denotes the identity operator. We solve (6) in the flaw by the Galerkin technique.

P is approximated by the first N elements of a complete vector function series defined as

$$\mathbf{P} \approx \mathbf{P}^{(N)} = \sum_{n=1}^{N} v_n \mathbf{p}_n \tag{7}$$

where the $\mathbf{p}_n$'s are linearly independent and have support in the anomalous region Ω_A. We build an equation system for the unknown v_n's by substituting (7) into (6) and testing the equation by the same basis functions $\mathbf{p}_m$, $m=1,\ldots,N$, using the reaction scalar product.

Following the usual procedure, we arrive at an equation system

$$\underline{\underline{A}}\,\underline{v} = \underline{b} \tag{8}$$

where

$$A_{mn} = \left\langle \eta^{-1}\mathbf{p}_n - \mathcal{E}\mathbf{p}_n, \mathbf{p}_m \right\rangle, \tag{9}$$

$$b_m = \left\langle \mathbf{E}^{(u)}, \mathbf{p}_m \right\rangle. \tag{10}$$

To calculate the signal, we first need to determine the unknowns $\underline{v}$ in (8). After substituting $\mathbf{P}$ from (7) into the signal expression (4) the order of the summation and the scalar product is changed. Then, the signal can be calculated by

$$\zeta = \underline{v}^T \cdot \underline{b}. \tag{11}$$

3.2. Sensitivity calculation

The flaw is represented by a set of appropriate parameters

$$\eta = \eta(\underline{\alpha}); \qquad \underline{\alpha} = [\alpha_1, \ldots, \alpha_j, \ldots]^T \tag{12}$$

where vector $\underline{\alpha}$ contains these flaw parameters. We assume that representation (12) can describe any allowable flaw function. The objective function of the optimization (5) then takes the form

$$\Phi(\underline{\alpha}) = \sum_{p=1}^{L_{obs}} w_p \left| \zeta_p(\underline{\alpha}) - \zeta_{obs,p} \right|^2. \tag{13}$$

We may try to minimize (13) by a stochastic or a first-order minimization technique, or a combination of these two. When, however, the speed of the inversion is essential, we cannot avoid involving a first-order minimization component. Therefore, the fast and reliable computation of sensitivity information is a necessity.

In most first order minimization algorithms, the iteration step $\Delta\underline{\alpha}$, which gives the next flaw estimate, is selected in the opposite direction to the gradient of the error function with respect to the parameter vector

$$\Delta\underline{\alpha} = -\gamma \left[\frac{d\Phi(\underline{\alpha})}{d\underline{\alpha}} \right]^T \tag{14}$$

where γ is a constant that depends on which algorithm is used and may vary during the iterations. Taking the gradient of the error function (13) gives

$$\frac{d\Phi(\underline{\alpha})}{d\underline{\alpha}} = 2\,\mathrm{Re}\left\{ \sum_{p=1}^{L_{obs}} w_p \left[\zeta_p(\underline{\alpha}) - \zeta_{obs,p} \right]^* \frac{d\zeta_p(\underline{\alpha})}{d\underline{\alpha}} \right\} \tag{15}$$

where * denotes complex conjugate. The gradient of the objective function contains the gradients of the computed signal at every observation point. This can be calculated by performing the integral

$$\frac{d\zeta(\underline{\alpha})}{d\underline{\alpha}} = \int_{\Omega} \mathbf{E}^2 \frac{d\eta}{d\underline{\alpha}} d\Omega \tag{16}$$

where $\mathbf{E}$ is the total perturbed electric field generated by the probe coils. The proof of (16) is given in [4].

3.3. Discretization of the model

The basic idea of the discretization is that the anomalous region is subdivided into small sub-regions called *flaw cells* (also called voxels) whose size and shape can be different. Hence, the flaw function can be expressed as

$$\eta(\underline{c})=\sum_{I=1}^{L_C} c_I H_I \tag{17}$$

where the *cell function* H_I is 1 in cell I and 0 elsewhere and L_C denotes the number of cells.

We approximate the current dipole density as

$$\mathbf{P}_{f,r}=\sum_{I=1}^{L_C}\sum_{i=1}^{3} v_{f,r,Ii}\hat{\boldsymbol{p}}_{Ii}\,; \qquad \begin{array}{l} f=1,\ldots,L_{freq} \\ r=1,\ldots,L_{pos} \end{array} \tag{18}$$

where indices f and r stand for the frequency and the probe position, respectively. $\hat{\boldsymbol{p}}_{Ii}$'s ($i$=1,2,3) are three mutually orthogonal vector functions whose supports are cell I corresponding to Ω_I.

The coefficients of the algebraic equation system (8) then become

$$A_{f,Kk,Ii}=\delta_{KI}\frac{V_K}{c_K}-\int_{\Omega_K}\boldsymbol{E}_f(\hat{\mathrm{p}}_{Ii})\cdot\hat{\mathrm{p}}_{Kk}\,d\Omega\,, \tag{19}$$

$$b_{f,r,Kk}=\int_{\Omega_K}\mathbf{E}_{f,r}^{(u)}\cdot\hat{p}_{Kk}\,d\Omega\,. \tag{20}$$

where V_K is the volume of cell Ω_K, and δ_{KI} is 1 for K=I and 0 otherwise. The coefficients (19) and (20) are evaluated by a 3D electromagnetic field simulator that can compute the arbitrary unflawed field $\mathbf{E}_{f,r}^{(u)}$ and the electric field $\boldsymbol{E}_f(\hat{\boldsymbol{p}}_{Ii})$ induced by the current dipole $\hat{\boldsymbol{p}}_{Ii}$. Note that the system matrix coefficients (19) do not depend on the geometry of the probe coils or the probe positions because these coefficients are basically reactions between dipole current density pulses (18). This is not true of course for the right-hand-side coefficients because these are reactions between the probe coils and the dipole current density pulses.

The total system to be solved can be summarized as

$$\underline{\underline{A}}_f\,\underline{v}_{f,r}=\underline{b}_{f,r}\,, \tag{21}$$

where an entire scan requires that we solve at L_{freq} frequencies and L_{pos} positions

$$\zeta_{f,r}=\underline{v}_{f,r}^T\cdot\underline{b}_{f,r}\,; \qquad \begin{array}{l} f=1,\ldots,L_{freq} \\ r=1,\ldots,L_{pos} \end{array} \tag{22}$$

Note that at a given frequency we need to solve a linear equation system with the same system matrix but with different right hand sides. This speeds up the evaluation of an entire signal scan because the time-consuming LU decomposition is performed only once and the solutions for the different right hand sides can be obtained by simple back-substitutions.

We would like to point out some merits of this discrete model. First, note that in (19) only the diagonal elements depend on the flaw function thus the η-independent part of the matrix can be calculated for the whole anomalous region before entering the inversion loop. Next, note that a flaw that is a sub-region of the anomalous region can be modeled by using the same pre-calculated data set because the corresponding coefficient set is simply a subset of the coefficient set of the whole anomalous region. Finally, since the discrete system can be derived from a variational formulation, the solution is stationary and the system matrix coefficients whose modulus is smaller than a given small positive threshold can be neglected without significantly affecting the accuracy of the calculation. Therefore we can utilize sparse matrix procedures which also speeds up the solution considerably.

Finally, we have proved in [5] that the sensitivity is calculated by

$$\frac{\partial \zeta_{f,r}}{\partial c_I} = \frac{V_I}{c_I^2} \sum_{i=1}^{3} v_{f,r,Ii}^2 \, . \tag{23}$$

It is a very simple formula and allows us to calculate the sensitivity from the same reaction data set that is constructed for the signal calculation.

4. Definition of the Inverse Problem

4.1. Measurement data

The impedance changes due to the presence of the studied crack are assumed to be the measurement data. Axial cracks are considered that are located below the scanning line. The impedance data are given along a 20 mm scanning line with the resolution of 0.5 or 1.0 mm. In the following, two types of measurement data are used for the examples. In the first inversion example real experimental data provided with the E'NDE Benchmark Problem #3 are used. For these data the resolution is 1.0 mm. In the succeeding examples simulated data are used where the impedance change is calculated by the above described method and random noise is added to the simulated impedance changes. The noise is a random value between zero and 25% of the calculated data. If the calculated data is smaller than the average of the whole scan, the noise is assumed to be a random value between zero and 25% of the average value. For the simulated data the resolution is 0.5 mm.

4.2. Crack parameters

During the reconstruction procedure a crack is searched for whose parameters are chosen from the possible set of crack parameters that are described in this subsection. The shape of the crack is assumed to be a trapezoid described by five parameters as shown in Fig. 2 (actually four parameters are required for the description of a general trapezoid, the fifth free parameter is used for the description of the location of the crack along the axial direction that is arbitrary). The thickness of the crack is either 0.1 or 0.2 mm. If the thickness is 0.1 mm its conductivity might differ from zero. Ten discrete conductivity values are used for the representation of the scale expanding from zero conductivity to the conductivity of the host material (σ_u). The discrete conductivity values are chosen in the way to obtain similar magnitude in the change of the measurement signal amplitude if the conductivity of a crack is changed from a discrete value to the next one. Consequently, the conductivity differences are smaller between two discrete values in the low conductivity region than in the high conductivity region. The goal of the reconstruction procedure is to

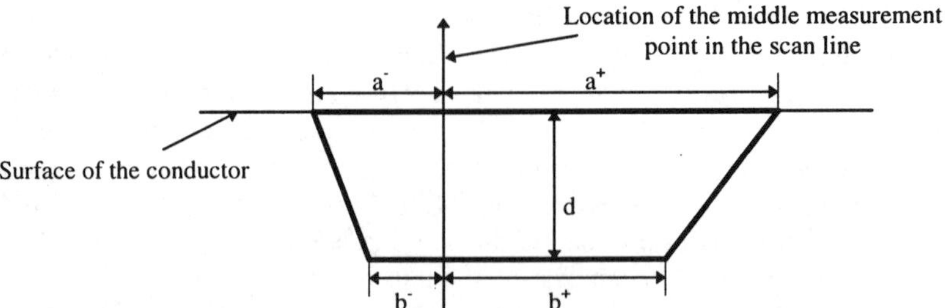

Fig. 2. Parameters of a trapezoidal shape crack.

find a crack described by the listed parameters that gives the smallest objective function (5). This task is attempted by an optimisation procedure described in the following.

5. Reconstruction Procedure

5.1. Search for a good initial shape

Our inverse problem of shape *and* conductivity reconstruction is strongly ill-posed (note that the increase of the conductivity of the crack, for instance, results in similar changes in the probe signal as the decrease of the size, especially the depth, of the crack). Consequently it is thought that the objective function has many local minima. This is specially true for our discrete crack model. (The crack model is discrete because - due to the method used for the solution of the direct problem - the crack shape can be changed only by adding or removing finite number of voxels to the original crack shape, similarly the conductivity and the thickness are also changed along a discrete scale.) One way of trying to avoid the termination of the process in a local minimum is to start the optimisation close to the global minimum. In the same time, by starting the optimisation from an initial point that is close to the optimum, we might speed the procedure up significantly.

A nearly optimal initial point is found by a simple search in a pre-calculated data set containing the signal of representative crack prototypes. Using the data set, not just a good shape but a good location is also found. The data set used for the reconstruction in our examples contains the signal of rectangular shape cracks with different sizes. The resolution of the data set in length and depth parameters are 0.8 mm and 10%, respectively. The thickness or conductivity parameters mapped in the data set are 0.2 mm, 0.1 mm void and 0.1 mm with the conductivity represented by the middle of our ten grade scale $\sigma=0.28\sigma_u$.

In the case of very noisy and unreliable measurement data, using the described simple search we can further increase the possibility of finding the global minimum by selecting several starting points. During the search in the data set not just the best fitting shape but shapes giving objective function values very close to the best one are also considered. If there are significant shape differences between the crack prototypes giving similar objective function values (like, for example, the objective function values of two crack prototypes are close to each other, although their depth and thickness or conductivity parameters are significantly different), it is worth starting individual local minimisers from these significantly different initial points. Analysing the results of the local minimisations, the global minimum can be found with higher probability. This statement is true even if some of the local minimisers are terminated in a local minimum. Since the local minimisations are independent calculations, the reconstruction time is not increased significantly if this parallel calculations are supported by computer hardware (for example more than one computers are used).

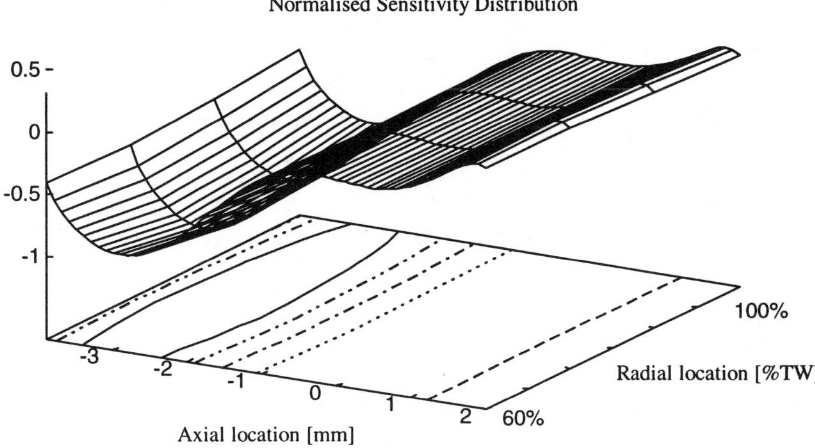

Fig. 3. Normalised sensitivity distribution.

5.2. Utilisation of the sensitivity information

The sensitivity distribution, that is the sensitivity (15) when the flaw parameters are assumed to be the conductivity of the voxels, might be calculated for any crack assumption. As an example in Fig. 3 the normalised sensitivity distribution is shown for an 0.1 mm thick OD40% (that is the depth of the crack is 40% of the tube wall thickness and the crack is located at the outer surface of the tube) crack with the shape parameters (see Fig. 2) $a^+=b^+=2$ mm, $a^-=b^-=-4$ mm when the original crack (that is the crack providing the measurement values $\{\zeta_{obs,p} | p = 1,...,L_{obs}\}$) is a 0.1 mm thick OD40% crack with the shape parameters $a^+=b^+=4$ mm, $a^-=b^-=-2$ mm. The conductivity of both cracks is zero.

The sensitivity distribution is used to calculate the sensitivities with respect to the optimisation parameters [2], [5]. As an example, the integral of the sensitivity along the perimeter of the crack gives the gradient of the objective function with respect to the change of the size of the crack [2]. Unfortunately it is difficult to use this data in an extent that is predicted by the theory. It is hard, for example, to use the numerically calculated sensitivity for the setting of the parameters of the optimisation (e.g. the step size in the case of steepest descent algorithm) as it is described in [1]. One of the reason of the difficulties is that - due to numerical errors coming from the discretization of the crack volume - the calculated sensitivity is only roughly approximated at the perimeter of the crack. The other reason is that, even if the sensitivity was highly reliable, the crack size cannot be modified as it is required by the theory because the crack shape can be changed by only adding or removing integer number of voxels. Despite the difficulties described, the sensitivity information is very important and useful to design a high performance optimisation. If it is available, the use of it is essential for crack reconstruction.

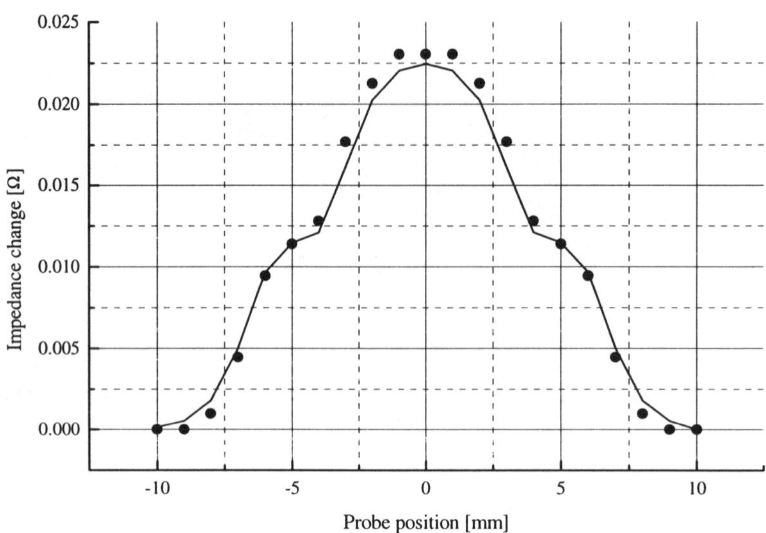

Fig. 4. Absolute value of the impedance change of the probe scanning above a 10 mm long 0.2 mm thick OD40% EDM notch. Comparison of the experimental (dots) and numerical results (line).

5.3. Optimisation procedure

The objective function is minimised by the steepest descent algorithm using the sensitivity information. Due to the uncertainties and difficulties coming from the discretization, the algorithm cannot be purely applied. Sometimes it is possible that the update of the optimisation step dictated by the theory is not leading the optimisation to the required direction. Thus, the applied algorithm has been modified to escape from such problems by continuing the procedure not necessarily to the steepest descent direction if the objective function is not decreased in an iteration step.

6. Numerical Examples

6.1. Comparison of the results of the forward solver with experimental data

The results of the forward solution is compared to the real experimental data provided for the E'NDE Benchmark Problem #3. Accordingly the impedance change due to a 0.2 mm thick 10 mm long OD40% EDM notch has been calculated and compared to the measurement data. The result of the comparison is shown in Fig. 4.

6.2. Reconstruction of rectangular shape EDM notch from real experimental data

Using the experimental data of the E'NDE Benchmark Problem #3 shown in Section 6.1, the designed algorithm is tested. In the search for the initial shape, a 10.4 mm long 0.2 mm thick OD40% void is found. This result is already very close to the final result, so the optimisation was almost unnecessary in this case. After a couple of iterations the selected shape has been accepted as the result of the reconstruction.

Table II. Results of the reconstruction of a trapezoidal OD30% crack.

	Original crack	*Selected initial estimate*	*Reconstruction result*
Shape:			
a^- [mm]	1.4	0.6	1.5
a^+ [mm]	3.4	2.6	3.7
b^- [mm]	0.2	0.6	0.2
b^+ [mm]	2.6	2.6	2.5
d [%TW]	30	30	30
thickness [mm]	0.1	0.1	0.1
conductivity [$1/\sigma_u$]	0	0	0
Optimisation result			
Number of function calls			166
RMS error of the result [%]			16.9

Table III. Results of the reconstruction of a trapezoidal ID30% crack.

	Original crack	*Selected initial estimate*	*Reconstruction result*
Shape:			
a^- [mm]	2.4	2.8	2.6
a^+ [mm]	2.4	2.8	2.6
b^- [mm]	1.2	2.8	1.1
b^+ [mm]	1.6	2.8	1.6
d [%TW]	0	30	30
thickness [mm]	0.1	0.1	0.1
conductivity [$1/\sigma_u$]	0.19	0.28	0.19
Optimisation result			
Number of function calls			187
RMS error of the result [%]			14.2

6.3. Reconstruction of trapezoidal shape void crack

In this example an 0.1 mm thick OD30% trapezoidal shape crack with zero conductivity is reconstructed. The measurement data were simulated by the forward solver and 25% noise was added to the obtained data. In Table II the results of the reconstruction procedure are listed (see also Fig. 2).

6.4. Reconstruction of trapezoidal shape crack with finite conductivity

In this example an 0.1 mm thick ID30% (ID stands for cracks that are located at the inner surface of the tube wall) trapezoidal shape crack that conductivity is 19% of the host material is considered. The measurement data were simulated by the forward solver with the addition of 25% noise. In Table III the results of the reconstruction procedure are listed.

7. Conclusions

An inversion algorithm is presented for the reconstruction of the shape and conductivity/thickness of cracks in tube specimens using eddy current impedance data. The reconstruction is based on deterministic optimisation. In the optimisation procedure the sensitivity information is used. The forward problem is solved by a rapid solution method using a pre-calculated data set. The sensitivity is easily calculated by the application of the same data set. In order to speed up the optimisation and avoid the termination in a local minimum, the initial estimate of the optimisation is obtained by searching crack prototypes analysed prior to the optimisation and stored in a data set. Numerical results are presented for the demonstration of the performance of the developed reconstruction procedure.

References

[1] S.J. Norton and J.R. Bowler, "Theory of eddy current inversion," *J. Appl. Phys.*, **73** (2), pp.501-512, 1993.

[2] J.R. Bowler, S.J. Norton and D.J. Harrison, "Eddy current interaction with an ideal crack II. The inverse problem," *J. Appl. Phys.*, **75** (12), pp.8138-8144, 1994.

[3] J. Pavo and K. Miya, "Reconstruction of crack shape by optimization using eddy current field measurement," *IEEE Trans. Magn.*, **30** (5), pp.3407-3410, 1994.

[4] Z. Badics, H. Komatsu, Y. Matsumoto, and K. Aoki, "Inversion scheme based on optimization for 3D eddy current flaw reconstruction problems," *J. Nondestructive Evaluation*, **17** (2), pp.67-78, 1998.

[5] Z. Badics, J. Pávó, H. Komatsu, S. Kojima, Y. Matsumoto and K. Aoki, "Fast flaw reconstruction from 3D eddy current data," *IEEE Trans. Magn.*, **34** (5), pp.2823-2828, 1998.

[6] T. Takagi, et al, "ECT research activity in JSAEM - benchmark models of eddy current testing for steam generator tube - (Part 1)," in *Nondestructive Testing of Materials,* R. Collins, et al, Eds., Amsterdam: IOS Press, pp.253-264, 1995.

Sizing of 3-D Surface Cracks by Using Hall Element Probe

Dorian Minkov and Tetsuo Shoji
Research Institute for Fracture Technology, Tohoku University, Aoba-Ku, Sendai, Miyagi 980, Japan

ABSTRACT. The finite size of a Hall element measuring leakage magnetic field in the vicinity of surface cracks in magnetic materials is taken into account when sizing such cracks. An analytical expression is derived for the dependence of the Hall voltage on the average value of the z-component of the intensity of the leakage field of a crack over the volume of the active region of the Hall element. An existing method for crack sizing is modified for the case when the magnetic field measurements are performed by a Hall element probe. Computations are performed for sizing of different surface cracks with complex cross-sections and unknown shapes. The computed results show that precise sizing of the investigated surface cracks can be achieved by using the modified method. It is confirmed that the finite size of the Hall element has to be taken into account for precise sizing of surface cracks.

1. INTRODUCTION

A new electrical and magnetic method for sizing of surface flaws was developed in [1]-[3]. The method can be used when e.g. magnetic field is applied inside a magnetic sample which results in bending of the lines of the intensity of the applied field in the vicinity of a flaw and generating of leakage magnetic flux outside the sample. This method was improved aiming at precise sizing of surface cracks with complex cross-sections and unknown shapes [4]. In the improved method, the volume of the investigated crack is represented more accurately than in [1], and the z-component of the intensity of the leakage field is measured at several points located along three lines which are parallel to the direction of the applied field and positioned above the crack, at the same distance from the surface of the sample. The improved method utilises an improved algorithm for crack sizing which contains no more than four regressions for the distribution of the z-component of the intensity of the leakage field at the measurement points. A drawback of the work performed in [4] is that the magnetic sensor measuring leakage magnetic field is considered to have a negligibly small size.

In practice, Hall element probes are commonly used when measuring magnetic leakage fields for detecting and characterising defects in magnetic materials. In many cases, the size of the semiconductor Hall element is commensurate with the size of the crack, which indicates that the finite size of the Hall element might have to be considered in crack characterisation methods [5]. In this paper, the finite size of the Hall element is taken into account when crack sizing is performed by the improved method [4]. An analytical expression is derived for the dependence of the Hall voltage as a function of the average value of the z-component of the intensity of the leakage field of a crack over the volume of the active region of the Hall element. Correspondingly, the improved method [4] is modified for crack sizing based on measuring Hall voltages. Computations performed for sizing of different surface cracks show that the improved method allows precise sizing of all of the investigated cracks.

2. ANALYTICAL EXPRESSION FOR THE HALL VOLTAGE

In the improved method, the volume of the investigated crack is represented as a sum of the volumes of N - 1 constituent cracks with parallelepiped shape and variable depth d such as the constituent crack shown in Fig. 1, where (x0y) is the surface plane of the sample and the applied magnetic field is directed along the x axis. The constituent cracks are chosen to have the same length $2l=2l_0/(N-1)$ and width 2a, where $2l_0$ and 2a are the length and the width of the investigated crack. Based on the 'dipole model' of a crack [6], the z-component of the intensity of the leakage magnetic field at a point with co-ordinates (x, y, z) above the surface of the constituent crack from Fig. 1 can be expressed as [4]:

$$H_c(x,y,z)=c\left\{\begin{aligned}&-\sum_{j=1}^{p}\left[\ln\frac{y_{2j}+(x_1^2+y_{2j}^2+z_{1j}^2)^{0.5}}{y_{1j}+(x_1^2+y_{1j}^2+z_{1j}^2)^{0.5}}\right]+\ln\frac{y_3+(x_1^2+y_3^2+z^2)^{0.5}}{y_4+(x_1^2+y_4^2+z^2)^{0.5}}\\&+\sum_{j=1}^{p}\left[\ln\frac{y_{2j}+(x_2^2+y_{2j}^2+z_{1j}^2)^{0.5}}{y_{1j}+(x_2^2+y_{1j}^2+z_{1j}^2)^{0.5}}\right]-\ln\frac{y_3+(x_2^2+y_3^2+z^2)^{0.5}}{y_4+(x_2^2+y_4^2+z^2)^{0.5}}\end{aligned}\right\} \quad (1)$$

where $y_{1j}=-l-y+(j-1)2l/p$, $y_{2j}=-l-y+j2l/p$, $y_3=l-y$, $y_4=-l-y$, $x_1=x+a$, $x_2=x-a$, $z_{1j}=z+d_i(p+0.5-j)/p+d_{i+1}(j-0.5)/p$, $p\rightarrow\infty$, $c=m/4\pi\mu_o$, and m is the surface density of magnetic charge at the crack walls, which is proportional to the intensity of the applied magnetic field and depends on the material of the sample.

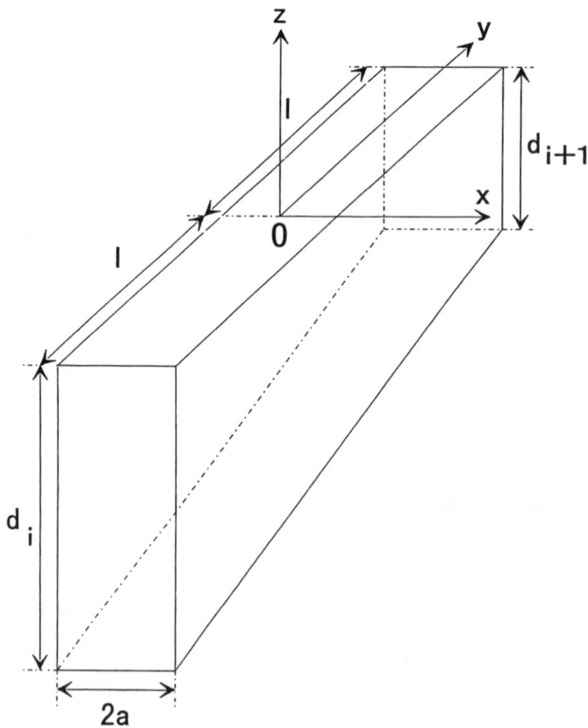

Figure 1. Sketch of a constituent crack with variable depth d. The applied magnetic field is directed along the x axis and (x0y) is the surface plane of the sample.

The z-component of the intensity of the leakage field of the investigated crack at the measurement point (x,y,z) is a sum of the z-components of the intensity of the leakage field of its N-1 constituent cracks at the same point:

$$H(x,y,z)=\sum_{k=1}^{N-1} H_{ck}(x,y,z) \tag{2}$$

A schematic drawing of a conventional semiconductor Hall element, including the orientation of its axes with respect to the axes of the investigated crack, is shown in Fig. 2. The active layer with thickness d_a is conductive and deposited on a semi-insulating substrate. Electrical contacts are located at the four opposite corners of the square surface of the structure. Electrical current I is applied between two opposite contacts, and Hall voltage V_H, proportional to the average value of the z-component of the intensity of the magnetic field, is generated between the other two opposite contacts. Assuming that the electrical current carriers flow only within the right-angular parallelepiped active region of the Hall element, the following analytical expression for the Hall voltage as a function of the average value of the z-component of the intensity of the leakage field of the investigated crack over the volume of the active region is derived:

$$V_H=\frac{\mu_o I}{qnd_a}\frac{\sum_{i_1=1}^{N_1}\sum_{i_2=1}^{N_1}\sum_{i_3=1}^{N_2} H_{i_1,i_2,i_3}}{N_1^{\,2}N_2}=\frac{\mu_o I}{qnd_a}\overline{H}_a=c_1\overline{H}_a \tag{3}$$

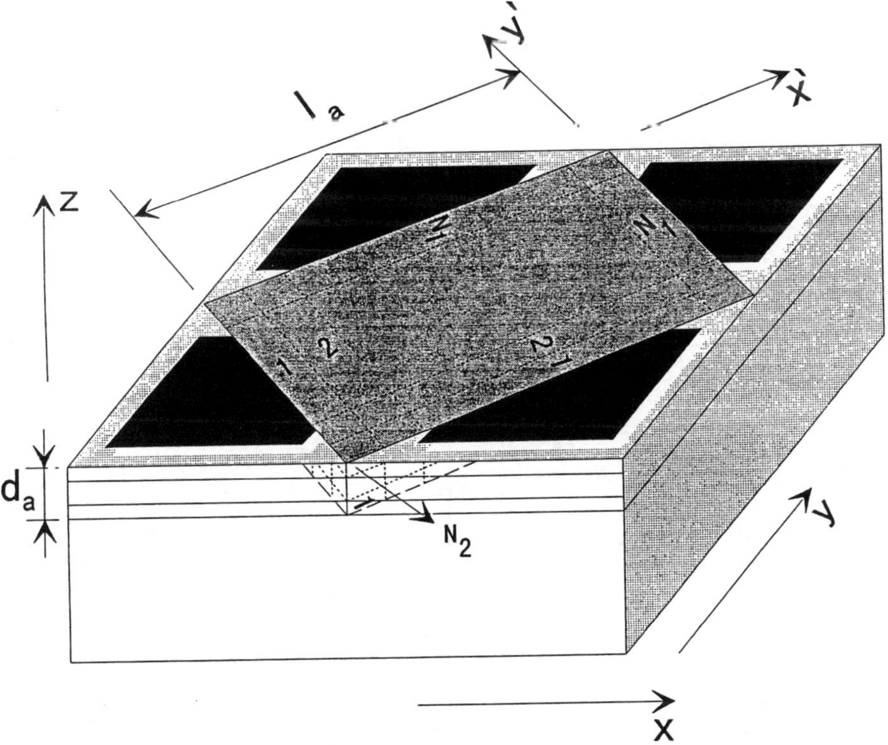

Figure 2. Schematic drawing of a conventional semiconductor Hall element. The four contacts are presented in black colour, the surface of the active region is in grey, and the sizes of the active area are $l_a \times l_a \times d_a$.

where μ_o is the vacuum permeability, q is the absolute value of the electron charge, n is the electron concentration in the active layer which is assumed to be n-type, the volume of the active region is represented as a sum of its constituent right-angular parallelepipeds, N_1 and N_2 are the numbers of such constituent parallelepipeds along the axes x and z, H_{i_1,i_2,i_3} is the z-component of the intensity of the leakage magnetic field of the investigated crack in the center of the constituent parallelepiped with number i_1,i_2,i_3, $\overline{H}_a$ is the average value of the z-component of the intensity of the leakage field of the investigated crack over the volume of its active region, and $c_1=\mu_0 I/qnd_a$ [Ω.m] is a constant.

3. UTILIZATION OF THE IMPROVED METHOD FOR CRACK SIZING

Following the improved method [4], measurements of the Hall voltage are performed, by a Hall element probe, along three lines which are parallel to the direction x of the applied field and positioned at the same distance above the surface of the sample. The central line is projected at the centre of the investigated crack, while the other two lines are projected at a distance $2l_0/6$ from the two opposite ends of the crack- Fig. 3.

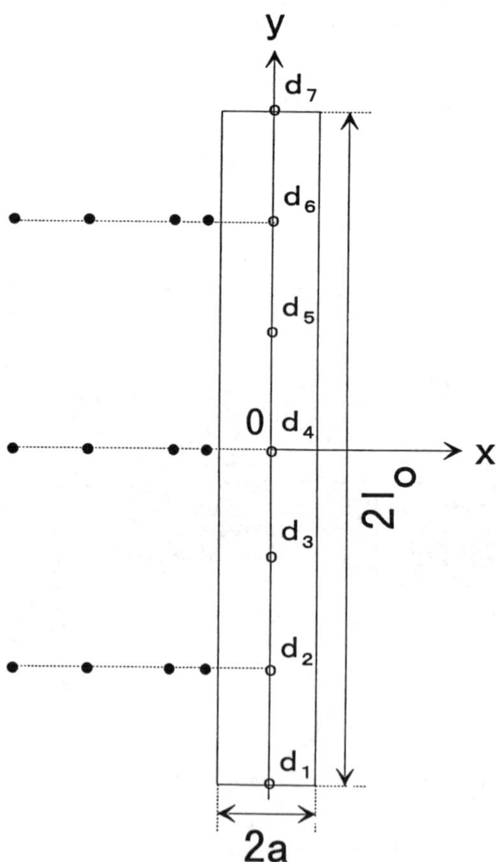

Figure 3. Surface view of an investigated crack. Projections of the measurement points are illustrated by (•) and located along three measurement lines (- - -), while the depth profile points of the crack for N=7 are illustrated by (o).

Measurements of the Hall voltage are performed at only four points of the geometrical centre of the active region of the Hall element located along each one of the three parallel lines (Fig. 3). The Hall sensor is oriented in such a way, with respect to the investigated crack, that the directions of the corresponding axes x, y, and z from Fig. 1 and Fig. 2 coincide. The distribution of the z-component of the intensity of the leakage field along each one of these three lines is symmetrical relative to the long axis y which passes through the centre of the crack (Fig. 4). Accordingly, the four measurement points are chosen to be located at distances $-3x_0$, $-2x_0$, $-x_0$, and $-0.7x_0$ from the centre of the investigated crack where $-x_0$ is the distance between the centre of the crack and the location of the maximum value of the z-component of the intensity of the leakage field along the x axis.

The improved algorithm for crack sizing [4] is shown in Fig. 5. The algorithm contains no more than four regressions for the distribution of the Hall voltage at the measurement points when Hall element probes are used. Each regression reduces to minimisation of the RMS error between the measured Hall voltage and the corresponding variable theoretical Hall voltage with respect to the depth profile of the crack $d_1, d_2, \ldots, d_N$, as well as a and c, where N is the number of equidistant points from the cross-section of the crack in the plane (yz). The subsequent regressions of the algorithm are performed to compute the depth profile of the crack for N = 4, 5, 7, and 10, as well as a and c. The utilisation of subsequent regressions for increasing N, which corresponds to more detailed depth profiles, aims at providing better initial approximation of the depth profile for the next regression which decreases the possibility for finishing the next minimisation at a local minimum. The algorithm can end when the RMS error falls below a small number eps.

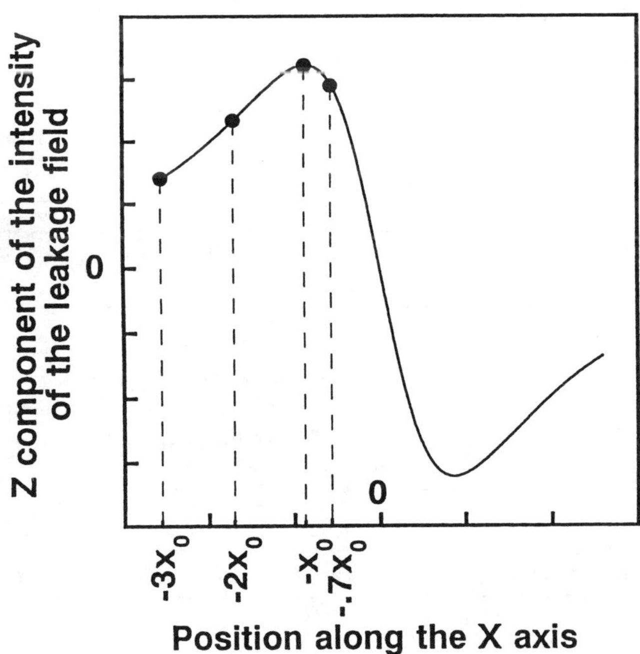

Figure 4. Distribution of the z-component of the intensity of the leakage field along a measurement line and location of the measurement points (●).

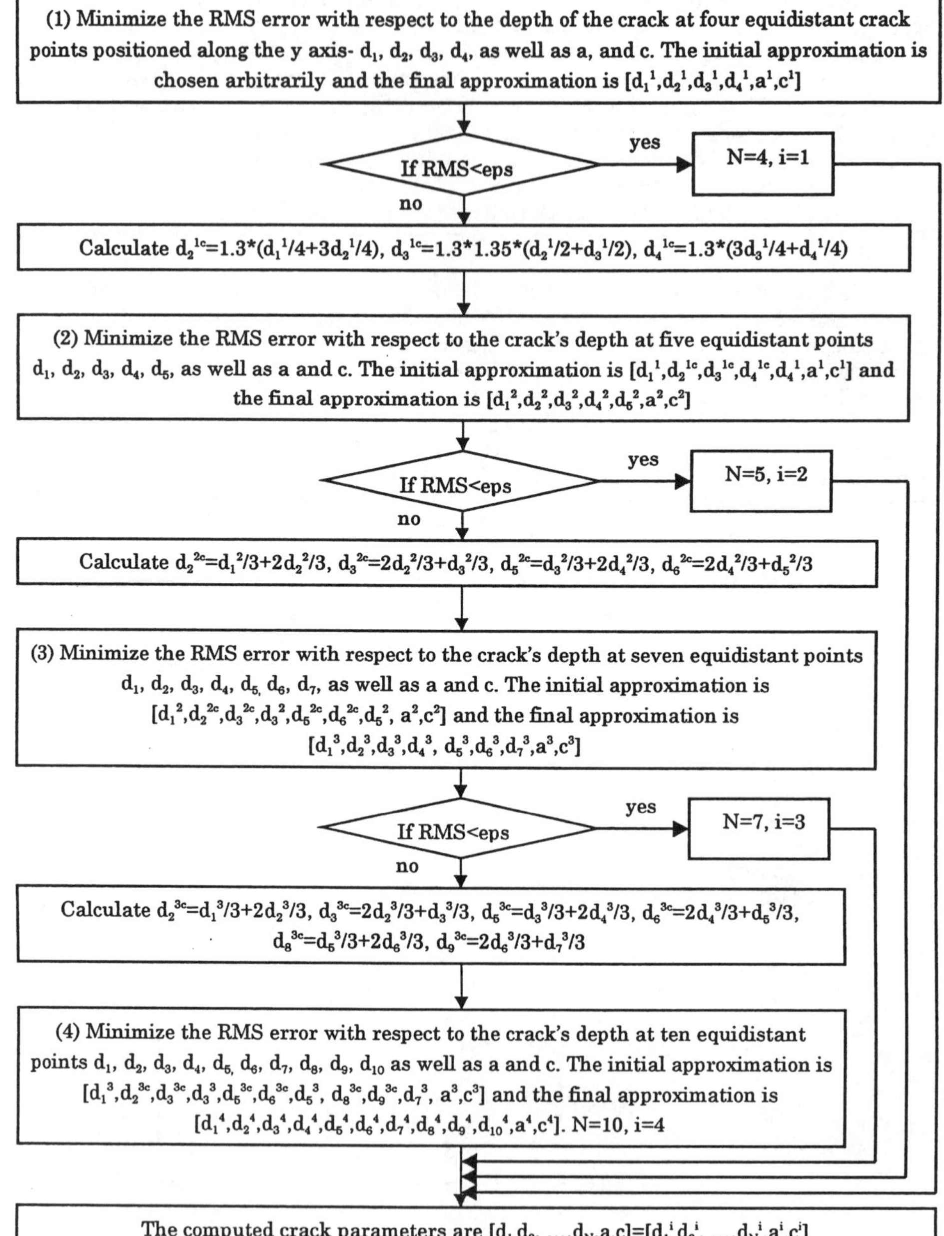

Figure 5. The improved algorithm for crack sizing, containing no more than four regressions.

4. COMPUTED RESULTS

Computations are performed for sizing of cracks with length $2l_0 = 10$ mm, width $2a = 0.1$ mm, depth $d = 2$ mm, and cross-sections in the (yz) plane shown in Fig. 6. Case symbols are introduced for simplicity while Case 1, Case 2, and Case 3 refer to the cracks with cross-sections shown respectively in Fig. 6a, Fig. 6b, and Fig. 6c. It is considered that the sizes of the active region of the Hall element are 125×125×6 μm which are typical values for some commercial Hall element probes, and the measurement distance above the surface of the sample is $z = 0.3$ mm which is similar to the distance between the active layer and the top plastic surface of the Toshiba THS124 Hall element probe.

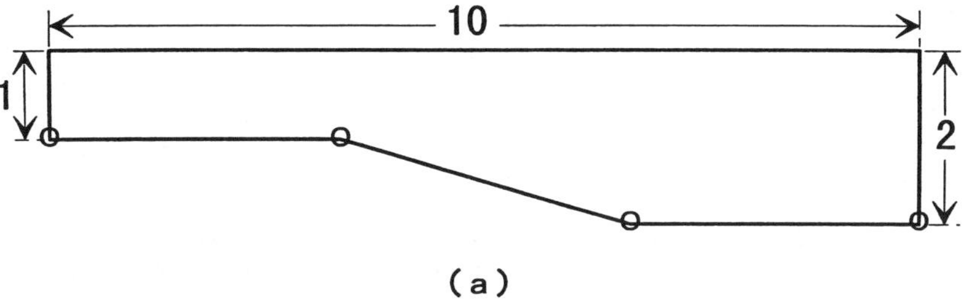

(a)

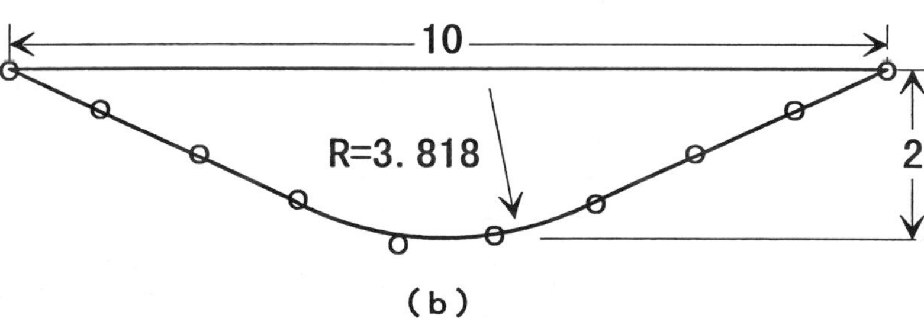

(b)

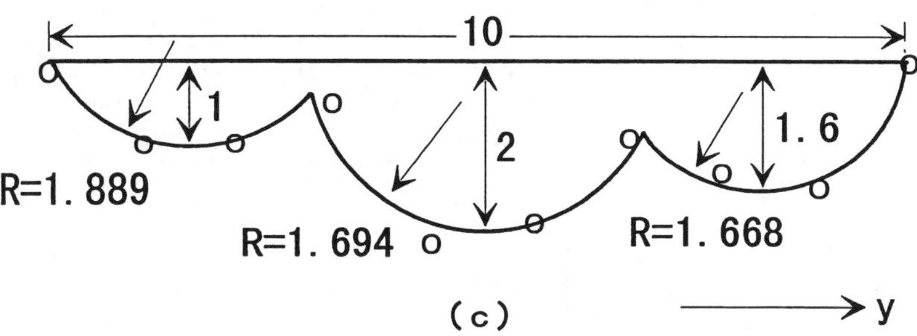

(c)

Figure 6. Cross-sections in the (yz) plane of the investigated cracks (---), and computed results for the depth profiles of these cracks (o). The dimensions are given in mm.

The following assumptions are made in the computations:

- The length of the investigated crack is measured independently, i.e. the value of l_0 is known,
- The width of the crack 2a is unknown,
- The proportionality parameter in (1) has a value of $c = 10^6$ A/m in all the field distribution measurements performed, while the measured field distributions are simulated by replacing this value of c in (1),
- The parameter c is assumed to be unknown when variable theoretical field distributions are generated using (1), taking into account that c is proportional to the intensity of the applied magnetic field and depends on the material of the sample,
- The measured Hall voltages are simulated using (1) - (3) for N = 102, p = 3, $N_1 = 18$, and $N_2 = 3$,
- The variable theoretical Hall voltages are generated using (1) - (3) for p = 8, $N_1 = 4$, and $N_2 = 1$,
- The constant c_1 from (3) is always considered to be known and $c_1 = 1.822*10^{-6}$ [Ω.m].

The computed results for the depth profiles of the cracks from Case 1- Case 3 are illustrated in Fig. 6. The computed data for the depth profile, a, and c, as well as their true values, are presented in Table 1 for the cracks from Case 1- Case 3.

Table 1. Computed data for the depth profiles, a, and c, of the investigated cracks, and their true values. The set of true values of the parameters of the investigated crack is designated by TSP.

Crack Parameter	CASE 1		CASE 2		CASE 3	
	Computed Value	True Value	Computed Value	True Value	Computed Value	True Value
d_1(mm)	0.998	1.000	0.025	0.000	0.203	0.000
d_2(mm)	1.001	1.000	0.541	0.569	0.977	0.921
d_3(mm)	2.000	2.000	1.069	1.099	0.941	0.921
d_4(mm)	2.001	2.000	1.617	1.629	0.467	1.080
d_5(mm)			2.046	1.960	2.168	1.917
d_6(mm)			1.948	1.960	1.910	1.917
d_7(mm)			1.642	1.629	0.951	1.080
d_8(mm)			1.063	1.099	1.398	1.510
d_9(mm)			0.542	0.569	1.609	1.510
d_{10}(mm)			0.037	0.000	0.000	0.000
$a*10^{-2}$ (mm)	5.219	5.000	5.216	5.000	5.236	5.000
$c*10^6$ (A/m)	0.958	1.000	0.959	1.000	0.955	1.000

5. DISCUSSION

The results from Fig. 5 and Table 1 show that precise sizing of surface cracks with complex cross-sections and unknown shapes can be achieved by using the improved method [4] when the z-component of the intensity of the leakage field is measured by a Hall element probe.

The computations for crack sizing are performed for two initial approximations of the first minimisation of the new sizing algorithm correspondingly TSP*1.5 and TSP/1.5 for each one of the investigated cracks. The final approximations of the first minimisation are the same, for both initial approximations, for all of the investigated cracks. This is an indication that the crack sizing results are the same within a wide range of initial approximations of the crack parameters.

The largest error at the computation of the depth profile is made for the crack from Case 3 in comparison with the cracks from Case 1 and Case 2. This result can be explained considering that the crack from Case 3 has the most complex cross-section, and correspondingly a precise representation of its volume requires using larger number N-1 of its constituent cracks.

A comparison between the computed results for the depth profile of the investigated cracks shows that increased complexity of the cross-section of a crack requires computation of its depth profile at a larger number N of equidistant depth profile points, which is in accordance with the previous paragraph.

The RMS error at the end of each subsequent minimisation decreases, for all of the investigated cracks, which means that an increase of the number N of depth profile points leads inevitably to more precise crack sizing. Correspondingly, it can be speculated that increase of N above ten would result in further improvement of the accuracy of crack sizing.

It turns out that precise sizing of the investigated cracks for z = 0.3 mm can be achieved by using $N_1 \geq 4$, and $N_2 \geq 1$ when theoretical Hall voltages are generated. This is an indication that the finite size of the Hall element has to be taken into account for precise sizing of surface cracks.

REFERENCES

[1] D. Minkov and T. Shoji, *NDT&E International*, 31 (1998) 317-324.

[2] D. Minkov and T. Shoji, Electromagnetic Nondestructive Evaluation (II), IOS Press, Amsterdam, 1998, pp. 271-279.

[3] D. Minkov and T. Shoji, The 75th JSME Spring Annual Meeting, Tokyo, 1998, pp. 606-609.

[4] D. Minkov and T. Shoji, The ASNT Fall Conference and Quality Testing Show, Nashville, 1998, pp. 180-182.

[5] W. Lord and R. Srinivasan, Review of Progress in Quantitative Nondestructive Evaluation, Plenum Press, 1984, v. 3B, pp. 855-862.

[6] S. Mukae, M. Katoh and K. Nishio, *Nondestructive Inspection*, 37 (1988) 885-893.

Electromagnetic Nondestructive Evaluation (III)
D. Lesselier and A. Razek (Eds.)
IOS Press, 1999

Remote Field Eddy Current Technique Applied to Steam Generator Tubes

Hiroyuki FUKUTOMI, Toshiyuki TAKAGI and Tatsuya AIZAWA
Institute of Fluid Science, Tohoku University
Katahira 2-1-1, Aoba-ku, Sendai 980-8577, JAPAN

Abstract. The remote field eddy current (RFEC) technique is used to nondestructively evaluate ferromagnetic tubes. Evidently, this technique is especially attractive when the tubes must be inspected from the inside as well as original eddy current testing (ECT) is. Unlike the impedance plane analysis form of ECT, the technique is a through-transmission effect that reduces problems such as lift-off normally associated with ECT. In the inspection of steam generator tubes, the real issue is to detect the minute cracks grow up from the outside. However, using ECT, it is considered infeasible to find them accurately because of the limitations of penetration of eddy currents. This paper describes the results of numerical simulation for the RFEC technique applied to non-magnetic steam generator tubes. Two models are compared: axisymmetric and three-dimensional finite elements, and the $\boldsymbol{A}$-ϕ method is used to compute the RFEC signal.

1. Introduction

The remote field eddy current (RFEC) technique is a well established nondestructive evaluation method for detecting corrosion and wall losses in ferromagnetic conducting tubes. It is used particularly in heat exchanger and small diameter pipes since it detects internal and external discontinuities with approximately equal sensitivity and the probes fit small space requirements. The remote field phenomenon is based on the through wall indirect coupling energy path that exists for a low frequency field generated in a straight tube [1]. This indirect path contains two effective transits of the pipe wall. One transit is from the inside out, the other from the outside back in, making it an attractive NDE method for wall metal losses.

The typical configuration consists of two internal solenoids, one exciter coil and one detector. At low frequencies, where the wall thickness to skin depth ratio is of the order of one, the field diffuses rapidly out through the tube wall with some attenuation and phase shift. This external field is guided along the outer surface and then rediffuses back into the tube, where it combines with the rapidly attenuated direct field. The region where the direct and indirect fields are of comparable magnitude and can interfere destructively is termed the transition zone. At approximately 2 to 2.5 outside diameters away from the exciter, the rediffusing external field will dominate the internal field at the detector. This region is termed the remote field. On the other hand, the region that is close to the exciter is termed near field.

Research has recently been done to study RFEC phenomena and its applications by using analytical solutions [2] or axisymmetric finite element modeling [3]. The RFEC problems require larger domains and finer meshes because electromagnetic fields far from exciting coils are the most important to predict the phenomena and signals, and the RFEC signal changes due to defects are fairly smaller than those of ECT in general. For the reduction of computer resources, an approach has been proposed in which small important parts including the defects,

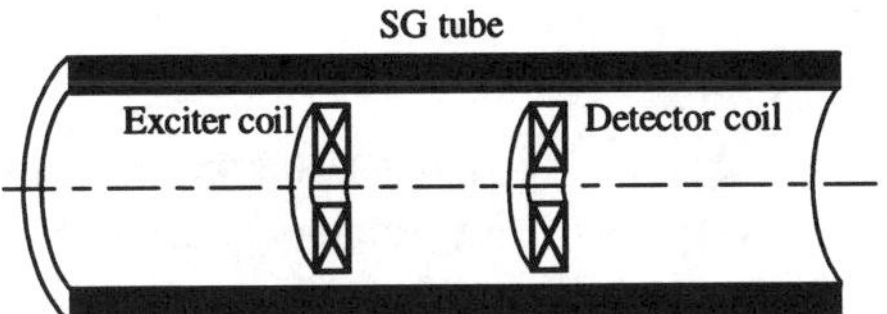

Fig. 1: RFEC probe inside a SG tube. The probe consists of an exciter and a detector, which have the same dimensions.

Table 1: Configurations of components in the RFEC model

coils -	outer diameter	18 mm
	inner diameter	8 mm
	thickness	2 mm
	number of turns	500
	current	57.32 mA
	frequencies	100kHz to 500kHz
tube -	outer diameter	22.24 mm
	inner diameter	19.7 mm
	thickness	1.27 mm
	relative permeability	1
	conductivity	1×10^6 S/m

basically three-dimensional parts, are under consideration, and some data from the corresponding axisymmetric problems are used as the boundary conditions [4]. To authors' knowledge, there has been no report of success in three-dimensional numerical simulation, because of the necessity of very high accuracy.

Using ECT, the steam generator tubes, which are non-magnetic in nuclear power plants, are inspected for the secure operations of the plants. Although minute cracks such as fretting corrosion, stress corrosion cracking, and intergranular attacking on the outer surfaces of the tubes are problematic in this inspection, it is difficult to detect them because of the shallow skin depth or the lack of penetration.

Studies for the development of ECT for the steam generator tubes have been performed on such topics as measurement apparatus and probe design, numerical modeling, and signal processing. In the numerical modeling, high numerical accuracy in various numerical methods is demonstrated in quantitatively predicting eddy current signals for the realistic steam generator tube geometry [5-8]. Recently RFEC phenomena in non-magnetic tubes have been demonstrated experimentally and numerially [9,10], and nodal and edge-based elements are used in the numerical simulations

To work out the problem that ECT in SG tube inspection has difficulty in the detectability of the outer cracks, RFEC is considered in this paper, applied to non-magnetic tubes. The RFEC signals related to the defects are given by using axisymmetric and three-dimensional finite elements and the A-ϕ method. For the purpose of improvement of the detectability for the outer defects, parametric studies are performed to detect the outer cracks in the SG tubes and its applicability is discussed.

2. Numerical Simulation for RFEC techniques

To test the validity of the numerical simulation codes which uses finite elements and the A-ϕ method, and to investigate RFEC phenomena in a non-magnetic tube, these codes are applied to the problem as shown in Fig. 1. Assuming the straight tube is fabricated from nickel-based superalloy from which realistic SG tubes are made, the probe consists of an exciter and a detector coil, which have the same dimensions. For simplification, nearby conducting material such as support plates, rigs or bent parts are not modeled here. Table 1 summarizes the important dimensions, material properties, and test conditions.

Electromagnetic induction phenomena associated with RFEC as well as ECT are described by the quasi-static form of Maxwell's equations because of the negligibly small displacement current term. Therefore the governing equation of the A-ϕ method is given:

$$\nabla\times\frac{1}{\mu}\nabla\times \boldsymbol{A}=\boldsymbol{J}_s-\sigma(j\omega\boldsymbol{A}+\nabla\phi)\ , \tag{1}$$

where A and ϕ are a magnetic vector potential and an electric scalar potential in complex vector form. In order to solve the above equation, two types of finite elements are used: first order quadrilateral nodal elements for axisymmetric models; first order hexahedral edge-based elements for three-dimensional models [6,8]. Introducing edge-based finite elements makes it possible to reduce computational costs because of the gauge transformation ϕ=0 in (1).

After obtaining the solution of the governing equation, a RFEC signal, voltage V induced in the detector, is therefore

$$V=-j\omega\int_c \boldsymbol{A}dl\ , \tag{2}$$

where this integration is carried out in the coil alone with a line C.

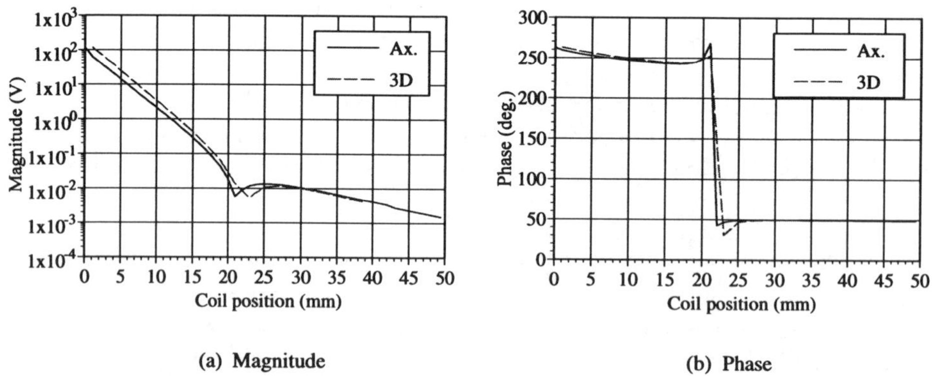

(a) Magnitude (b) Phase

Fig. 2: RFEC signals as a function of the distance between the exciter and detector at 200kHz. The first line (Ax.) and second line (3D) represent the results in axisymmetric and three-dimensional problems. (a) On the sides of the valley in the transition zone between the near and remote field zones, the tendencies that signal traces decrease are different. (b) In the transition zone, there are sharp drops in phase.

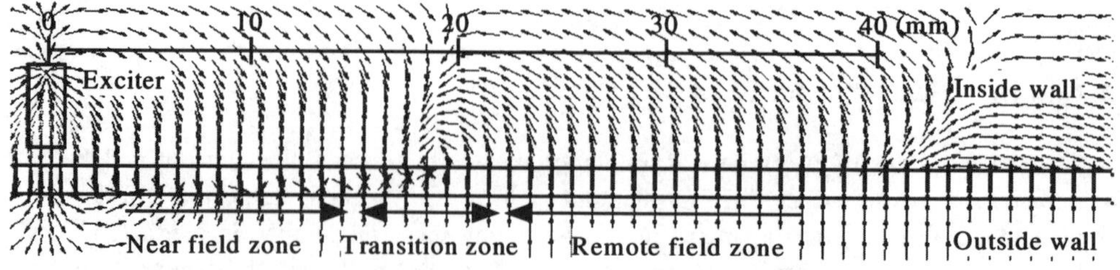

Fig. 3: Axisymmetric finite element prediction of Poynting vector real energy directions under the RFEC phenomena in the non-magnetic tube.

3. Results and Discussion

Predictions of the internal field were taken as the detector was pulled away from the stationary exciter at 200 kHz (skin depth: 1.13mm). By the four-way symmetry, only a quarter of the volume is discretized in three-dimensional problems, and table 2 summarizes the computational costs. Fig. 2 shows the solutions in the axisymmetric and three-dimensional problems. Since these results are in excellent agreement, the computational code for the three-dimensional problems are considered useful. At this frequency whose skin depth corresponds to the tube wall thickness, the log magnitude falls rapidly with a near straight line in the direct field (0 to 0.9 diameters from the exciter), and dips are observed; the phase promptly drops in the same field. This can be explained by the fact that the decline of magnetic flux directly transmited from the exciter is much faster than that outside the tube in the axial direction. Therefore, indirect magnetic flux is dominant in the remote field zone inside the tube. To comprehend the RFEC phenomena, it is also worth mentioning that a Poynting vector real energy plot shows that the transition zone occurs where outwardly directed energy from the exciter meets inwardly directed energy from the outside tube in Fig. 3. Although it is known that the distance between an exciter and a transition zone, in general, is 2 to 2.5 diameters, it appears in a diameter from the exciter in this case.

The signal traces associated with a circling circumferential outer defect, which are non-through, 2mm wide and 20% deep through the wall, are shown in Fig.4, which are computed by using axisymmetric finite elements. In this figure, the indications occur when the exciter and detector pass the defects. A circumferential defect gives us two approximately same high peaks in magnitude with the exciter and detector passages. In this model, the same solution is obtained by using nodal finite elements. The voltage changes due to the defect mentioned above are plotted in Fig. 5, following impedance plane analysis form of eddy current testing. The maximum value and its phase changes due to the defect are defined as:

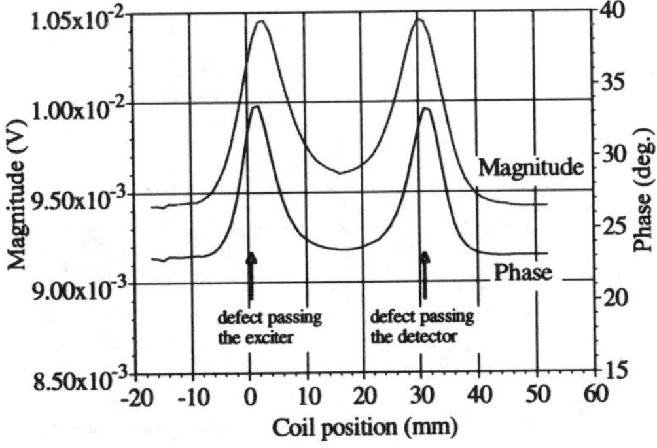

Fig. 4: Signal traces associated with a circumferential outer defect 2mm wide and 40% deep through the wall at 200kHz with coil interval 30mm. The height of the two peaks are the same both in magnitude and phase. Instead of the probe motion, the defect moves in the numerical simulation.

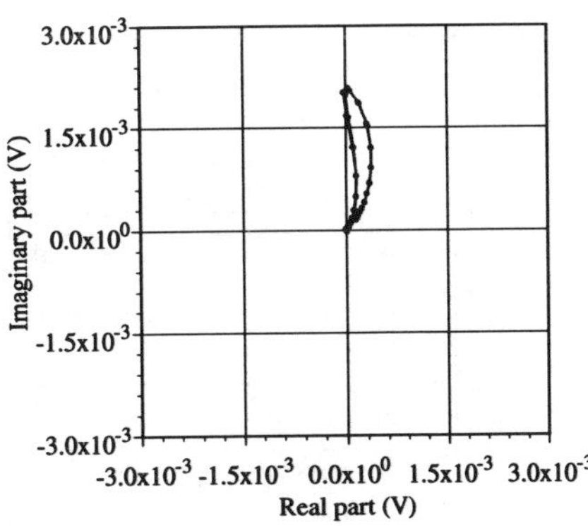

Fig. 5: Trajectories of the voltage changes due to the defect in a complex plane. The defect considered is the same as in Fig.4.

$$|V| = \sqrt{\Delta V_x^2 + \Delta V_y^2} \,, \tag{3}$$

$$\angle V = \tan^{-1}\left(\Delta V_y / \Delta V_x\right) \times 180 / \pi \,, \tag{4}$$

where ΔV_x and ΔV_y are the real and imaginary parts of the changes, respectively.

To detect outer defects (ODs) with higher sensitivity, maximum changes due to two 40% deep - 1mm wide defects (an inter defect (ID) and an OD) were predicted with various coil intervals and frequencies. Fig. 6 shows the ratio of the OD to ID in magnitude and differences of the phase between them. It can be seen that the signal changes of the OD are discernibly higher than those of the ID when detector is in and around the transition zone. (Saying exactly, volume of two defects is slightly different but the difference can be neglected in this comparison.) Irrespective of preferable high sensitivity, due to noises linked to lift-off variations or magnetic inhomogeneities, the use of the transition zone is generally avoided. However SG tubes are non-magnetic and have smooth surfaces inside, and the velocity effect by the probe motion system can be negligible. Therefore putting a detector in the transition zone is expected to be useful in terms of detection of small outer cracks. As long as the tube wall is thicker, the same characteristics are observed with lower frequency where the wall thickness to skin depth ratio is of the order of one.

The RFEC signals obtained by means of the axisymmetric and the three dimensional models are compared at 200kHz for defects with various depths and for two locations of the detecor: in typical remote field zone and in the transition zone. The defects are 2 mm wide, and 20%, 40%, 60%, 80% and 100% deep. Fig. 7 shows maximum changes in

Table 2: Computational costs in three-dimensional problem

elements	22,176
nodes	25,810
edges	73,669
unknowns	59,637
computational memory (MB) [a]	200
computational time(s) [b]	1,302

[a] The computation is carried out with double precision.
[b] The computer is from Visual Technology with DEC alpha A21164, whose clock frequency is 533 MHz. The acceleration factors of the ICCG method are 1 in the nodal elements and 1.02 in edge-based elements. For the convergence the tolerance is chosen as $\varepsilon = 10^{-7}$

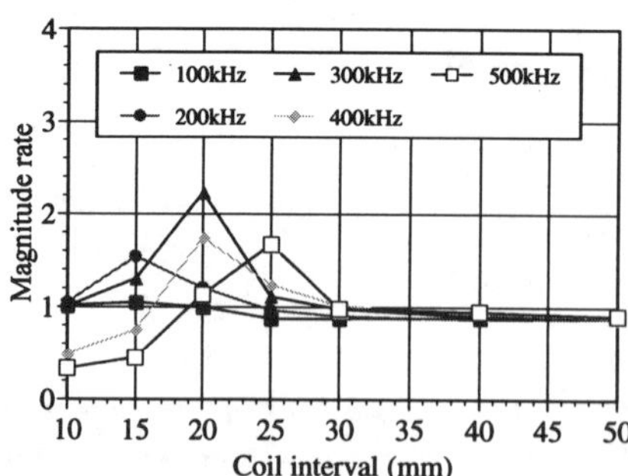

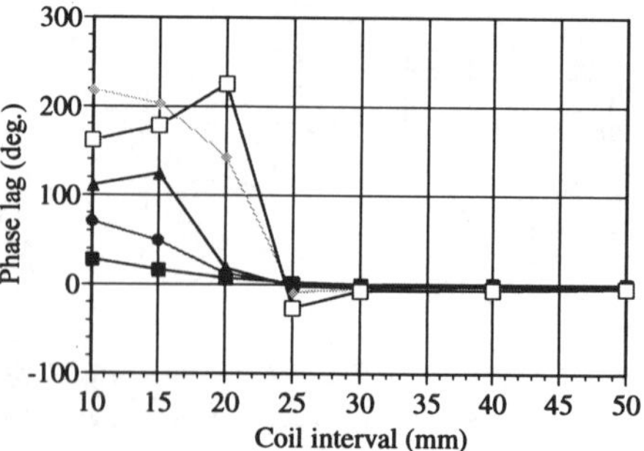

Fig. 6: Magnitude rate and phase lag of maximum signals of a ID and OD which are 2mm wide and 40% deep, S_i and S_o, with the various frequencies and coil intervals. The rate and lag are defined as $| S_o | / | S_i |$ and arg(S_o)-arg(S_i), respectively. The OD is detected with higher sensitivity when the detector is in the transition zone, and then the phase lag appears between the maximum signals of the ID and OD.

magnitude due to the defects as a function of the defect depth. The results demonstrate good agreement and it is distinctive that correlation between the changes and depths is approximately linear although the changes due to outer defects from typical ECT exponentially decline, not in direct proportion to their depths. The signal changes due to the ODs are larger than those of the IDs in Fig.7(a), different from in Fig.7(b) As for the phase, while the changes reflect the depths in Fig.7(a), the changes approximately remain constant in Fig.7(b). This means setting the detector in the transition zone makes it possible to know defect depths from acquired signals as well as typical ECT. The same tendency is observed with pit-type defects as shown in Fig.8.

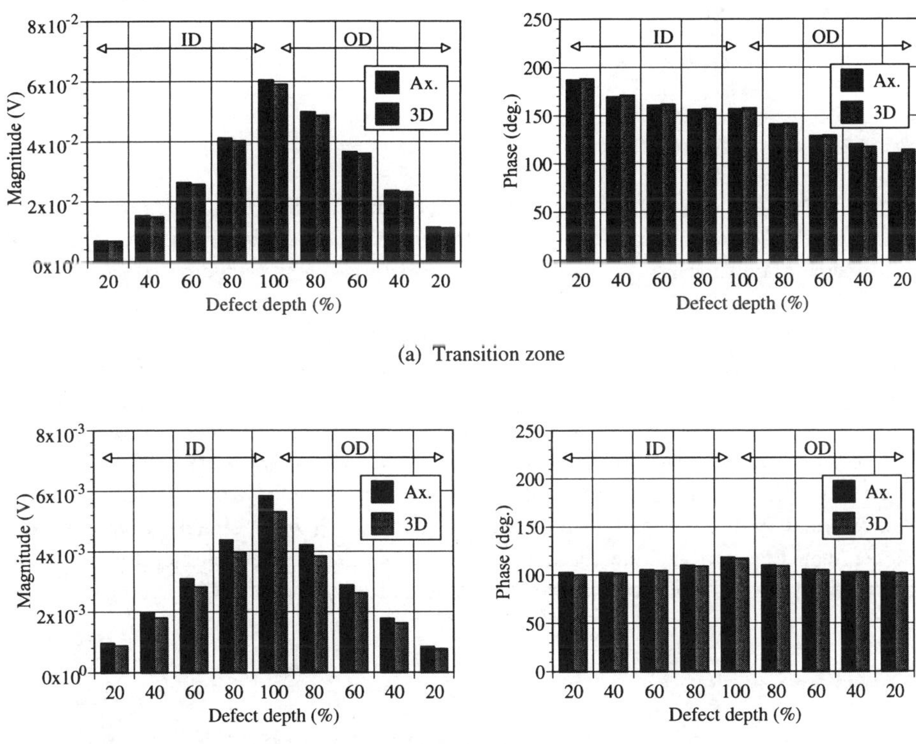

Fig. 7: RFEC signals as a function of the detect depth due to the circumferential IDs, a through detect and ODs, 2mm wide, at 200kHz with coil intervals 15mm (transition zone) and 30mm (remote field zone). (a) There is an advantage that ODs are detected more sensitive than IDs are, and the phase lag is useful to know defect depths. (b) With approximately equal sensitivity to the depths, the phase does not depend on the depth, and the phase corresponding to the maximum magnitude remains constant.

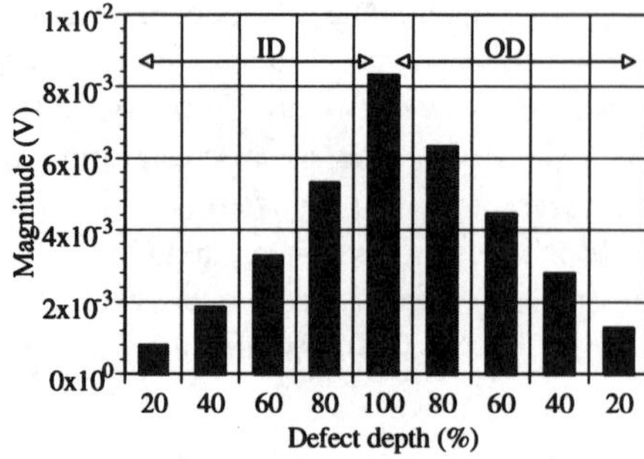

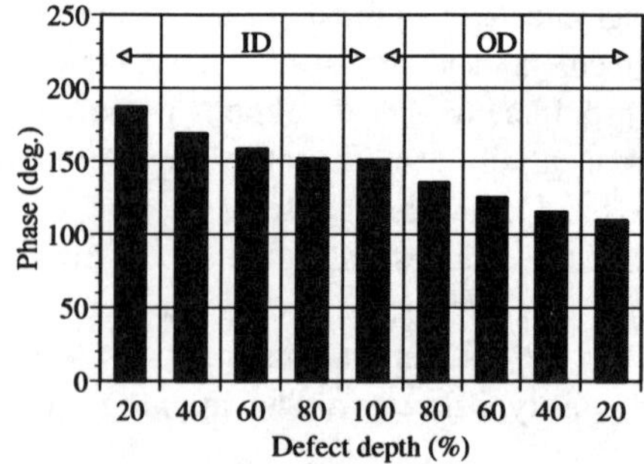

Fig. 8: RFEC signals as a function of the detect depth due to inner, through, and outer pits, 2 mm long and wide, at 200kHz with coil interval 15mm. The same characteristics are observed on condition that the detector is fixed in the transition zone as the case of the circumferential defects.

4. Summary

For the improvement in detectability in the outer cracks, the advantages of the RFEC techniques which stems from the through-transmission effect was used for the inspection of non-magnetic tubes. The A-ϕ method was used with nodal and edge-based finite elements in the axisymmetric and three-dimensional problems, to evaluate the detectability. Both predictions were compared each other for the purpose that the accuracy of the numerical simulation was tested. Since they are in excellent agreement, the use of the simulation is possible and, what is more, valid to study the phenomena and performance under the test. From the results of the numerical simulation, the common RFEC techniques are expected to detect outer cracks as sensitively as inner cracks when they are applied to the non-magnetic tubes such as steam generator tubes in nuclear power plants. Moreover, this paper demonstrates that, when putting a detector in a transition zone, signals to an outer defect is higher than those to an inner defect with the same depth and volume, and the depths are classified by their signal phase information as it is usually done in ECT.

Acknowledgments

This study was supported in part by the Research Committee on Advanced Eddy Current Testing Technology of the Japan Society of Applied Electromagnetics and Mechanics through a grant from 5 PWR utilities and Nuclear Engineering Ltd. The authors would like to thank Prof. Masahiko Nishikawa of Osaka University, and Prof. Hiroshi Hoshikawa and Dr. Kiyoshi Koyama of Nihon University for their advice and encouragement on this work.

References

[1] T. R. Schmidt, The Remote Field Eddy Current Inspection Technique, *Materials Evaluation*, **42-2**, pp.225-230 (1984).

[2] S. M. Haugland, Fundamental Analysis of the Remote-Field Eddy-current Effect, *IEEE Trans. Magn.*, **32-4**, pp.3195-3211 (1996).

[3] W. Lord, Y. S. Sun, S. S. Udpa and S. Nath, A Finite Element Study of the Remote Field Eddy Current

Field, *IEEE Trans. Magn.*, **24-1**, pp.435-438 (1988).
[4] H. Y. Lin and Y. S. Sun, Application of "Zoom-in" Technique in 3D Remote Field Eddy Current Effect Computation, *IEEE Trans. Magn.*, **26-2**, pp.881-884 (1990).
[5] T. Takagi, M. Hashimoto, H. Fukutomi, M. Kurokawa, K. Miya, H. Tsuboi, M. Tanaka, J. Tani, T. Serizawa, Y. Harada, E. Okano and R. Murakami, Benchmark Models of Eddy Current Testing for Steam Generator Tube: Experiment and Numerical analysis, *Intern. J. of Applied Electromagnetics in Materials*, **5**, pp.149-162 (1994).
[6] H. Fukutomi, T. Takagi, J. Tani and M. Hashimoto, Numerical Evaluation of ECT Impedance Signal due to Minute Cracks, *IEEE Trans. Magn.*, **33-2**, pp.2123-2126 (1997).
[7] T. Takagi, M. Uesaka and K. Miya, Electromagnetic NDE Research Activities in JSAEM, *Electromagnetic Nondestructive Evaluation*, IOS Press, pp.9-16 (1997).
[8] T. Takagi, H. Huang, H. Fukutomi and J. Tani, Numerical Evaluation of Correlation between Crack Size
[9] and Eddy Current Testing Signal by a Very Fast Simulator, *IEEE Trans. Magn.*, in press.
[10] M. Isobe, R. Iwata and M. Nishikawa, High Sensitive Remote Field Eddy Current Testing by Using Dual Exciting Coils, *Nondestructive Testing of Material*, IOS Press, pp.145-152 (1997).
[11] H. Fukutomi, T. Takagi, J. Tani and M, Nishikawa, Numerical Simulation for the Remote Field Eddy Current Technique Applied to Non-magnetic Materials, *IEEE Trans. Magn.*, submitted.

Benchmark Problems

Electromagnetic Nondestructive Evaluation (III)
D. Lesselier and A. Razek (Eds.)
IOS Press, 1999

Numerical Analysis of Eddy Current Non Destructive Testing (JSAEM Benchmark Problem #2- Flat Plates with Cracks)

Marco Angeli, Ermanno Cardelli
Institute of Energetics, University of Perugia, Via G. Duranti 1/A-4, 06125 Perugia, Italy

Ryszard Sikora
Department of Electrical Engineering, Technical University of Szczecin, Poland

Abstract
In this paper we show the numerical solution of the second JSAEM - E'NDE benchmark problem. The problem deals with the impedance change of a pancake coil when it moves on a square plate with different crack depths. We have solved the problem by means of a Finite Difference approach in the frequency domain. In the paper we give the main features of this numerical approach, and some information about the hardware board used and about the needs of memory allocation and of CPU time of the software used.

1 Introduction

One of the activities in the research field about the electromagnetic non destructive evaluation is the discussion dealing with the efficiency of different numerical methods in solving both forward and inverse problems. Recently six benchmark problems have been discussed and proposed at the Research Committee on Advanced Eddy Current Testing Technology of the Japan Society of Applied Electromagnetics and Mechanics. We continue our activity of analysis and numerical evaluation of these benchmark problems and we present here the results obtained for the problem #2 – Flat Plates with cracks.

2 Brief Description of the Problem

A pancake type coil is placed above a flat plate with a crack and moves parallel to the x-axis (see fig. 1a), placed along the crack length direction. The probe coil has axisymmetric shape and is made of 140 turns. It is supplied with a current of an amplitude of 1/140 A (see fig 1b). The plate is made of Inconel 600 and has an electric conductivity $\sigma = 1.0^6\ \Omega^{-1}m^{-1}$ and a relative magnetic permeability $\mu_r = 1$.

The solution of the problem requires the determination of the impedance change of the probe. This parameter is evaluated by subtracting the values obtained for the plate without crack from the values obtained for the plate with crack. The impedance change must be calculated either at the frequency of 150 kHz or of 300 kHz, and for lift-offs of 0.5 and 1 mm at different coil locations, starting from the centre of the plate x=0 (and of the crack) to the distance of 10 mm along the crack direction. Three different types of crack having rectangular shape and length of 10 mm are taken into account:

- Full depth crack height 1.25 mm width 0.2 mm
- Inner defect crack height 0.5 mm width 0.21 mm
- Outer defect crack height 0.5 mm width 0.21 mm

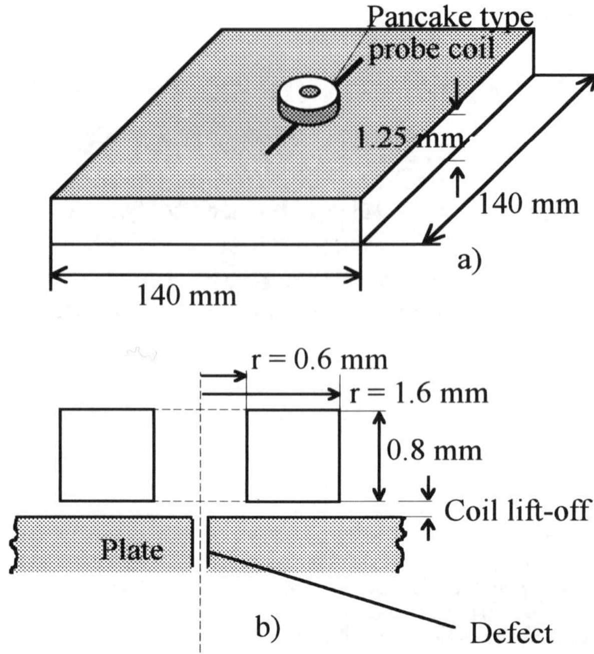

Fig. 1 - Description of the benchmark geometry

3 Numerical Approach

The numerical method used for the eddy current analysis is based on a finite difference frequency domain algorithm, obtained using the spatial discretization proposed by Yee [1].

We have adopted here a variable spatial step in order to obtain a more accurate discretization in those regions where high variations of the values of the electromagnetic field are expected or where the eddy current analysis is particularly important, for example in the proximity of the coil and of the plate.

In addition this variable step grid allows us to well simulate a far boundary with relatively few cells.

On the boundary we have imposed the conditions $\dot{E}_t = 0$ and $\dot{B}_n = 0$.

The domain of the problem can be decomposed in cells containing only one kind of material, and the components of the electric field and of the magnetic induction used in the discretization are continuous over the surfaces separating two cells.

Using this technique the Maxwell equations, written in integral form:

$$j\omega \iint_S \varepsilon \dot{\mathbf{E}} \cdot d\mathbf{S} + \iint_S \dot{\mathbf{J}} \cdot d\mathbf{S} = \oint_l \dot{\mathbf{H}} \cdot d\mathbf{l} \tag{1}$$

$$-j\omega\mu \iint_S \dot{\mathbf{H}} \cdot d\mathbf{S} = \oint_l \dot{\mathbf{E}} \cdot d\mathbf{l} \tag{2}$$

where the dot notation indicates the phasor, can be approximated by the following set of algebraic equations:

$$\left(j\omega\varepsilon\,\dot{E}_{x(i+\frac{1}{2},j,k)}+\dot{J}_{x(i+\frac{1}{2},j,k)}\right)\Delta y_j\Delta z_k =$$
$$\left(\dot{H}_{z(i+\frac{1}{2},j+\frac{1}{2},k)}-\dot{H}_{z(i+\frac{1}{2},j-\frac{1}{2},k)}\right)\Delta z_k-\left(\dot{H}_{y(i+\frac{1}{2},j,k+\frac{1}{2})}-\dot{H}_{y(i+\frac{1}{2},j,k-\frac{1}{2})}\right)\Delta y_j \tag{3}$$

$$\left(j\omega\varepsilon\,\dot{E}_{y(i,j+\frac{1}{2},k)}+\dot{J}_{y(i,j+\frac{1}{2},k)}\right)\Delta x_i\Delta z_k =$$
$$\left(\dot{H}_{x(i,j+\frac{1}{2},k+\frac{1}{2})}-\dot{H}_{x(i,j+\frac{1}{2},k-\frac{1}{2})}\right)\Delta x_i-\left(\dot{H}_{z(i+\frac{1}{2},j+\frac{1}{2},k)}-\dot{H}_{z(i-\frac{1}{2},j+\frac{1}{2},k)}\right)\Delta z_k \tag{4}$$

$$\left(j\omega\varepsilon\,\dot{E}_{z(i,j,k+\frac{1}{2})}+\dot{J}_{z(i,j,k+\frac{1}{2})}\right)\Delta x_i\Delta y_j =$$
$$\left(\dot{H}_{y(i+\frac{1}{2},j,k+\frac{1}{2})}-\dot{H}_{y(i-\frac{1}{2},j,k+\frac{1}{2})}\right)\Delta y_j-\left(\dot{H}_{x(i,j+\frac{1}{2},k+\frac{1}{2})}-\dot{H}_{x(i,j+\frac{1}{2},k+\frac{1}{2})}\right)\Delta x_i \tag{5}$$

$$-j\omega\mu\,\dot{H}_{x(i,j+\frac{1}{2},k+\frac{1}{2})}\,\Delta y_{j+\frac{1}{2}}\Delta z_{k+\frac{1}{2}} =$$
$$\left(\dot{E}_{z(i,j+1,k+\frac{1}{2})}-\dot{E}_{z(i,j,k+\frac{1}{2})}\right)\Delta z_{k+\frac{1}{2}}-\left(\dot{E}_{y(i,j+\frac{1}{2},k+1)}-\dot{E}_{y(i,j+\frac{1}{2},k)}\right)\Delta y_{j+\frac{1}{2}} \tag{6}$$

$$-j\omega\mu\,\dot{H}_{y(i+\frac{1}{2},j,k+\frac{1}{2})}\,\Delta x_{i+\frac{1}{2}}\Delta z_{k+\frac{1}{2}} =$$
$$\left(\dot{E}_{x(i+\frac{1}{2},j,k+1)}-\dot{E}_{x(i+\frac{1}{2},j,k)}\right)\Delta x_{i+\frac{1}{2}}\quad\left(\dot{E}_{z(i+1,j,k+\frac{1}{2})}-\dot{E}_{z(i,j,k+\frac{1}{2})}\right)\Delta z_{k+\frac{1}{2}} \tag{7}$$

$$-j\omega\mu\,\dot{H}_{z(i+\frac{1}{2},j+\frac{1}{2},k)}\,\Delta x_{i+\frac{1}{2}}\Delta y_{j+\frac{1}{2}} =$$
$$\left(\dot{E}_{y(i+1,j+\frac{1}{2},k)}-\dot{E}_{y(i,j+\frac{1}{2},k)}\right)\Delta y_{j+\frac{1}{2}}-\left(\dot{E}_{x(i+\frac{1}{2},j+1,k)}-\dot{E}_{x(i+\frac{1}{2},j,k)}\right)\Delta x_{i+\frac{1}{2}} \tag{8}$$

where

$$\Delta x_i = \frac{x(i+1)-x(i-1)}{2} \tag{9}$$

and

$$\Delta x_{i+\frac{1}{2}} = x(i+1)-x(i) \tag{10}$$

In (1) the term $\dot{J}$ may be supposed to be known in some region of the space, for example in the current driven coils where, taking into account the fact that the diameter of each turn is lower than the penetration depth, $\dot{J}$ has an amplitude uniformly distributed in space and direction tangential to the coil.

The current density is supposed to be uniformly distributed in the coil, so it may be expressed by the relations:

$$\dot{J}_x = -\left|\dot{\mathbf{J}}\right|\frac{y-y_c}{r} \tag{11}$$

$$\dot{J}_y = \left|\dot{\mathbf{J}}\right|\frac{x-x_c}{r} \tag{12}$$

where (x_c, y_c) is the centre of the coil and $r = \sqrt{(x-x_c)^2 + (y-y_c)^2}$.

So the mean values of the current density are given by the relations:

$$\dot{J}_x = -|\dot{\mathbf{J}}| \frac{\left. \sqrt{(x-x_c)^2 + (y-y_c)^2} \right|_{y_1}^{y_2}}{y_2 - y_1} \tag{13}$$

$$\dot{J}_y = |\dot{\mathbf{J}}| \frac{\left. \sqrt{(x-x_c)^2 + (y-y_c)^2} \right|_{x_1}^{x_2}}{x_2 - x_1} \tag{14}$$

where x_1, x_2, y_1, y_2 are the extrema of the rectangular surface centred in a grid node.

These expressions may be used as impressed values of the current density in the numerical simulation and ensure that the relation $\operatorname{div} \dot{\mathbf{J}} = 0$ is respected also in the discretized model.

In the other regions we have $\dot{\mathbf{J}} = \sigma(\dot{\mathbf{E}} + \dot{\mathbf{E}}^*)$, where $\dot{\mathbf{E}}^*$ is the electromotive force applied between two nodes of the grid, and may be known in the case of impressed voltage, or equal to zero.

We have solved this set of algebraic equations by means of the Bi-Conjugate Gradient method [2]. More traditional iterative methods, such as Jacobi or Gauss-Seidel or Successive Over-relaxation, have in fact produced numerical instabilities.

Current driven coils can be simulated by imposing, in every node taken by the coil, the current density.

Once solved the system of algebraic equations, the equivalent resistance and reactance of the system have been evaluated by means of the complex expression of the electric power:

$$\iiint_{\Gamma} \dot{\mathbf{E}}^* \cdot \breve{\mathbf{J}} \, d\Gamma = RI^2 + jXI^2 \tag{15}$$

where $\breve{\mathbf{J}}$ is the conjugate of $\dot{\mathbf{J}}$ and I is the rms value of the current.

4 Results

We have used as hardware board a PC with Pentium II INTEL® 300 MHz Processor with 128 MB RAM. The numerical code is written in C++ programming language and it is compiled at 32 bit in Windows environment. In Table 1 the main features of the computer program used are given. We show here, in Fig. 2 and 3, the results obtained for the full depth defect, in Fig. 4 and 5 the results for the 40% depth inner defect and in Fig. 6 and 7 the results for the 40% depth outer defect.

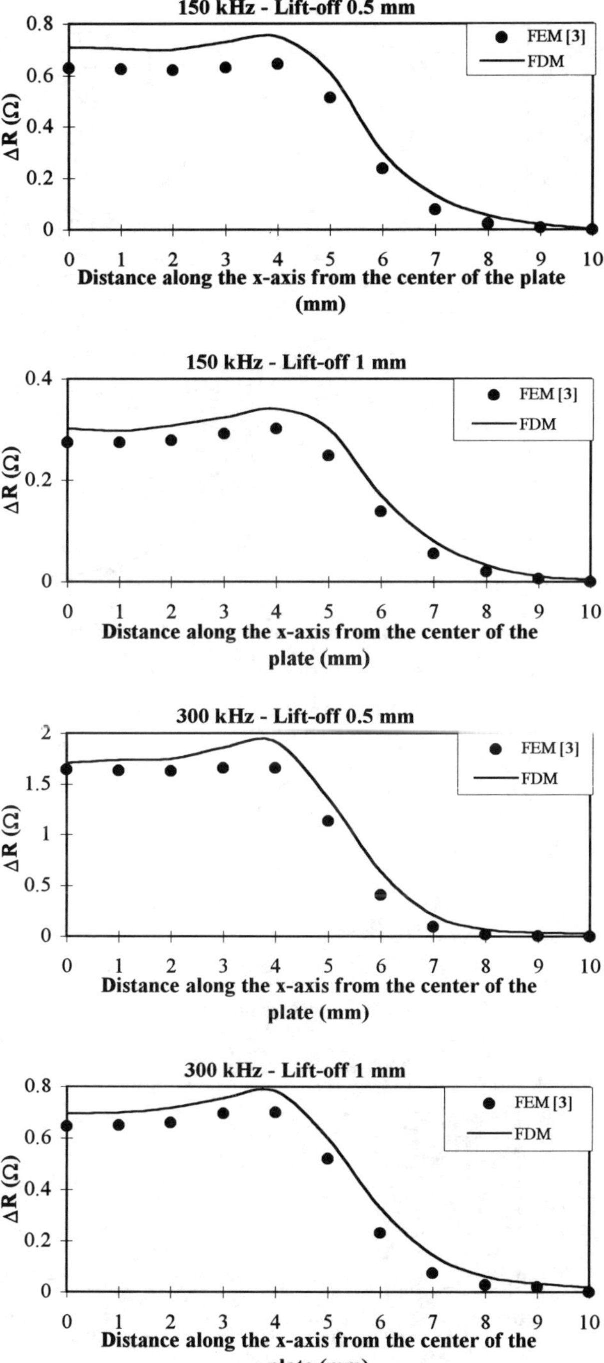

Fig. 2 - Resistance changes for the full depth crack

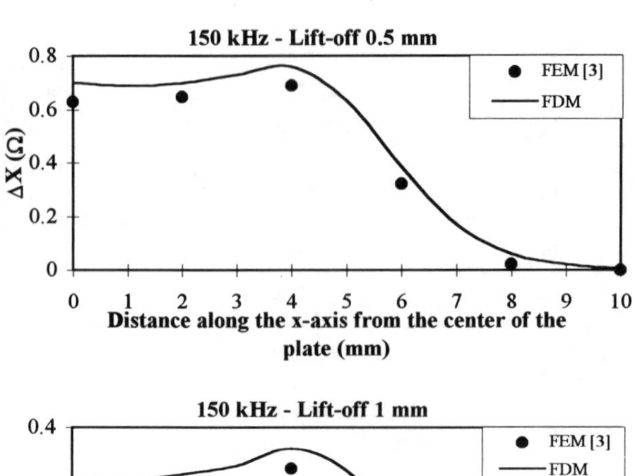

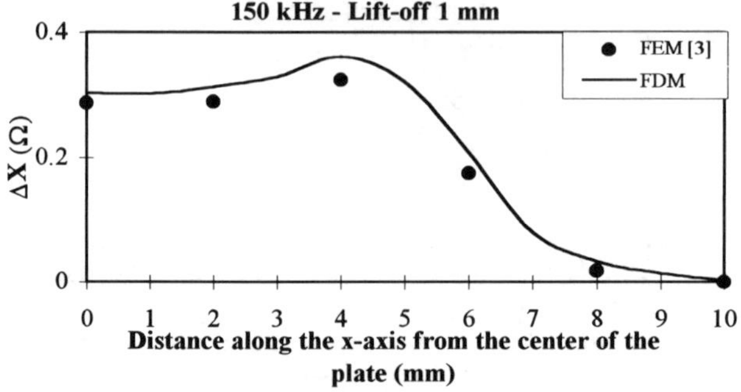

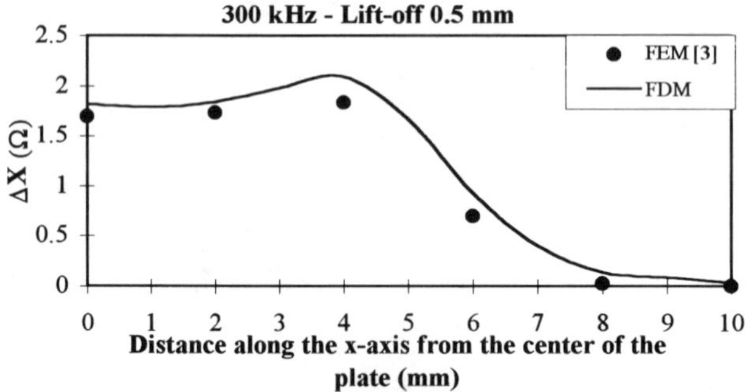

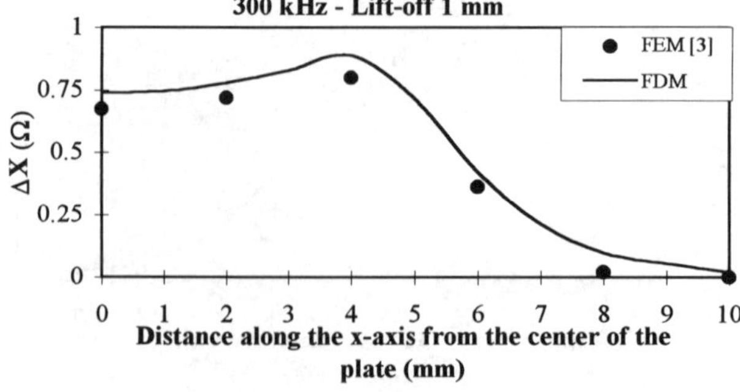

Fig. 3 - Reactance changes for the full depth crack

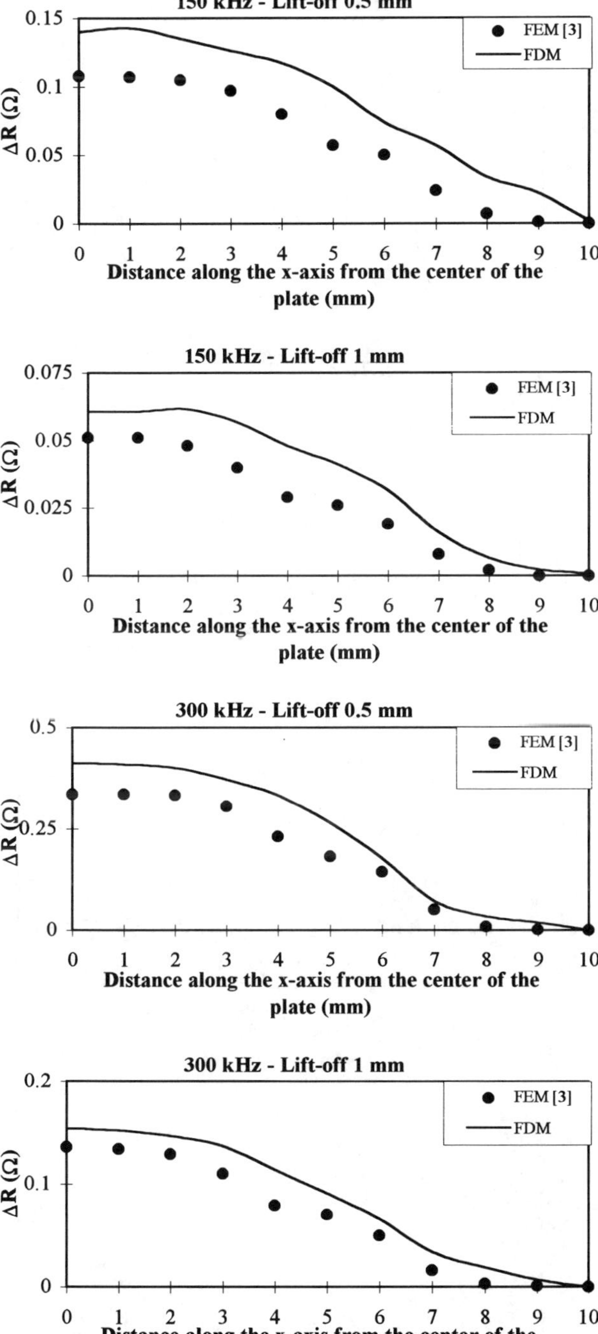

Fig. 4 - Resistance changes for the 40% depth inner crack

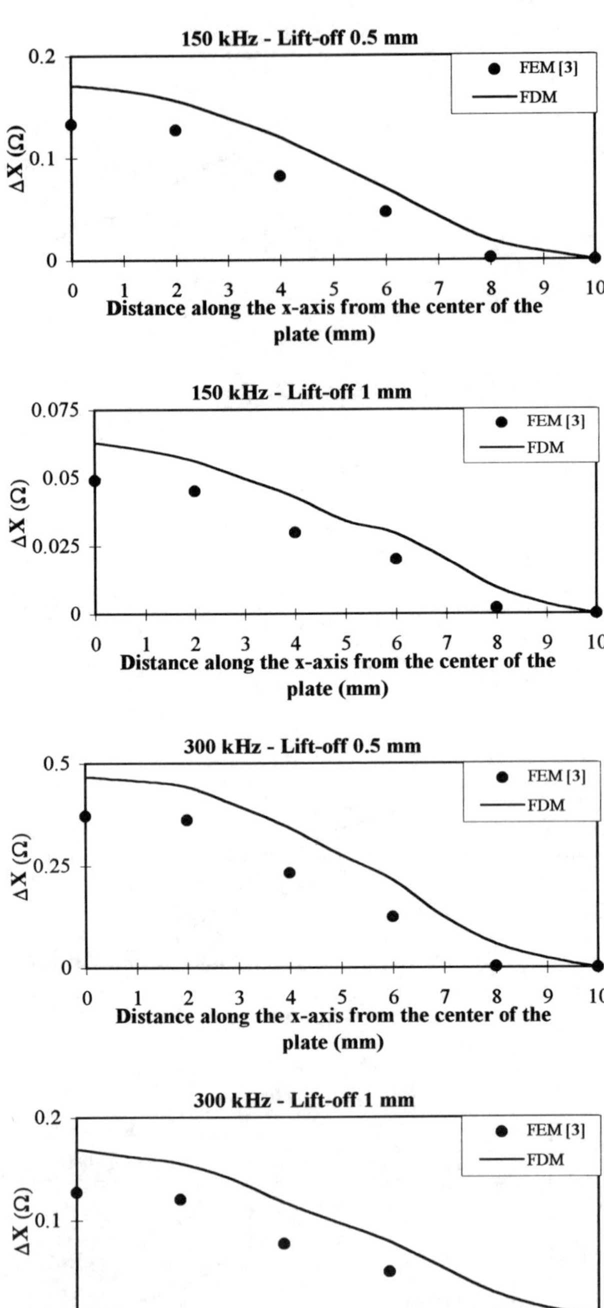

Fig. 5 - Reactance changes for the 40% depth inner crack

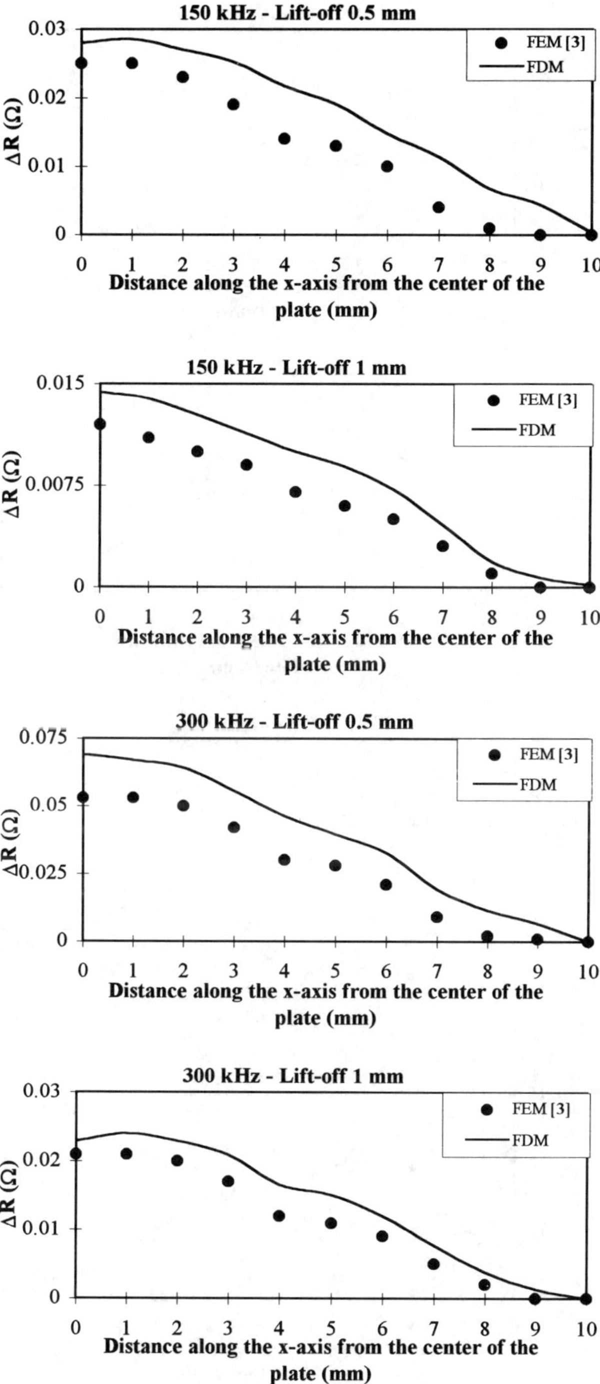

Fig. 6 - Resistance changes for the 40% depth outer crack

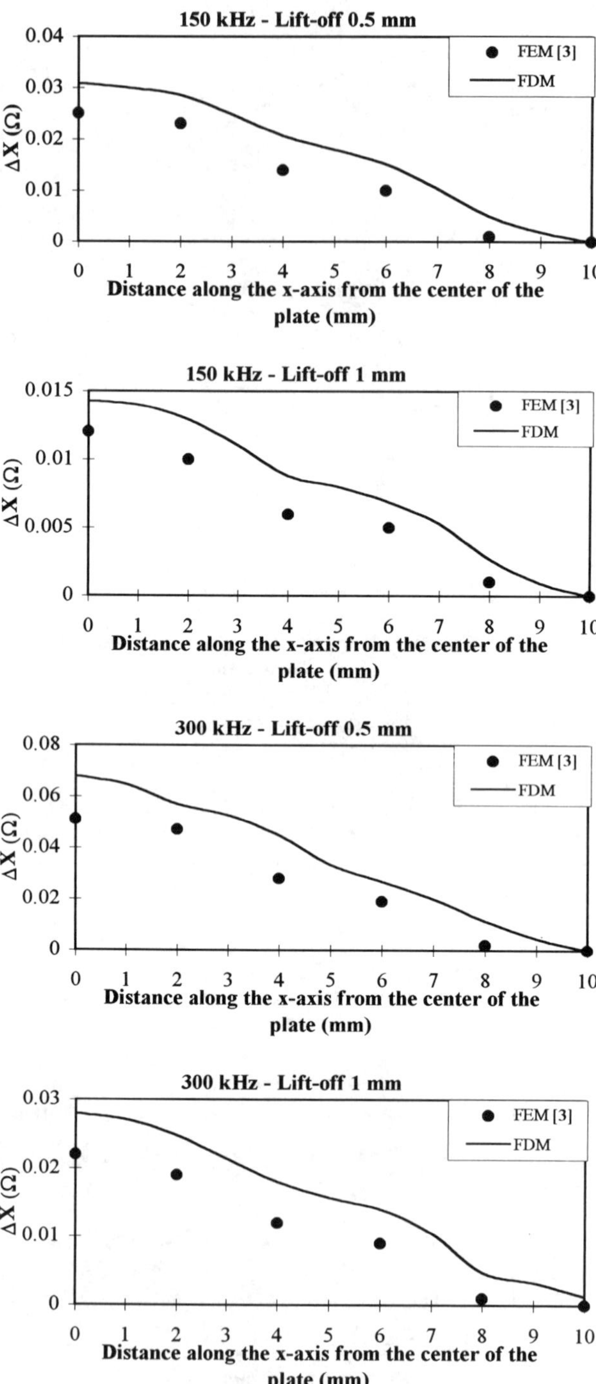

Fig. 7 - Reactance changes for the 40% depth outer crack

Table 1
Description of computer program

No.	Item		Specification
1	Code name		
2	Formulation		4. FDFDM (Finite Difference Frequency Domain Method)
3	Governing equations	In conductor	$j\omega\iint_{\Gamma}\varepsilon\dot{\mathbf{E}}\,d\Gamma+\iint_{\Gamma}\sigma\left(\dot{\mathbf{E}}+\dot{\mathbf{E}}^{*}\right)d\Gamma=\oint_{\gamma}\left(\nabla\times\dot{\mathbf{H}}\right)\cdot d\gamma$ or $j\omega\iint_{\Gamma}\varepsilon\dot{\mathbf{E}}\,d\Gamma+\iint_{\Gamma}\dot{\mathbf{J}}\,d\Gamma=\oint_{\gamma}\left(\nabla\times\dot{\mathbf{H}}\right)\cdot d\gamma$ $-j\omega\iint_{\Gamma}\mu\dot{\mathbf{H}}\,d\Gamma=\oint_{\gamma}\left(\nabla\times\dot{\mathbf{E}}\right)\cdot d\gamma$
		In vacuum	$j\omega\iint_{\Gamma}\varepsilon\dot{\mathbf{E}}\,d\Gamma=\oint_{\gamma}\left(\nabla\times\dot{\mathbf{H}}\right)\cdot d\gamma$ $-j\omega\iint_{\Gamma}\mu\dot{\mathbf{H}}\,d\Gamma=\oint_{\gamma}\left(\nabla\times\dot{\mathbf{E}}\right)\cdot d\gamma$
4	Solution variables	In conductor	$\dot{\mathbf{E}},\dot{\mathbf{H}}$
		In vacuum	$\dot{\mathbf{E}},\dot{\mathbf{H}}$
5	Gauge condition		1. Not imposed
6	Treatment of exciting current		2. Elements for coil
7	Impedance calculation method	equation	$X=\mathrm{Im}\left(\frac{\iiint_{\Gamma}\dot{\mathbf{E}}^{*}\cdot\breve{\mathbf{J}}\,d\Gamma}{I_{C}^{2}}\right)$ $R=\mathrm{Re}\left(\frac{\iiint_{\Gamma}\dot{\mathbf{E}}^{*}\cdot\breve{\mathbf{J}}\,d\Gamma}{I_{C}^{2}}\right)$
		description	
8	Solution method for equation		4. BiCG (BiConjugate Gradient)
	Convergence criterion for interaction method		Number of iteration = number of unknown
9	Element type		5. rectangle
			1. Nodal element (4 nodes)
10	Number of elements		147000
11	Number of nodes		158202
12	Number of unknowns		849241

13	Computer	Name	PC Intel Pentium II 300 MHz
		Speed	128 MIPS 15 MFLOPS
		Main memory (MB)	256
		Used Memory (MB)	60
		Precision of data (bits)	32

		CPU time (s)	total	1400
			solving linear equations	1260

14	References on computer program	

References

[1] K. S. Yee, Numerical solution of initial boundary value problems involving Maxwell' s equations in isotropic media. *IEEE Trans. Antennas Propagat.*, vol. AP-14, pp. 302-307, May 1966.

[2] R. Barrett et al., *Templates for the solutions of linear systems: building blocks for iterative methods.* SIAM, 1994.

[3] H. Huang, Private communication.

[4] T. Takagi *et al.*, ECT research activity in JSAEM -benchmark models of eddy current testing for steam generator tube - (Part 1), *Nondestructive Testing of Materials*, Studies in Applied Electromagnetics and Mechanics, Vol. 8, IOS press, 1995, pp. 253-264.

[5] T. Takagi *et al.*, ECT research activity in JSAEM -benchmark models of eddy current testing for steam generator tube - (Part 2), *Nondestructive Testing of Materials*, Studies in Applied Electromagnetics and Mechanics, Vol. 8, IOS press, 1995, pp. 312-320.

[6] A. Kameari, Solution of axisymmetric conductor with a hole by FEM using edge-element, COMPEL, Vol. 9, 1990, pp. 230-232.

[7] J. Shimone, K. Maeda and Y. Harada, Impedance measurement of unique shape cracks by pancake type coil, *Applied Electromagnetics*, JSAEM Studies in Applied Electromagnetics, Vol. 5, 1996, pp. 12-17.

Electromagnetic Nondestructive Evaluation (III)
D. Lesselier and A. Razek (Eds.)
IOS Press, 1999

Benchmark Test on Improved ECT Probe Based on FLUXSET Sensor

Antal GASPARICS, Csaba S. DARÓCZI, Gábor VÉRTESY
Research Institute for Technical Physics and Materials Sciences,
H-1525 Budapest-114, P.O. Box 49, Hungary
József PÁVÓ
Department of Electromagnetic Theory, Technical University of Budapest,
H-1521 Budapest, Egry J. u. 18., Hungary

Abstract. An improved ECT probe based on Fluxset sensor [1] and measurement system is presented in this paper. The basic experimental setup was published during the ENDE '97 workshop [2]. Based on the previous experiments a new and improved probe has been developed and tested in order to obtain information about the performance of the measurement system. The signal/noise ratio of the output signal was determined in case of different type of cracks and in case of lift-off and tilting errors which were proposed in ENDE'97 [3].

1. Introduction

The aim of our development was the 15% and 10% OD and ID type crack detection in 1.25 mm thin INCONEL 600 plate. We want to determine as many proposed parameters [3] of benchmark test as possible in case of very small cracks. The quality of the measurement method is characterised by the signal/noise ratio of the output data, the spatial distortion and the required electrical power for the detection. The usually investigated cracks have 40% (or higher) depth [4,5,6,7,8], but these cracks are very large from signal/noise point of view. The influence of the surface roughness and any other external disturbances on the output signal cause worse signal/noise ratio in case of smaller cracks because of the lower perturbation of eddy currents due to the presence of the crack itself [2]. This fact was the reason why 15% and 10% cracks were chosen for our benchmark investigations.

2. Probe geometry

The size of the plate specimen (80 mm width and 80 mm long) requires a smaller probe than the pervious one which was presented in ENDE'97 [2] because of the side (or edge) effect (the influence of the edge of the plate specimen). The basic structure of the measuring head was not changed - it consists of one exciting coil and one Fluxset sensor (see Fig.1). The most important difference between the new and the old head is the size of the exciting coil. The sensor was also improved, Fluxset #6 type sensor was applied in the measurement.

The sensor is located in the middle of the probe geometry under the exciting coil parallel to the surface. The lift-off (the distance between the surface and the component) of the exciting coil is 2.0 mm, the lift-off of the sensor core is approx. 0.5 mm. The sensor detects the magnetic field in y axial direction. There is no parallel component of the magnetic field of the exciting coil in the middle, when there is no crack in the plate. It means that the sensor measures the perturbation of the magnetic field due to the distortion of eddy current

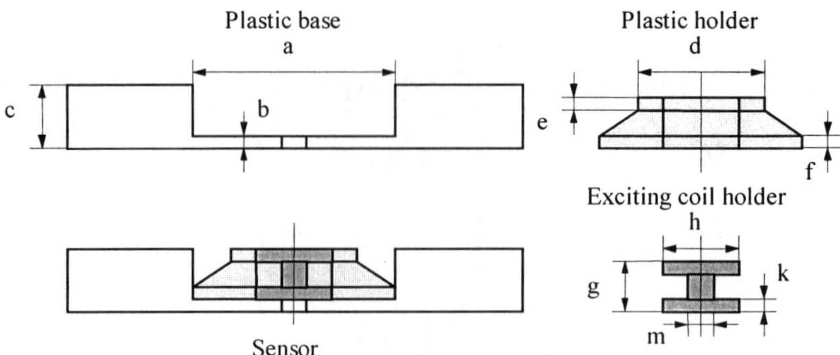

Fig.1 The geometry of the probe

Table I. Geometrical data of the probe

Parameter	Symbol	Size [mm]
Holder hole	a	24.5
Coil holder lift-off	b	1.5
Base thickness	c	6.5
Diameter	d	16.0
Thickness	e	3.0
Thickness	f	0.5
Coil height	g	6.5
Coil outer diameter	h	10.0
Coil inner lift-off	k	0.5
Coil inner diameter	m	4.8

distribution in the plate [9]. The outer diameter of the exciting coil is 10 mm which is more than two times smaller than it was in the pervious configuration. All other important geometrical data can be found in Table I.

3. The applied sensor

The main target of the ECT application is the solution of the inverse problem: the reconstruction of the crack or some of its important parameters such as depth and the shape. For this reason is important to get as much information about the crack as it is possible. The spatial distortion of the crack detection may have influence on the quality of the output data and on the signal/noise ratio of the measurement of course. The Fluxset based ECT system detects the magnetic field perturbation. The sensor has a large size (10 mm length) relatively to the investigated crack sizes and, of course, relatively to the distribution of the magnetic field perturbation.

In Fig.2 the geometry of the applied Fluxset #6 type sensor used in the investigated probe is shown. The spatial resolution of the sensor was tested by a bobbin coil arrangement (Fig.2) at 1 and 10 kHz frequencies. The two exciting bobbin coils generate the magnetic field in opposite direction regarding to each other. The results of this test and the calculated magnetic field of the exciting coils can be seen in Fig.3. We can say, that the spatial resolution of the Fluxset sensor is much higher than its lengths. (We were not able to determine the resolution due to the small error.)

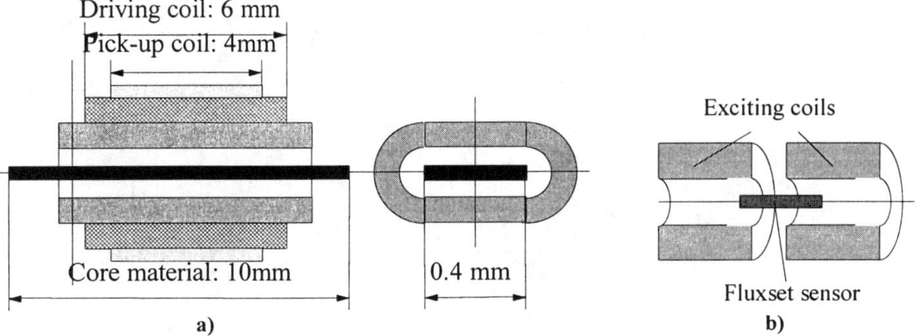

Fig.2 a) Fluxset sensor geometry, b) Bobbin type arrangement

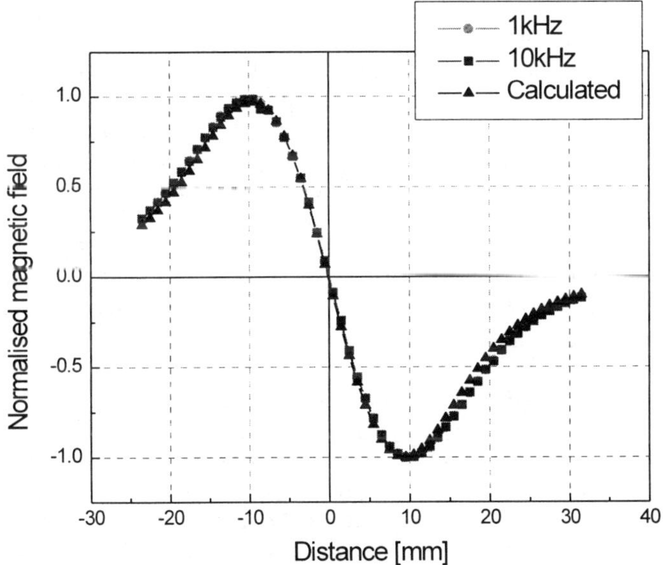

Fig.3 Results of the spatial resolution test of the Fluxset sensor

4. The measuring system

One of the most important advantages of the Fluxset sensor is the simplicity of the required measuring system (Fig.4). The output of the sensor is the phase of the induced voltage pulses in the pick-up coil and no special instruments are necessary for the phase demodulation [2]. A computer was applied to control the measurement and for the data acquisition. The digital signal generator produced the signal for the exciting coil in order to induce eddy currents and the digital signal processing was made by a digital oscilloscope. The driving frequency of the improved sensor was 180.6 kHz, the frequency of the excitation signal was 20 kHz and the signal had sinusoidal shape (single frequency method). The highest frequency of the excitation is the half of the driving frequency - theoretically that means 90.3 kHz. An eight

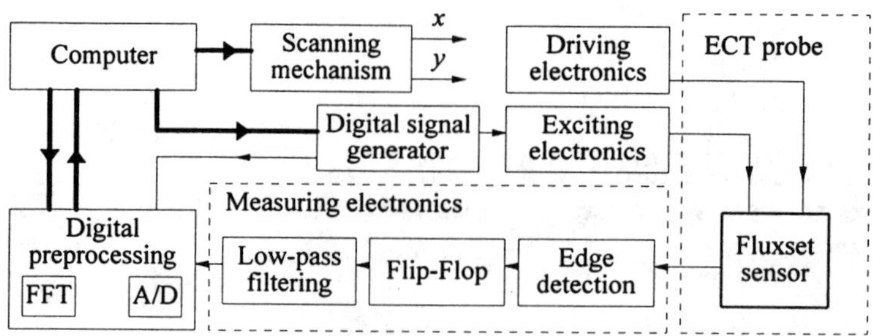

Fig.4 Measuring system

order low pass analogue filter filtered out the driving frequency components from the pick-up coil signal.

5. The results of the scans

The probe was tested on an 80 x 80 mm plate specimen made of INCONEL 600. The geometry of the probe arrangement is shown in Fig.5. The sensor was perpendicular position (y direction) relatively to the direction of the crack (x direction). The thickness of the plate is 1.25 mm, two cracks were investigated with 187.5 μm (15%) and 125 μm (10%) depth from the both side of the plates (OD and ID types). The length of the cracks was 9 mm, the thickness was approx. 0.2 mm [10]. The origin of the co-ordinate system was set in the middle of the geometry. The surface scans were made in x direction 1, 3, 5 and 7 mm far from the *x* axis.

The output of the surface scan can be seen in Fig.6. The comparison of the results from the measurement (Fig. 6) and from the calculation of the magnetic field perturbation (Fig. 7) shows the probe has enough high spatial resolution for crack detection in contrast with the size of the exciting coil and the length of the sensor.

All scans made by the same probe and measuring electronics was tuned in the same way. The step in *x* direction was 0.5 mm (Fig. 8). The probe is very sensitive to the positioning error of the sensor core relatively to the exciting coil. Sensor position was set by plastic screws only which had not high enough stability and solidity. The surface scans are comparable with each other in spite of the fact that the sensor slip has strong influence on the results. The comparisons can bee seen in Fig. 9 in case of each scan lines in case for the four studied situations.

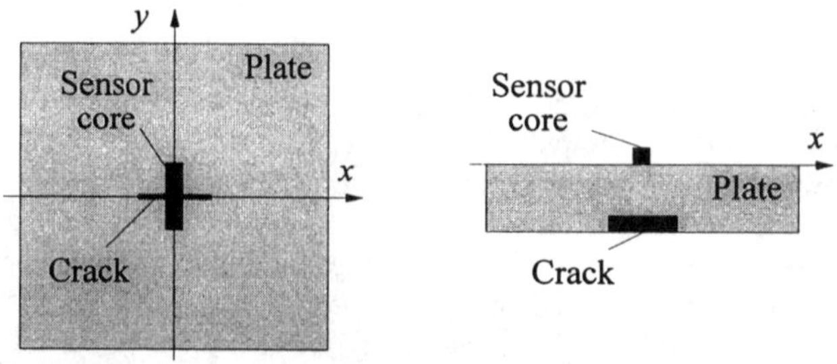

Fig.5 Probe arrangement

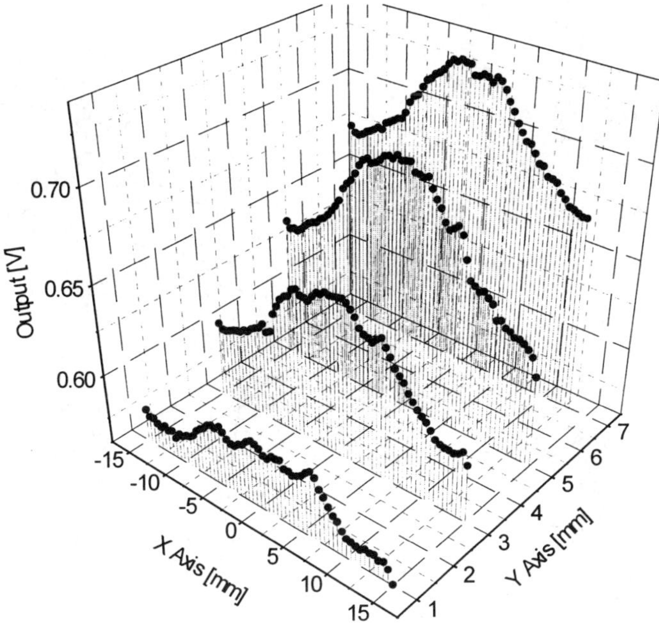

Fig. 6 Surface scan results of OD15% type crack

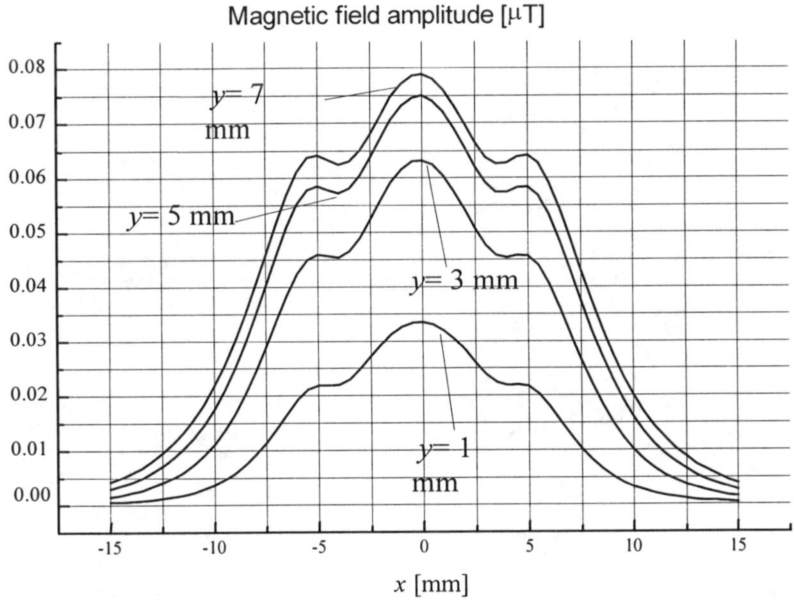

Fig. 7 Absolute value of the *y* component of the magnetic field at the middle of the sensor while the probe is scanning above a 9 mm long OD15% crack along different *y*=constant scanning lines (see Fig.5).

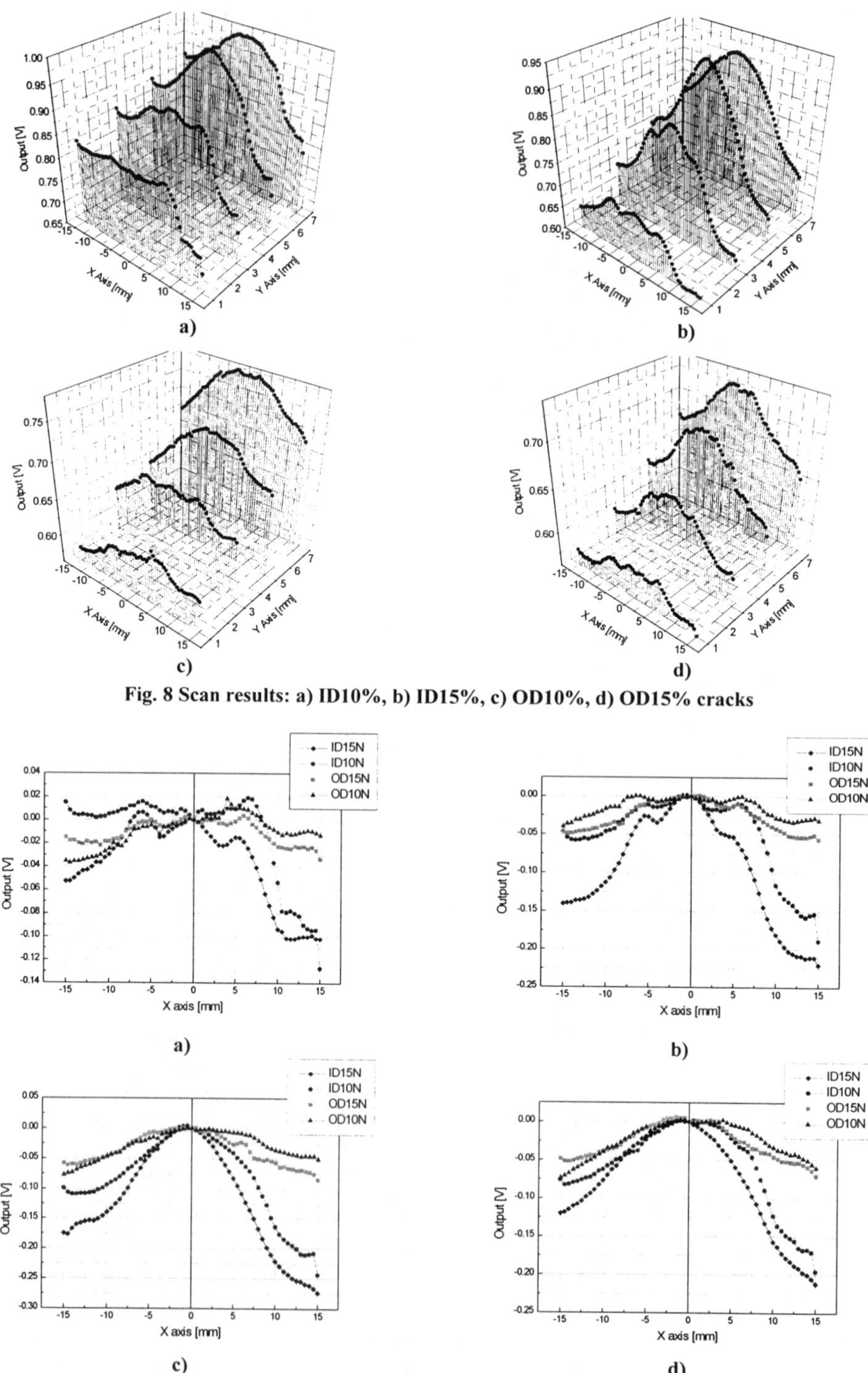

Fig. 8 Scan results: a) ID10%, b) ID15%, c) OD10%, d) OD15% cracks

Fig. 9 Comparison of scan lines at a) 1 mm, b) 3 mm, c) 5 mm, d) 7 mm far from the x axis without offset

6. Benchmark definitions

There are several aspects of the qualification of the performance of different ECT probes. There are also several techniques and methods in ECT measurements. The performance of our probe was identified by the determination of the signal/noise ratio that is proposed in [3]. It should be noted, that there are also several numerical or physical techniques for improving the signal/noise ratio (such as dual or multiple frequency methods [11,12], noise filtering and optimisation and statistical techniques). Only the raw output signal of the whole measurement system is used for our benchmark calculations presented in this paper. Single frequency type excitation was used for inducing eddy currents in the plate specimen.

There are several sources of the noise in our measurement system, for example: positioning error, surface roughness, noise of the sensing element, the noise of the analogue signal processing, the noise of the signal discretisation, the error from FFT algorithm etc. The different type of noises can be classified into two main groups: time dependent noises and position dependent noises. The filtering of the time dependent noises is easier than the filtering of the position dependent noise. The FFT algorithm also has influence on the signal/noise ratio. We applied 250 kHz sampling frequency and FFT was made on time-windows included of 10,000 points (that means the time-window has 40 ms lengths that means 800 periods of the excitation cycle). The S/N ratio, which is based on the time dependent noises, is investigated in this paper.

7. Benchmark results

The following parameters of the output signal proposed in [3] were identified:

- The amplitude of the time dependent noise {Noise}
- The difference between the maximum and the minimum value of the signal {Peak to peak}
- The signal/noise ratio {S/N as $20\log_{10}$("peak to peak"/"noise")}

It can be well seen in Fig. 6 that scan lines 3 and 5 mm far from the *x* axis hold the most importation information about the crack - so the data of these lines play a significant role in the comparison.

7.1. Qualification of the depth resolution

The depth resolution means the boundary of the smallest crack which can be detected in case of a minimal necessary signal/noise ratio for the inversion. The results of the calculated parameters based on the presented data (Fig. 9) can be found in Table II. The results show, the S/N value is about 32dB in case of OD10% crack and about 46dB in case if ID15% crack. The results show also the difference between ID and OD type detection. It is roughly 8-10 dB from S/N point of view (in case of the described specimen).

7.2. Qualification of the noise suppression

The noise suppression means the ability of the system to suppress the external noise sources. We investigated the influence of the lift-off and the tilting error [3] on the output of the measuring system in case of OD15% crack detection.

Table II. Benchmark results of different type cracks detection

Parameter:	Noise [V]	Peak to peak [V]	S/N [dB]
ID15 - 1mm	0.00079	0.10834	42.74
ID15 - 3mm	0.00029	0.22257	57.70
ID15 - 5mm	0.00245	0.27822	41.10
ID15 - 7mm	0.00104	0.21445	46.29
ID10 - 1mm	0.00129	0.14645	41.10
ID10 - 3mm	0.00620	0.18936	29.70
ID10 - 5mm	0.00096	0.24483	48.13
ID10 - 7mm	0.00115	0.19778	44.71
OD15 - 1mm	0.00066	0.03684	34.94
OD15 - 3mm	0.00047	0.05768	41.78
OD15 - 5mm	0.00105	0.08673	38.34
OD15 - 7mm	0.00087	0.07634	38.86
OD10 - 1mm	0.00118	0.05528	33.41
OD10 - 3mm	0.00099	0.04057	32.25
OD10 - 5mm	0.00228	0.07798	30.68
OD10 - 7mm	0.00142	0.07606	34.58

Two situations were investigated: a) The probe has no tilting error, but it has 0.5 mm lift-off error (OD15 L); b) the probe has 3° tilting error (around y axis) and 1 mm lift-off error in the same time (OD15 LT). (It was impossible to apply tilting error only, because of the contact of the plain surfaces.) The results of the scans can be found in Fig. 10 and Fig. 11.

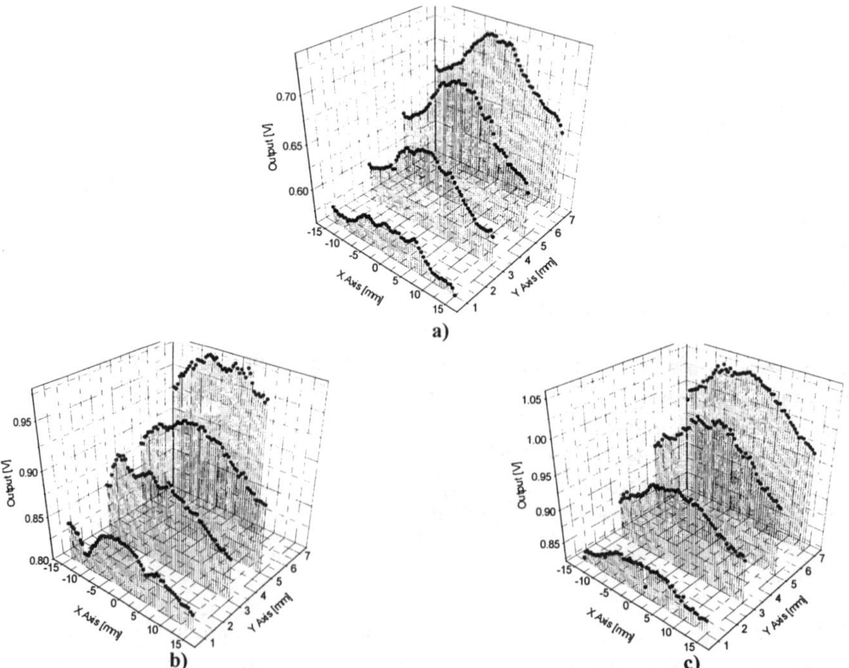

Fig. 10 Scan lines about OD15% crack: a) without lift-off, b) 0.5 mm lift-off (OD15 L), c) 1mm lift-off and 3° tilting error (OD15 LT)

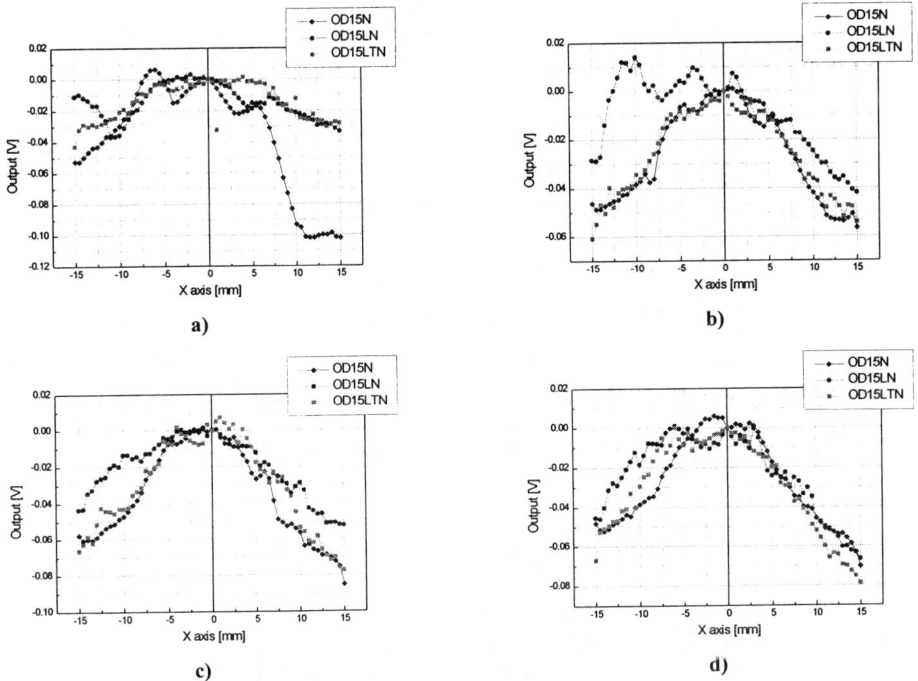

Fig. 11 Comparison of scan lines about OD15% crack a) 1 mm, b) 3 mm, c) 5 mm, d) 7 mm far from *x* axis (without offset)

The lift-off and tilting errors have influence on the signal/noise ratio also. The shapes of the scan lines are also changed. We have to mention that the symmetry of the exciting magnetic field is also changed due to the tilting error and this fact may cause a large magnetic field in the middle point of the sensor core.

The calculated benchmark parameters of the noise suppression can be found in Table III. The S/N ratio is about 39 dB without lift-off and tilting. The 0.5 mm lift-off decreased the S/N ratio by 4 dB roughly, and there is no difference between the S/N ratio of the lift-off only and lift-off and tilting error scans.

The 0.5 mm lift-off does not cause significant change in the measurement from the excitation point of view, because. the default lift-off of the excitation coil is 2.0 mm (Table I). The presented probe is not sensitive to the tilting error. This deficiency causes a constant increase of the magnetic field in the middle of the core material, but the output of the probe is the changing of the magnetic field. The tilting defect is negligible practically from S/N ratio point of view while the offset of the magnetic field amplitude does not saturate the measuring electronics. Nevertheless, the tilting error has influence on the shape of the scan lines, that may cause problems during the solution of the inverse problem. That means the changing of the S/N ratio due to the tilting does not characterise the quality of the measuring system in case of the presented probe.

Table III. Results of lift-off and tilting error tests

Parameter:	Noise [V]	Peak to peak [V]	S/N [dB]
OD15 - 1	0.00066	0.03684	34.94
OD15 - 3	0.00047	0.05768	41.78
OD15 - 5	0.00105	0.08673	38.34
OD15 - 7	0.00087	0.07634	38.86
OD15 - 1 L	0.00097	0.03984	32.27
OD15 - 3 L	0.00317	0.05608	24.95
OD15 - 5 L	0.00187	0.05319	29.08
OD15 - 7 L	0.00297	0.06754	27.14
OD15 - 1 LT	0.00247	0.04426	25.07
OD15 - 3 LT	0.00211	0.06070	29.18
OD15 - 5 LT	0.00414	0.08470	26.22
OD15 - 7 LT	0.00173	0.07898	33.19

8. Conclusions

Benchmark tests and the improved Fluxset sensor based ECT probe were presented in this paper. The target of our investigation was to determine the relationship between the quality of the measuring system and the signal/noise ratio. The objects of the ECT measurement were 15% and 10% ID and OD type cracks in 1.25 INCONEL 600 plate. The influence of the lift-off and the tilting error on the signal/noise ratio was also determined, and the efficiency of S/N parameter like a quality factor was investigated. The raw output signal of the measurement was used for the calculations. The obtained S/N values can bee seen in Fig.12.

The results show that there is a strong relationship between the depth of the crack and the S/N values. The difference between the ID and OD type crack was also significant from the S/N factor point of view. The S/N ratio can be used for qualification of the measurement systems and this factor is useful for the determination of the detectable crack depth boundary - taking into consideration the requirements of the inverse problem solution.

The results show also that the presented Fluxset based probe is not sensitive to the lift-off noise and the tilting error with respect to the noise of the output signal. The S/N factor is not characteristic for the quality of the measurement system because there is no relationship between the distortion of the shape of the scan lines and the noise of these data.

The presented improved Fluxset type ECT probe was able to detect 125 μm deep crack in the metallic plate with 1.25 mm thickness from the other surface (OD10%) at the level of 32 dB S/N ratio. The spatial resolution of the applied Fluxset sensor and the probe were also tested experimentally by bobbin type coil arrangement and analytically by the comparison of the calculated and measured values of the surface scan. We found that the spatial resolution of the sensor is much higher than its length and the spatial distortion of the sensor was not possible to be determined due to the small error.

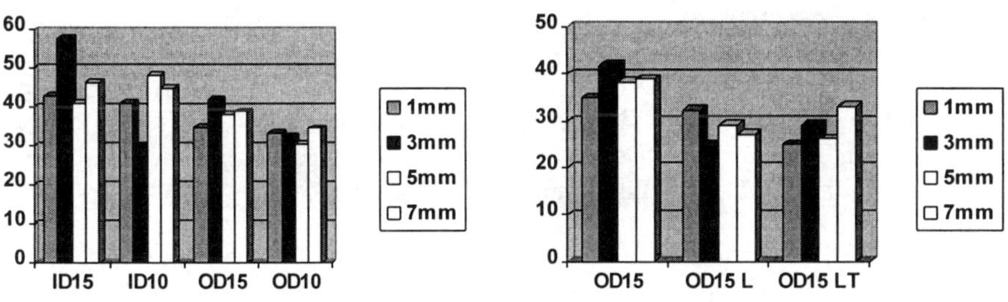

a) b)

Fig. 12 S/N ratio [dB] of the scan lines 1,3,5, and 7 mm far from the *x* axis: a) in case of different cracks, b) without lift-off (OD15), in case of lift-off (OD15 L), tilting and lift-off (OD 15 LT)

9. Outlook

The problem of the eddy current testing can be divided into two main parts: the measurement and the inverse problem solution. We have to distinguish these two jobs from the benchmark tests point of view. Some significant parameters have influence on the quality of the inverse problem solution and characterise the quality of the measurement output at the same time. The S/N ratio is one of these important properties of the measurement and it can be easily determined in case of all kinds of measurements. The spatial resolution (or distortion) may have also the same importance in the combination of the S/N factor from the inversion point of view.

The driving frequency of the applied Fluxset sensor was 180.6 kHz. The highest frequency of the excitation is the half of the driving frequency - theoretically that means 90.3 kHz. The pick-up signal of the sensor at this driving frequency can be seen in Fig. 13.

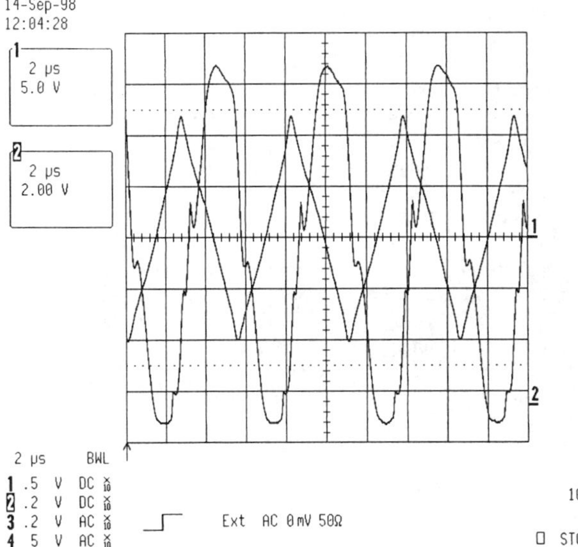

Fig. 13 Oscilloscope plot of the sensor pick-up and driving signal at 180.6 kHz frequency

10. Acknowledgements

This work was supported by the EU INCO-COPERNICUS Project ERBIC-15-CT-960703 by the Hungarian Scientific Research Fund through grant T-023559 and by National Committee for Technological Development (OMFB) Hugarian-Italyan Intergovermental S&T cooperation program trhought grant I-39/98.

References

[1] G. Vértesy, J. Szöllösy, L.K. Varga, A. Lovas, Electronic Horizon, **53** (1992) 102

[2] A. Gasparics, Cs.S. Daróczi, G. Vértesy and J. Pávó, *" Improvement of ECT Probes Based on FluxSet Type Magnetic Field Sensor"*, in *Studies in Applied Electromagnetics and Mechanics 14, Electromagnetic Nondestructive Evaluation (II)*, eds. R. Albanese, G. Rubinacci, T. Takagi and S.S. Udpa, IOS Press, Amsterdam pp.146-151, 1998.

[3] József Pávó, Antal Gasparics, Csaba S. Daróczi, Gábor Vértesy, *"Proposal for Benchmark Problem Qualifying - Some Aspects of the Performance of ECT Probes"*, in *Studies in Applied Electromagnetics and Mechanics 14, Electromagnetic Nondestructive Evaluation (II)*, eds. R. Albanese, G. Rubinacci, T. Takagi and S.S. Udpa, IOS Press, Amsterdam pp.337-342, 1998.

[5] M. Yan, M. Afzal, S. S. Udpa, S. Mandayam, Y. Sun, L. Udpa, P. Sacks, *"Iterative Algorithms for Electromagnetic NDE Signal Inversion"*, in *Studies in Applied Electromagnetics and Mechanics 14, Electromagnetic Nondestructive Evaluation (II)*, eds. R. Albanese, G. Rubinacci, T. Takagi and S.S. Udpa, IOS Press, Amsterdam pp.287-296, 1998.

[6] Radu C. Popa, Kenzo Miya, *"A Data Processing and Neural Network Approach for the Inverse Problem in ECT"*, in *Studies in Applied Electromagnetics and Mechanics 14, Electromagnetic Nondestructive Evaluation (II)*, eds. R. Albanese, G. Rubinacci, T. Takagi and S.S. Udpa, IOS Press, Amsterdam pp.297-304, 1998.

[7] Hiroyuki Fukutomi, Toshiyuki Takagi, Junji Tani, Fumio Kojima, *"Crack Shape Characterization in Eddy Current Testing"*, in *Studies in Applied Electromagnetics and Mechanics 14, Electromagnetic Nondestructive Evaluation (II)*, eds. R. Albanese, G. Rubinacci, T. Takagi and S.S. Udpa, IOS Press, Amsterdam pp.305-312, 1998.

[8] Toshiyuki Takagi, Hiroyuki Fukutomi, Mitsuo Hashimoto, Yoshikatsu Yoshida, Kenzo Miya, Hajime Tsuboi, Yutaka Harada, Junri Shimone, *"ECT Research Activities in JSAEM - Benchmark Models of Eddy Current Testing for Steam Generator Tube - (Part I)"*, in *Studies in Applied Electromagnetics and Mechanics 8, Nondestructive Testing of Materials*, eds. R. Collins, W.D. Dover, J.R. Bowler, K. Miya, IOS Press, Amsterdam pp.253-264, 1995.

[9] József Pávó, Imre Sebestyén, Gábor Vértesy, Csaba S. Daróczi, Kenzo Miya, *"Calculation of the Interaction of Fluxset Type ECT Probe and Crack Located in Metallic Plate"*, in *Studies in Applied Electromagnetics and Mechanics 8, Nondestructive Testing of Materials*, eds. R. Collins, W.D. Dover, J.R. Bowler, K. Miya, IOS Press, Amsterdam pp.211-222, 1995.

[10] Csaba S. Daróczi, János Szöllösy, Gábor Vértesy, József Pávó, Kenzo Miya, *"Electromagnetic NDT Material Testing by Magnetic Field Sensor"*, in *Studies in Applied Electromagnetics and Mechanics 8, Nondestructive Testing of Materials*, eds. R. Collins, W.D. Dover, J.R. Bowler, K. Miya, IOS Press, Amsterdam pp.75-86, 1995.

[11] Csaba S. Daróczi, Antal Gasparics, *"Dual-Frequency Eddy Current NDT Measuring Technique Improved Depth Sensitivity in Metallic Plates"*, in *Studies in Applied Electromagnetics and Mechanics 13, Non-Linear Electromagnetic Systems - Advanced Techniques and Mathematical Methods*, eds. V. Kose, J. Sievert, IOS Press, Amsterdam pp.189-192, 1998.

[12] Csaba S. Daróczi, Antal Gasparics, *"Depth Sensitive Dual-Frequency Eddy Current NDT Measuring Technique"*, in *Studies in Applied Electromagnetics and Mechanics 14, Electromagnetic Nondestructive Evaluation (II)*, eds. R. Albanese, G. Rubinacci, T. Takagi and S.S. Udpa, IOS Press, Amsterdam pp.17-23, 1998.

Author Index

CARDIFF UNIVERSITY
PRIFYSGOL CAERDYDD